OPTICAL INFORMATION
PROCESSING AND
HOLOGRAPHY

WILEY SERIES IN PURE AND APPLIED OPTICS

Advisory Editor

Stanley S. Ballard, University of Florida

Optical Information Processing and Holography

W. THOMAS CATHEY

University of Colorado
Denver, Colorado

John Wiley & Sons, New York / London / Sydney / Toronto

Library of Congress Cataloging in Publication Data:

Cathey, W. Thomas, 1937–
    Optical information processing and holography.

    (Wiley series in pure and applied optics)
    "A Wiley-Interscience publication."
    Includes bibliographical references.
    1. Optical data processing. 2. Optical pattern
    recognition. 3. Holography. I. Title.

TA1630.C37   621.36   73-14604
ISBN 0-471-14078-3

Printed in the United States of America

10 9 8 7 6 5 4 3 2

To Susan and Cheryl

Preface

Holography, pattern recognition, and optical information processing are fields of increasing importance to engineers and scientists. Many are learning through self education, and this book should prove useful to them. The book evolved, however, from notes used in a one-semester class on optical information processing and holography for first-year graduate students in engineering and physics and for well-prepared undergraduates. The course was taught at the Boulder campus of the University of Colorado to primarily younger, full-time students and at the Denver campus to older, part-time students who may have had experience in the field. The necessity of providing a text for such a variety of students has resulted in a book that is larger than anticipated, but contains both basic discussions and detailed descriptions of applications.

The book deals mainly with information conveyed by spatial rather than temporal modulation. The discussion of spatial information processing draws from two areas—information or communication theory and electromagnetic theory. Many of the concepts are borrowed from communication theory, and the existence of information on a wavefront of necessity introduces the problems of propagation and diffraction.

The discussion of spatial information carried on a wavefront and of the means of recording and recovering the information is the basis of holography and spatial filtering. In Chapter 1 scalar diffraction theory is applied to moving fields, emphasizing the propagation of an angular spectrum of plane waves. Chapter 2 introduces the mathematical tools to be used in the processing of spatial information—two-dimensional transforms, convolution, and sampling. Chapter 3 goes into the details of wavefront recording, reconstruction, and modulation. The concept of coherence is introduced in Chapter 4. The degree of coherence required in holography and information processing is discussed, and the coherence properties of many radiating sources are described. In Chapter 5 the imaging and Fourier-transforming properties of a lens are derived from the diffraction formulas.

The general properties of some detectors, recorders, and spatial modulators are presented in Chapter 6. The detection mechanisms and the effects of nonlinearities and finite resolution of recording materials are discussed as they apply to wavefront recording, reconstruction, and modulation.

In communication theory, several useful concepts and techniques have been developed in conjunction with the detection and processing of temporal information. Almost all of these concepts apply with little modification to the detection and processing of spatial information. Detection theory is applied to detection of spatial signals (pattern recognition); spectral analysis and filtering theory are used to analyze and modify the spatial frequency distribution of a spatial signal; and, perhaps the most significant of all, transfer functions are found for the elements operating on the spatial information, enabling us to use the "black box" approach to describe many of the operations.

Chapter 7 discusses spatial filtering and pattern recognition. The first general area covered is that of straightforward spatial filtering by altering the amplitude and phase of the spatial spectrum with separate transparencies. The more powerful technique of complex filtering, made practical by the use of wavefront recording to form a filter, is shown to be useful in pattern recognition. Many of the problems encountered in pattern recognition are discussed in detail.

An analysis of imaging systems appears in Chapter 8. The concept of an angular spectrum of plane waves is used to determine the transfer function of an imaging system in terms of spatial frequencies, and the linear system approach to imaging is discussed.

Chapter 9 deals with holography in general and with types of holograms. The concepts of wavefront recording and reconstruction are examined in Chapter 3, but a more complete description of the holographic process is given in Chapter 9. The topics of magnification and the associated aberrations are covered and some techniques for making holograms with incoherent illumination are presented. An important section of this chapter deals with the effect of the recording medium on the quality of a reconstructed image. Other hologram-generating techniques, such as the computer, are also treated in Chapter 9. Computer generation of spatial modulators is also important in spatial filtering, but because much of the work in this field has been in relation to holography, the subject is placed in Chapter 9.

Applications of holography are dealt with in Chapter 10. This includes holographic interferometry and pulsed laser holography. The application of the concept of holography to wavelengths other than optical ones is discussed and the advantage of using the concept of holography in other fields is pointed out. For example, reconstruction of an image through a phase-distorting medium is an interesting way to describe the operation of adaptive antenna arrays. The vast amount of information that can be stored on a hologram has led to the use of holograms as memory elements in such applications as computer memory and data storage. The techniques used in these applications are discussed briefly. One of the most tantalizing applications is the production of three-dimensional television. Some of the difficulties and some of the proposed solutions are discussed in the last section.

The topics are presented in a logical sequence. No effort was made to preserve the historical order of the developments, but extensive references are given to assist the student in finding background material. The list is not all inclusive, however, and omission of any reference is not intended to reflect on the value of that work. The problem sets are intended to further illustrate the concepts presented in the respective chapters. Hopefully, some of the problems will also encourage reading of the original literature.

One comment concerning notation is in order. The form $e^{-i(\omega t - \gamma z)}$, where γ is the propagation constant, is used to represent a wave traveling in the positive z direction. This notation is generally encountered in diffraction theory and in books on optics. The time variable will usually be dropped, leaving $e^{i\gamma z}$ to represent the wave traveling in a positive direction rather than $e^{-j\gamma z}$ as seen in some engineering texts. To reconcile the formulas, i is to be replaced by $-j$.

I acknowledge the helpful suggestions of my students, the extensive comments of Helmut Lotsch, who carefully read the entire manuscript, and the patience of Judy Price who typed most of the manuscript.

W. THOMAS CATHEY

Denver, Colorado
May 1973

Contents

**OPTICAL INFORMATION
PROCESSING AND
HOLOGRAPHY**

1

Wavefront Propagation and Scalar Diffraction

In introducing scalar diffraction theory, the phenomenogical arguments of Huygens are given to provide an intuitive understanding of the concepts involved. However, the mathematical relations describing scalar diffraction are derived by the use of an ensemble of plane waves which can be considered as being *equivalent* to the actual wave distribution. The resulting equation is examined for possible simplifications through approximations, leading to a definition of the Fresnel and Fraunhofer diffraction regions. The final section introduces the concept of a transfer function description of diffraction.

1.1 Introduction to Scalar Diffraction

To exactly describe the diffraction of a wave by an aperture, the polarization of the wave must be known. That is, the vector representation of the wave must be used [1-1]. However, the vectorial representation of the wave is neglected in this chapter because (1) the mathematical development is more complicated and (2) the scalar theory gives acceptable results for the cases that we consider. In general, the scalar theory is acceptable if the detail of the diffracting structure and the distance between the aperture and the plane in which the field is to be determined are large with respect to a wavelength. Two examples in which these approximations do not hold are microwave diffraction by small apertures and optical diffraction by a closely spaced grating [1-2].

An expression relating the wave in one plane to the wave in another can be derived using Green's theorem. A summary of this procedure is given by Goodman [1-3]. We shall use a different approach for arriving at the

1

diffraction formula which involves following, from one plane to the next, the angular spectral components of the wavefront. That is, the distribution in one plane is written in terms of an ensemble of plane waves which, when added, in both amplitude and phase, are equivalent to the distribution. The plane waves are allowed to propagate to another reference plane and are added to find the new distribution in the second reference plane. This approach is similar to that of Ratcliffe [1-4]. Before attacking the diffraction problem, let us consider the description of a propagating plane wave.

Propagation of plane waves. A uniform plane wave traveling in an arbitrary direction can be represented by

$$|U| Re \exp [-i(\omega t - \alpha x - \beta y - \gamma z + \epsilon)] \tag{1-1}$$

where $|U|$ represents the magnitude of the wave, ϵ denotes the phase, and *Re* denotes real part. We will normally allow a complex U to include both amplitude and phase. The power density of the wave is given by UU^*. The variable U therefore represents neither the electric nor the magnetic field but could be associated with either, depending on the definition of U. That is, $U = E/\sqrt{\eta}$ or $U = \sqrt{\eta}\, H$ where η is the intrinsic impedance of the medium, E represents the electric field, and H represents the magnetic field. In our scalar work, a choice need not be made. The intensity of a wave is a measure of the energy per unit time per unit area *normal* to the direction of propagation. The variables α, β, and γ determine the direction in which the wave propagates, and are restricted by

$$\alpha^2 + \beta^2 + \gamma^2 = k^2 = \left(\frac{2\pi}{\lambda}\right)^2 \tag{1-2}$$

where λ is the wavelength. The values of α, β, and γ are related to the direction of propagation; that is

$$\alpha = k \cos\theta, \tag{1-3}$$

$$\beta = k \cos\xi, \tag{1-4}$$

$$\gamma = k \cos\chi \tag{1-5}$$

where θ, ξ, and χ, are respectively, the angles between the x, y, and z axes and the direction of propagation (normal to the wave). Note that the dimensions of α, β, and γ must be radians per unit length.

In writing a description of a propagating wave, we shall normally omit the temporal variable and the propagation constant for the z direction. The temporal variation is understood and the value of γ can be found from (1-2). The expression (1-1) can then be written as

$$U \exp \{ i2\pi[ux + vy] \} \tag{1-6}$$

where u and v are *spatial* frequencies having the dimensions of cycles per unit length:

$$u = \frac{\alpha}{2\pi} = \frac{\cos\theta}{\lambda}, \qquad v = \frac{\beta}{2\pi} = \frac{\cos\xi}{\lambda}. \qquad (1\text{-}7)$$

Figure 1-1 illustrates the parameters discussed for a plane wave propagating in the x and z directions. The solid lines are maxima and the dotted lines are minima of the tilted wave. Along the x axis the distance between maxima is given by $\lambda/\cos\theta$. That is, the spatial period in the x direction is $\lambda/\cos\theta$, and the spatial frequency is given by (1-7). A wave propagating partially in the negative x direction, such that the spatial period is again $\lambda/\cos\theta$, has the same magnitude spatial frequency, but the phase regresses $(\cos\theta)\lambda$ cycles per unit length rather than advancing by the same amount. This could be considered as a *negative* progression of spatial phase. Notice, however, that the real part of (1-6) is the same in either case. When two waves represented by $U\exp(i2\pi ux)$ and $U\exp(-i2\pi ux)$ are combined, the resulting amplitude distribution along the x axis is $2U\cos(2\pi ux)$, giving a fringe distribution having a spatial frequency of u cycles per unit length. Consequently, we see an intensity pattern described by $4U^2\cos^2(2\pi ux)$.

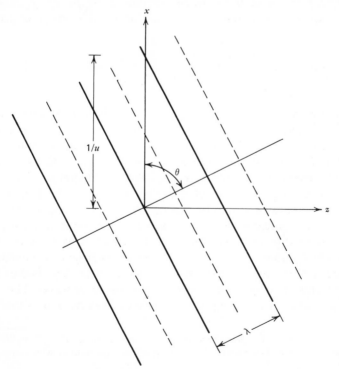

Figure 1-1   A plane wave propagating in the positive x and z directions.

Angular spectrum of plane waves. Let us temporarily leave the discussion of the propagation of a plane wave and consider the Fourier transform representation*

$$A_1\left(\frac{\cos\theta}{\lambda}, \frac{\cos\xi}{\lambda}\right) = \int\int U_1(x_1, y_1)\exp\left[-i2\pi\left(\frac{x_1\cos\theta}{\lambda} + \frac{y_1\cos\xi}{\lambda}\right)\right]dx_1\,dy_1$$

$$(1\text{-}8)$$

and the inverse transform

$$U_1(x,y) = \int\int A_1\left(\frac{\cos\theta}{\lambda}, \frac{\cos\xi}{\lambda}\right)\exp\left[i2\pi\left(\frac{x_1\cos\theta}{\lambda} + \frac{y_1\cos\xi}{\lambda}\right)\right]$$

$$\times d\left(\frac{\cos\theta}{\lambda}\right)d\left(\frac{\cos\xi}{\lambda}\right) \qquad (1\text{-}9)$$

where $U_1(x_1, y_1)$ represents the wave distribution in the $x_1 - y_1$ plane and $A_1[(\cos\theta)/\lambda, (\cos\xi)/\lambda]$ is the Fourier transform of U_1. We have seen that

$$\exp\left[i2\pi\left(\frac{x_1\cos\theta}{\lambda} + \frac{y_1\cos\xi}{\lambda}\right)\right] = \exp[i2\pi(ux_1 + vy_1)] \qquad (1\text{-}10)$$

represents a plane wave propagating partially in the x and y directions (the $i\omega t$ and $i\gamma z$ terms are dropped). Consequently, (1-9) says that the distribution U_1 can be considered as being made up of plane waves propagating in directions determined by $\cos\theta$ and $\cos\xi$ and having amplitudes and phases as described by A_1. We can therefore say that A_1 describes an angular spectrum of plane waves, just as the Fourier transform of a temporal distribution yields the frequency spectrum of the distribution. The integral of (1-9) extends from minus infinity to positive infinity. The only plane waves that are of interest to us are those having arguments of A_1 from $-1/\lambda$ to $1/\lambda$. Values of the argument outside these limits require that $|\cos| > 1$; that is, the angle is imaginary. Waves having such arguments are called *evanescent* or *inhomogeneous* waves. Their directions of propagation are along the positive or negative x axis and decay

---

*We could have used the Fourier transform pair $A_1(\cos\theta, \cos\xi)$ and $U_1(x_1/\lambda, y_1/\lambda)$. That is, the coordinates of the distribution U_1 could have been normalized with respect to the wavelength.

exponentially in the z direction. Investigation of (1-8) shows that these evanescent waves occur only when $\mathbf{U}$ has spatial frequency components with periods of less than a wavelength. Even when they occur, they decay rapidly with z so that they contribute to the field only very near the diffracting structure. In most of our work, we shall neglect the effects of evanescent waves.

The propagation of a homogenous plane wave is easy to describe—it remains an infinite, uniform plane wave if the medium is homogenous and isotropic. Only the phase changes. Consequently, we can describe a distribution $\mathbf{U}_1$ in plane $x_1 - y_1$ in terms of its angular spectrum of plane waves $\mathbf{A}_1$, and allow the plane wave components to propagate to plane $x_2 - y_2$ giving us $\mathbf{A}_2$ from which $\mathbf{U}_2(x_2, y_2)$ can be found. Figure 1-2 shows the propagation of one of the plane wave components. For simplicity, only the x and z axes are shown. As can be seen from the figure, the tilted waves (not propagating just in the z direction) travel a shorter distance than the nontilted wave. The nontilted wave travels a distance z giving a phase shift of $\exp(ikz)$. In general, the plane wave components travel a distance q giving the phase shift

$$\exp(ikq) = \exp(ikz\cos\chi). \qquad (1\text{-}11)$$

Because

$$\cos^2\chi + \cos^2\theta + \cos^2\xi = 1, \qquad (1\text{-}12)$$

$$\exp(ikq) = \exp\left(ikz\sqrt{1 - \cos^2\theta - \cos^2\xi}\,\right). \qquad (1\text{-}13)$$

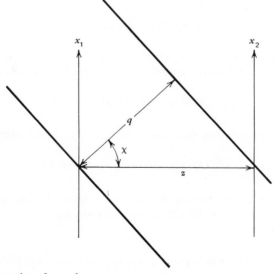

Figure 1-2 Propagation of one plane wave component.

Consequently, the relation between the angular spectrum of the distribution in plane 2 to angular spectrum of the distribution in plane 1 is

$$\mathbf{A}_2\left(\frac{\cos\theta}{\lambda},\frac{\cos\xi}{\lambda}\right)=\mathbf{A}_1\left(\frac{\cos\theta}{\lambda},\frac{\cos\xi}{\lambda}\right)\exp\left(ikz\sqrt{1-\cos^2\theta-\cos^2\xi}\right). \quad (1\text{-}14)$$

Now, (1-9) can be used to find $U_2(x_2,y_2)$. Because we shall consider only cases where $\cos\theta$ and $\cos\xi$ are small, (1-14) is approximated by

$$\mathbf{A}_2\left(\frac{\cos\theta}{\lambda},\frac{\cos\xi}{\lambda}\right)=\mathbf{A}_1\left(\frac{\cos\theta}{\lambda},\frac{\cos\xi}{\lambda}\right)\exp(ikz)\exp\left(-ikz\frac{\cos^2\theta}{2}\right)$$

$$\times\exp\left(-ikz\frac{\cos^2\xi}{2}\right) \quad (1\text{-}15)$$

or

$$\mathbf{A}_2(u,v)=\mathbf{A}_1(u,v)\exp(ikz)\exp(-i\pi\lambda zu^2)\exp(-i\pi\lambda zv^2). \quad (1\text{-}16)$$

The use of (1-9) results in

$$U_2(x_2,y_2)=e^{ikz}\mathfrak{F}^{-1}\left\{\mathbf{A}_1(u,v)\exp\left[-i\pi\lambda z(u^2+v^2)\right]\right\} \quad (1\text{-}17)$$

where $\mathfrak{F}^{-1}$ denotes the inverse Fourier transform. In Chapter 2 we show that

$$\mathfrak{F}^{-1}[G(u)H(u)]=g(x)\otimes h(x) \quad (1\text{-}18)$$

where the capital letters represent the Fourier transforms of the functions indicated by the lowercase symbols and $\otimes$ denotes convolution, that is,

$$g(x)\otimes h(x)\equiv\int g(w)h(x-w)\,dw. \quad (1\text{-}19)$$

By use of the convolution theorem (1-18), (1-17) can be written as

$$U_2(x_2,y_2)=e^{ikz}U_1(x_2,y_2)\otimes\mathfrak{F}^{-1}\left\{\exp\left[-i\pi\lambda z(u^2+v^2)\right]\right\} \quad (1\text{-}20)$$

where $U_1(x_2,y_2)$ is the inverse transform of $\mathbf{A}_1(u,v)$. The inverse Fourier transform in (1-20) is the two dimensional integral

$$\int_{-\infty}^{\infty}\int\exp\left[-i\pi\lambda z(u^2+v^2)\right]\exp\left[i2\pi(x_2u+y_2v)\right]\,du\,dv \quad (1\text{-}21)$$

which can be separated into two integrals. Because

$$\mathcal{F}^{-1}\left\{\exp\left[-\pi(ax)^2\right]\right\} = |a|^{-1}\exp\left[-\pi\left(\frac{u}{a}\right)^2\right], \qquad (1\text{-}22)$$

(1-21) becomes

$$\frac{1}{i\lambda z}\exp\left(i\pi\frac{x_2^2+y_2^2}{\lambda z}\right) \qquad (1\text{-}23)$$

after combining the results of the two separate transformations. Equation (1-20) can then be written as

$$U_2(x_2,y_2) = \frac{1}{i\lambda z}\exp(ikz)U_1(x_2,y_2)\otimes\exp\left(ik\frac{x_2^2+y_2^2}{2z}\right) \qquad (1\text{-}24)$$

$$= \frac{1}{i\lambda z}\exp(ikz)\int\int U_1(x_1,y_1)$$

$$\times\exp\left\{ik\left[\frac{(x_2-x_1)^2+(y_2-y_1)^2}{2z}\right]\right\}dx_1\,dy_1 \qquad (1\text{-}25)$$

which relates the wave in the x_2,y_2 plane to the wave in the x_1,y_1 plane. This is the diffraction formula for small angles. For a derivation of the diffraction formula using the angular spectrum but without the small angle approximation, see Appendix I of reference [1-5].

1.2   Huygens' Principle

A concise treatment of the history of the development of diffraction theory is given by Born and Wolf [1-6] and more detailed treatments of the theory are available for those interested [1-7 through 1-9]. Only a brief introduction is given here. The account begins in 1678 when Christian Huygens postulated that each point on a wavefront could be regarded as the source of a secondary spherical wave. The position and form of the wave at a later time is then described by the envelope of the secondary waves as shown in Fig. 1-3. Huygens' concepts were incorporated into mathematical form by Augustin Fresnel in 1818. Fresnel predicted diffraction patterns, later experimentally verified, which could not be explained by particle theory. This work led to the acceptance of the wave theory of light over the particle theory. The mathematical description was later further refined by Kirchhoff, Rayleigh, and Sommerfeld. The mathematical techniques have

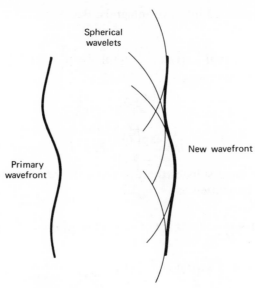

Spherical
wavelets

New wavefront

Primary
wavefront

Figure 1-3   The envelope of the secondary waves arising from the primary wave forms the new wave.

not yet been perfected, with most difficulty arising because of the boundary conditions. The boundary condition problems have been studied since 1888 by Maggi [1-10] up to the recent work of Rubinowicz, Wolf, and co-workers [1-11, 1-12].

We have obtained the expression (1-25) by considering the propagation of the plane wave components of the distribution U_1. The same equation can be developed using Huygens' assumption that each point of the distribution radiates a wavelet uniformly in all directions. Its wavefront is thus spherical and is described by $\exp(ikr)$. The total effect in the second plane is the superposition of the waves from all points in the first plane, being proportional to

$$\int\int U_1(x_1,y_1)\exp(ikr)\,dx_1\,dy_1 \qquad (1\text{-}26)$$

where $U_1(x_1,y_1)$ denotes the amplitude and phase of the wave radiated from the point (x_1,y_1), and the constant of proportionality has been shown to be $1/i\lambda z$. (See, for example [1-3].)

Reference to Fig. 1-4 shows that r is given as

$$r^2 = z^2 + (x_1 - x_2)^2 + (y_1 - y_2)^2, \qquad (1\text{-}27)$$

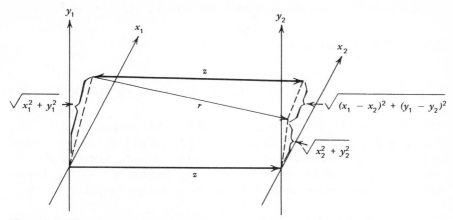

Figure 1-4 The relationship between the coordinates and the distance between a point on the x_1-y_1 plane and a point on the x_2-y_2 plane.

or equivalently as

$$r = z \left\{ 1 + \frac{(x_1 - x_2)^2 + (y_1 - y_2)^2}{z^2} \right\}^{1/2}. \qquad (1\text{-}28)$$

The binomial expansion can be used to obtain

$$r \cong z + \frac{1}{2z} \left[(x_1 - x_2)^2 + (y_1 - y_2)^2 \right] \qquad (1\text{-}29)$$

where the higher-order terms can be dropped as being negligible if the region of interest is restricted to be such that

$$z \gg |x_1 - x_2|, \qquad z \gg |y_1 - y_2|. \qquad (1\text{-}30)$$

Substituting the approximation (1-29) in (1-26) results in the same expression, (1-25), that was derived by the use of the angular spectrum. In this approach, the physical interpretation of the approximation is that only a portion of the spherical wave from each point is of concern. Consequently, a quadratic approximation to the spherical surface is adequate.

The diffraction formula. Equation (1-25) can be rewritten with the x_2 and y_2 terms outside of the integral as

$$U_2(x_2, y_2) = \frac{1}{i\lambda z} \exp(ikz) \exp\left(ik \frac{x_2^2 + y_2^2}{2z} \right)$$

$$\times \int \int U_1(x_1, y_1) \exp\left(ik \frac{x_1^2 + y_1^2}{2z} \right) \exp\left(-ik \frac{x_1 x_2 + y_1 y_2}{z} \right) dx_1\, dy_1. \qquad (1\text{-}31)$$

If the value of k is written as $2\pi/\lambda$ in the last exponential of (1-31),

$$U_2(x_2,y_2) = \frac{1}{i\lambda z} \exp(ikz) \exp\left(ik\frac{x_2^2+y_2^2}{2z}\right)$$

$$\times \int \int U_1(x_1,y_1) \exp\left(ik\frac{x_1^2+y_1^2}{2z}\right) \exp\left(-i2\pi\frac{x_1x_2+y_1y_2}{\lambda z}\right) dx_1\,dy_1 \quad (1\text{-}32)$$

takes the form of a *Fourier* transform. That is, the integral represents the Fourier transform of $U_1(x_1,y_1)\exp[ik(x_1^2+y_1^2)/2z]$ and the transformation is from the x_1,y_1 domain to the $x_2/\lambda z, y_2/\lambda z$ domain. Alternatively, the integral of (1-32) can be viewed as a *Fresnel* transform of $U_1(x_1,y_1)$. Let us further examine the significance of the term $\exp[ik(x_1^2+y_1^2)/2z]$.

1.3 Fresnel Region

In the inevitable effort to simplify the mathematics, the term

$$\exp\left(ik\frac{x_1^2+y_1^2}{2z}\right) \qquad\qquad (1\text{-}33)$$

of (1-32) comes under scrutiny. Obviously if $U(x_1,y_1)$ is nonzero over only a small region, then x_1^2 and y_1^2 can be small in comparison to λz, and the term (1-33) can be neglected. If this is not the case, by reason of either a large extent of U_1 or a small value of z, the x_2,y_2 plane is said to be in the Fresnel region of the distribution $U(x_1,y_1)$. The Fresnel region is then the region where z is large enough, with respect to the extent of U_1 and U_2, that the higher-order terms of (1-29) can be neglected but where λz is small enough with respect to the square of the extent of U_1 that the exponential term (1-33) cannot be neglected. This region is of importance in optical imaging and holography.

1.4 Fraunhofer Region

The point at which the term in (1-33) can be neglected is easier to define when U_1 is bounded by an aperture. The limits on the size of the aperture then give a direct means of comparing the maximum values of x_1^2 and x_2^2 with λz. Even so, the exponential term is not suddenly negligible, and different numbers have been given as the values of λz beyond which the exponential (1-33) can be ignored. Consideration of the integral of (1-32) shows that the exponential (1-33) is negligible if its phase change is negligible over the range of nonzero or nonnegligible values of U_1. It is assumed that this is so if the phase change over the range of integration is much less than 1 radian. If the maximum value of $x_1^2 + y_1^2$ is $(d/2)^2$, where d is the linear dimension of the maximum extent of $\mathbf{U}_1$, the restriction on the phase leads to

$$\frac{k(d/2)^2}{2z} = \frac{\pi d^2}{4\lambda z} \ll 1. \qquad (1\text{-}34)$$

This is frequently written as

$$z \gg \frac{d^2}{\lambda} \qquad (1\text{-}35)$$

where the ratio of $\pi/4$ is taken to be 1. The criterion is often that the allowable phase shift is much less then $\frac{1}{2}$ radian, leading to

$$z \gg \frac{2d^2}{\lambda}. \qquad (1\text{-}36)$$

In either case, the region in which (1-33) can be neglected is called the Fraunhofer region, or the far field. The Fraunhofer distribution is given by

$$U_2(x_2,y_2) = \frac{1}{i\lambda z} \exp(ikz) \exp\left(ik \frac{x_2^2 + y_2^2}{2z} \right)$$

$$\times \int \int U_1(x_1,y_1) \exp\left(-ik \frac{x_1 x_2 + y_1 y_2}{z} \right) dx_1 dy_1 \qquad (1\text{-}37)$$

or

$$U_2(x_2,y_2) = \frac{1}{i\lambda z} \exp(ikz) \exp\left(ik \frac{x_2^2 + y_2^2}{2z} \right) \mathbf{A}_1\left(\frac{x_2}{\lambda z}, \frac{y_2}{\lambda z} \right). \qquad (1\text{-}38)$$

If we recognize that, for small values of x/z and y/z, $\cos\theta = x_2/z, \cos\xi = y_2/z$, then (1-7) allows us to write

$$U_2(x_2,y_2) = \frac{1}{i\lambda z} \exp(ikz) \exp[i\pi\lambda z(u^2 + v^2)] \mathbf{A}_1(u,v). \qquad (1\text{-}39)$$

Either expression shows that the diffraction pattern in the Fraunhofer region is given by the Fourier transform of the distribution in the first plane. The multipliers show the $1/z$ decrease in amplitude ($1/z^2$ in power), the progressive phase shift due to propagation, and a quadratic phase factor.

It is interesting to put some numbers into (1-35) to determine where (1-33) can be neglected. If λ is 5×10^{-7} meter (in the visible region) and the aperture diameter is 0.5 cm, $z \gg 50$ meters. We shall see in Chapter 5 that there are methods by which we can reduce the value of z and still obtain a Fraunhofer diffraction pattern. These techniques involve the introduction of another phase term into (1-32) to cancel or reduce the effect of the term (1-33).

1.5   System Response and Transfer Function Description of Diffraction

The mathematical descriptions for diffraction can be expressed in terms of impulse responses or, alternately, in terms of transfer functions for space. In both approaches, the distribution in plane 1 is considered to be an input to a "system" and the distribution in plane 2 is the system output.

Consider first the relation (1-24), repeated below:

$$U_2(x_2,y_2) = \frac{1}{i\lambda z} \exp(ikz) U_1(x_2,y_2) \otimes \exp\left(ik\frac{x_2^2+y_2^2}{2z}\right). \quad (1\text{-}24)$$

The output U_2 is related to the input U_1 by the convolution of U_1 with the system's impulse response (the effect of a single point in space). Other terminology for the operation is that the input is convolved with the Green's function. If the input distribution is a point source, $U_1(x_1,y_1)$ can be represented by the delta function, $\delta(x_1,y_1)$ and U_2 becomes simply

$$\frac{1}{i\lambda z} \exp(ikz) \exp\left(ik\frac{x_2^2+y_2^2}{2z}\right) \quad (1\text{-}40)$$

which is a valid approximation to the impulse response of free space for regions near the z axis.

The results obtained in the discussion of the effect of propagation on the angular spectrum of a distribution can be described in terms of the spatial frequency transfer function. This gives the effect of the system on each component of the angular spectrum. For example, free space has the transfer function found in (1-15),

$$\exp(ikz) \exp\left(\frac{-ikz\cos^2\theta}{2}\right) \exp\left(\frac{-ikz\cos^2\xi}{2}\right) \quad (1\text{-}41)$$

which is the Fourier transform of the impulse response. If we write (1-41) in terms of spatial frequencies, we obtain

$$\exp(ikz) \exp(-i\pi\lambda z u^2) \exp(-i\pi\lambda z v^2). \quad (1\text{-}42)$$

That is, space simply shifts the phase of the spatial frequency components and does nothing to the amplitude. It is important to note that a description of the effect of propagation on the plane waves is also a description of the effect on the various spatial frequencies that make up the original distribution.

The effect of propagation of a wave through any optical system for which a transfer function can be obtained can be described in terms of the transfer function. The spatial frequency distribution of the output is equal to the spatial frequency spectrum at the input multiplied by the transfer function. More will be said of the spatial frequency transfer function later, but reference to Fig. 1-5 will illustrate the concepts discussed. The double arrow indicates the Fourier transform relations between the spatial distribution and the spatial frequency spectrum of the distribution. The transfer function H is the Fourier transform of the impulse response h.

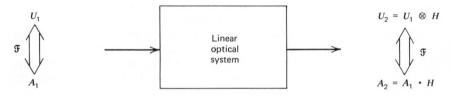

Figure 1-5 The linear system has an impulse response **h**. The Fourier transform of the impulse response yields the spatial frequency transfer function **H**.

PROBLEMS

1-1. Complete the square in the exponential of (1-21) and evaluate the integral to obtain (1-23).

1-2. Use the series representation for the expression of (1-33) to determine when it can be neglected in the evaluation of (1-32).

1-3. Show that the magnitude of the angular spectrum of the amplitude distribution of a wave is the same throughout the Fresnel region.

1-4. Consider the relations describing the effect of a spatial modulator such as a transparency having a transmittance $t(x,y)$, and find the relation between the spatial spectrum of the incident wave $A_1(u,v)$, the spatial spectrum of the transparency, and the spatial spectrum of the exit wave $A_2(u,v)$. Assume that the transparency has zero thickness.

1-5. Can a transfer function be found for the transparency of problem 4?

REFERENCES

1-1.   M. Born and E. Wolf, *Principles of Optics* (Macmillan Co., New York, 1964), Ch. 11.

1-2.   G. W. Stroke, "Etude Théorique et Expérimentale de Deux Aspects de la Diffraction par les Réseaux Optiques," *Rev. Opt.* **39**, 291–396 (1960).

1-3.   J. W. Goodman, *Introduction to Fourier Optics*, (McGraw-Hill Book Co., New York, 1968), Ch. 3.

1-4.   J. A. Ratcliffe, "Some Aspects of Diffraction Theory and Their Application to the Ionosphere," pp. 188–267 in *Reports on Progress in Physics*, vol. 19, A. C. Strickland, ed. (Physical Society, London, 1956).

1-5.   R. J. Collier, C. B. Burckhardt, and L. H. Lin, *Optical Holography* (Academic Press, New York, 1971).

1-6.   Ref. 1-1, Historical Introduction.

1-7.   B. B. Baker and E. T. Copson, *The Mathematical Theory of Huygens' Principle* (Claredon Press, Oxford, 1949).

1-8.   C. J. Bouwkamp, "Diffraction Theory," in *Reports on Progress in Physics*, vol. 17, A. C. Strickland, ed. (Physical Society, London, 1954).

1-9.   H. Hoenl, A. W. Maue, and K. Westpfahl, "Theorie der Bengung," pp. 218–573 in *Encyclopedia of Physics*, vol. XXV/1, S. Fluegge, ed. (Springer-Verlag, Berlin, 1961).

1-10. G. A. Maggi, "Sulla Propagazione Libera e Perturbata della Onde Luminose in un Mezzo Isotropo," *Ann. Math.* **16**, 21 (1888).

1-11. See, for example, A. Rubinowicz, "The Miyamoto-Wolf Diffraction Wave," Ch. V in *Progress in Optics*, vol. IV, E. Wolf, ed. (North Holland Publishing Co., Amsterdam, 1965).

1-12. E. W. Marchand and E. Wolf, "Diffraction at Small Apertures in Black Screens," *J. Opt. Soc. Am.* **59**, 79–90 (1969).

2

Fourier Transforms and Special Functions

Many of the topics treated require a working knowledge of Fourier transforms. The description of Fraunhofer diffraction is easier when Fourier transforms are used; Fresnel diffraction is described in terms of convolutions relating to the Fourier transform; and for the spatial filtering discussion, the use of Fourier transforms is indispensable. Consequently, before progressing to a discussion of imaging and spatial filtering or a more complete treatment of holography, Fourier transforms and their properties must be covered more thoroughly. This chapter is intended to be a review of Fourier transforms and a definition of the functions used in the remainder of the book. For those readers having a weak background in Fourier transforms, the use of a supplementary text is recommended [2-1, 2-2].

2.1 Two-Dimensional Fourier Transforms

Fourier transforms. The formulas relating a function to its transform are

$$\mathbf{G}(u) = \int_{-\infty}^{\infty} \mathbf{g}(x) e^{-i2\pi xu} dx \qquad (2\text{-}1)$$

and

$$\mathbf{g}(x) = \int_{-\infty}^{\infty} \mathbf{G}(u) e^{i2\pi xu} du \qquad (2\text{-}2)$$

where the $\mathbf{G}$ represents the transformed function. Notice that the 2π is incorporated into the exponential of both integrals rather than in the denominator of (2-2). Either is correct, but the listed formulas better suit our needs. The expression in (2-1) describes the direct Fourier transform and (2-2) represents the inverse Fourier transform. Quite often it is not necessary to distinguish between the two, but we shall normally do so. A common operational notation is

$$\mathbf{G}(u) = \mathcal{F}[\mathbf{g}(x)]$$

$$\mathbf{g}(x) = \mathcal{F}^{-1}[\mathbf{G}(u)]$$

where $\mathcal{F}$ denotes Fourier transformation and $\mathcal{F}^{-1}$ denotes inverse transformation.

Separable functions. The two-dimensional transform for rectangular coordinates is given by*

$$\mathbf{G}(u,v) = \int\int \mathbf{g}(x,y) e^{-i2\pi(xu+yv)}\, dx\, dy. \qquad (2\text{-}3)$$

If the function $\mathbf{g}(x,y)$ can be separated into a product—one a function of x only and the other a function of y only—the integration is simplified. Assume, for example, that $\mathbf{g}(x,y) = \mathbf{g}_x(x)\mathbf{g}_y(y)$. The integrals are now separable and

$$\mathcal{F}[\mathbf{g}(x,y)] = \mathcal{F}_x[\mathbf{g}_x(x)]\mathcal{F}_y[\mathbf{g}_y(y)]. \qquad (2\text{-}4)$$

The subscripts x and y on $\mathcal{F}$ are used only when it is desirable to stress that the integral involves only one dimension. Most functions with which we are concerned are separable, thus allowing us to perform two one-dimensional transformations rather than one two-dimensional one.

Many functions that we encounter are separable in polar coordinates. Most of them are, in addition, circularly symmetrical, so that

$$\mathbf{g}(r,\theta) = \mathbf{g}(r). \qquad (2\text{-}5)$$

A simple transformation of coordinates shows that (2-3) can be written in polar coordinates as

$$\mathbf{G}(\rho,\vartheta) = \int_0^{2\pi}\int_0^{\infty} r\mathbf{g}(r,\theta)\exp[-i2\pi r\rho(\cos\theta\cos\vartheta$$

$$+ \sin\theta\sin\vartheta)]\, dr\, d\theta \qquad (2\text{-}6)$$

where

$$x = r\cos\theta, \qquad y = r\sin\theta$$

$$u = \rho\cos\vartheta, \qquad v = \rho\sin\vartheta. \qquad (2\text{-}7)$$

*Unless otherwise indicated, all integrations are from $-\infty$ to ∞.

If **g** is not a function of θ, (2-6) can be written as

$$\mathbf{G}(\rho,\vartheta) = \int_0^\infty r\mathbf{g}(r) \int_0^{2\pi} \exp[-i2\pi r\rho\cos(\theta-\vartheta)] \, d\theta \, dr. \qquad (2\text{-}8)$$

We can use the identity involving the zero-order Bessel function,

$$J_0(a) = \frac{1}{2\pi} \int_0^{2\pi} \exp[-ia\cos(\theta-\vartheta)] \, d\theta, \qquad (2\text{-}9)$$

to obtain

$$\mathbf{G}(\rho) = 2\pi \int_0^\infty r\mathbf{g}(r) J_0(2\pi r\rho) \, dr. \qquad (2\text{-}10)$$

This is known as the Fourier-Bessel transform or the Hankel transform of zero order. Similarily,

$$\mathbf{g}(r) = 2\pi \int_0^\infty \rho\mathbf{G}(\rho) J_0(2\pi r\rho) \, d\rho. \qquad (2\text{-}11)$$

Transform theorems.   There are three fundamental relations that we shall use frequently. These are the linearity, similarity, and shift theorems. The linearity theorem states that

$$\mathscr{F}[a\mathbf{g}(x,y) + b\mathbf{h}(x,y)] = a\mathscr{F}[\mathbf{g}(x,y)] + b\mathscr{F}[\mathbf{h}(x,y)], \qquad (2\text{-}12)$$

or that the transform of a sum is the sum of the transforms. The similarity theorem is

$$\mathscr{F}[\mathbf{g}(ax,by)] = \frac{1}{|ab|} \mathbf{G}\left(\frac{u}{a}, \frac{v}{b}\right). \qquad (2\text{-}13)$$

Notice that a scale change in one domain causes an inverse change in the other. For example, a magnification of a wave distribution results in a shrinking of the extent of the spatial spectrum. The shift theorem can be expressed as

$$\mathscr{F}[\mathbf{g}(x-a,y-b)] = \mathbf{G}(u,v)e^{-i2\pi(au+bv)}. \qquad (2\text{-}14)$$

In terms of an angular spectrum, the shift theorem says that a shift in the position of a distribution results in a shift in the phase or tilt of the angular spectrum.

Complex functions.   It is often useful to relate the transform of a complex function to the transform of the conjugate of the same complex function. It can be shown (see problem 2-4 for one suggested procedure) that

$$\mathcal{F}\left[\mathbf{g}^*(x,y)\right]=\mathbf{G}^*(-u,-v) \tag{2-15}$$

where the superscript asterisk denotes the complex conjugate.

Special symbols.   There are some distributions that appear repeatedly but cannot be expressed as analytic functions. These have been assigned special symbols to facilitate their use.

One example is the Dirac delta function, represented by $\delta(x)$, which has infinite amplitude, zero width, and unit area. The delta function can be represented as the limit of any of several functions, but the most common is the limit, as $a \to \infty$, of rectangular pulse of width $1/a$ and height a. One of the more important properties of the delta function is the so-called sifting property. That is,

$$\int \delta(x)\mathbf{g}(x)\,dx = \mathbf{g}(0) \tag{2-16}$$

and

$$\int \delta(x \pm a)\mathbf{g}(x)\,dx = \mathbf{g}(\mp a). \tag{2-17}$$

The Fourier transform of $\delta(x-a)$ is therefore

$$\mathcal{F}\left[\delta(x-a)\right]=e^{-i2\pi a u}. \tag{2-18}$$

An additional property of the delta function worthy of note at this point is

$$\delta(ax)=\frac{1}{|a|}\delta(x). \tag{2-19}$$

The properties (2-16) through (2-19) are used extensively.

The rectangle function $\text{II}(x/a)$, or $\text{rect}(x/a)$, is defined as

$$\text{II}(x/a)=\begin{cases} 0 & |x|>\dfrac{a}{2} \\ \tfrac{1}{2} & |x|=\dfrac{a}{2} \\ 1 & |x|<\dfrac{a}{2} \end{cases} \tag{2-20}$$

See Fig. 2-1. The rectangle function can be shifted, expanded, or scaled. For example, $a\text{II}[(x-b)/c]$ is a rectangle of height a and width c, and is centered at $x=b$. One important use of the rectangle function is to select any segment of a function with any amplitude, and to reduce the rest of the function to zero. For example, $\text{II}(x-\tfrac{1}{2})\sin \pi x$ is shown in Fig. 2-2.

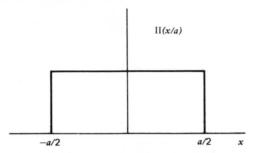

Figure 2-1 The rectangle function of width a and unit height, $\text{II}(x/a)$.

The Fourier transform of $\text{II}(x/a)$ is given by

$$\mathcal{F}\left[\text{II}\left(\frac{x}{a}\right)\right] = |a|\operatorname{sinc} au \qquad (2\text{-}21)$$

where

$$\operatorname{sinc} u = \frac{\sin \pi u}{\pi u}. \qquad (2\text{-}22)$$

The term sinc is sometimes defined without the π of (2-22), so that care must be taken to note the definition of sinc in different works. A sketch of sinc au is shown in Fig. 2-3.

Another useful symbol is $\Lambda(x/a)$ defined as

$$\Lambda\left(\frac{x}{a}\right) = \begin{cases} 0 & |x| > a \\ 1 - \left|\dfrac{x}{a}\right| & |x| < a \end{cases}. \qquad (2\text{-}23)$$

This function, shown in Fig. 2-4, defines a triangle of unit height and base

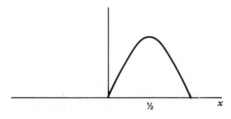

Figure 2-2 The function $\sin \pi x$ gated by $\text{II}(x - \tfrac{1}{2})$.

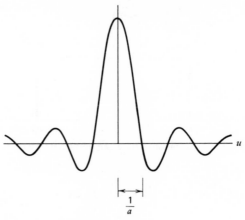

Figure 2-3   The function sinc au.

2a. The Fourier transform of $\Lambda(x/a)$ is

$$\mathfrak{F}\left[\Lambda\left(\frac{x}{a}\right)\right] = a\operatorname{sinc}^2(au). \qquad (2\text{-}24)$$

In circular coordinates we can use $II(r/2a)$, or $\operatorname{circ}(r/a)$, to denote a pillbox described by

$$\operatorname{circ}\left(\frac{r}{a}\right) = II\left(\frac{r}{2a}\right) = \begin{cases} 0 & r>a \\ \frac{1}{2} & r=a \\ 1 & r<a \end{cases} \qquad (2\text{-}25)$$

Notice the difference in the arguments of the circ and II functions. The

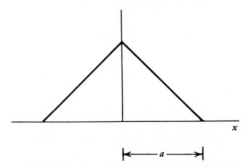

Figure 2-4   The triangle function $\Lambda(x/a)$.

transform of $II(r/2a)$ is

$$\mathfrak{F}\left[II\left(\frac{r}{2a}\right)\right]=|a|\frac{J_1(2\pi a\rho)}{\rho} \tag{2-26}$$

where J_1 is the Bessel function of the first kind, order 1, and ρ is the coordinate of the transform. A plot of $J_1(w)/w$ for positive w is given in Fig. 2-5.

Examples using Fraunhofer diffraction. We found in Chapter 1 that the far field or Fraunhofer diffraction pattern could be expressed in terms of a Fourier transform of the distribution immediately past the diffracting structure. The relation between the distribution in the distant x_2,y_2 plane and the one in the x_1,y_1 plane is

$$U_2(x_2,y_2)=\frac{1}{i\lambda z}\exp(ikz)\exp\left[\frac{ik(x_2^2+y_2^2)}{2z}\right]$$

$$\times\int\int U_1(x_1,y_1)\exp\left[\frac{-i2\pi(x_1x_2+y_1y_2)}{\lambda z}\right]dx_1\,dy_1. \tag{2-27}$$

The diffraction pattern of a rectangular aperture with uniform illumination

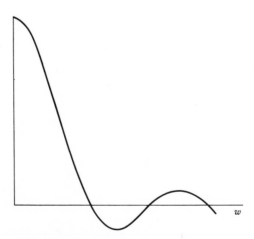

Figure 2-5 The function $J_1(w)/w$ for positive w.

can be found by transformation of the II function. If the multipliers outside the integral are neglected, $x_2/\lambda z$ is replaced by u, and $y_2/\lambda z$ is replaced by v, the expression (2-27) can be written as

$$U_2(u,v) \sim \mathfrak{F}\left[\Pi\left(\frac{x}{a}, \frac{y}{b}\right)\right] \qquad (2\text{-}28)$$

where a and b are the dimensions of the rectangle. Consequently, the amplitude of the diffraction pattern is proportional to

$$ab\, \mathrm{sinc}(au)\, \mathrm{sinc}(bv) \qquad (2\text{-}29)$$

and the intensity distribution is described by

$$[U_2(u,v)]^2 = \frac{a^2 b^2}{(\lambda z)^2}\, \mathrm{sinc}^2(au)\, \mathrm{sinc}^2(bv). \qquad (2\text{-}30)$$

Figure 2-6 shows a normalized cross section of the intensity of the Fraunhofer pattern in the plane $v=0$. Note that the nulls of the function fall at $u=m/a$ where m is an integer. That is, at

$$x_2 = m\frac{\lambda z}{a}. \qquad (2\text{-}31)$$

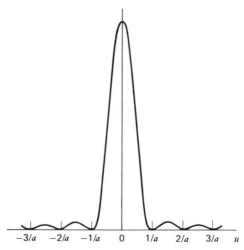

Figure 2-6 Cross section (at $v=0$) of the intensity diffraction pattern of a rectangular aperture.

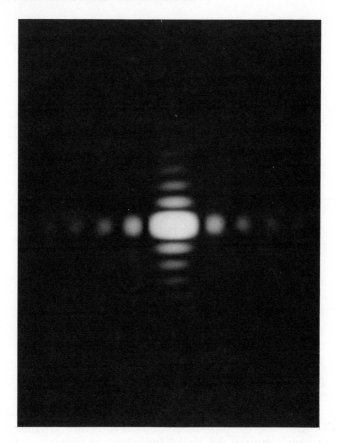

Figure 2-7 Photograph of the far-field intensity pattern caused by a rectangular aperture $(a/b = \frac{3}{5})$.

The angles, as seen from the diffracting aperture, at which the nulls appear are governed by

$$\frac{x_2}{z} = \frac{m\lambda}{a}. \tag{2-32}$$

Figure 2-7 is a photograph of the intensity pattern caused by a rectangle having a ratio of $a/b = \frac{3}{5}$.

If the wave is diffracted by a uniformly illuminated circular aperture of diameter d, the expression describing the Fraunhofer pattern, (2-27), can

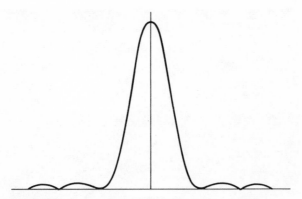

Figure 2-8   Cross section of the Airy pattern.

be written as

$$U_2(\rho,d) = \frac{1}{i\lambda z} \exp(ikz) \exp\left(\frac{ikr^2}{2z}\right) \mathcal{F}\left\{\Pi\left(\frac{r}{d}\right)\right\}$$

$$= \frac{d}{i2\lambda z\rho} \exp(ikz) \exp\left(\frac{ikr^2}{2z}\right) J_2(\pi\rho d). \qquad (2\text{-}33)$$

The intensity distribution associated with (2-33) is shown in a normalized plot in Fig. 2-8, and a photograph of the distribution is given in Fig. 2-9. This pattern is called the Airy pattern after the person who first described it. The nulls of this pattern are not equally spaced as is the case with a rectangular aperture, but are as given in Table 2-1. To find the actual location of a null, replace the spatial frequency coordinate ρ by $r_2/\lambda z$ and the first null appears at

$$r_2 = \frac{1.22\lambda z}{d}. \qquad (2\text{-}34)$$

The angular location of the first null is

$$\frac{r_2}{z} = \frac{1.22\lambda}{d}. \qquad (2\text{-}35)$$

2.2   Convolution and Correlation

One reason for an interest in convolution and correlation is that the Fourier transform of a product of functions results in the convolution of their transforms. This operation appears in many diffraction calculations.

Figure 2-9 Photograph of the diffraction pattern of a uniformly illuminated circular aperture (Airy pattern).

In addition, the operation of pattern recognition is often described in terms of correlation functions.

Convolution. The convolution of a function $\mathbf{g}(x)$ with $\mathbf{h}(x)$ is denoted by

$$\mathbf{p}(x) = \mathbf{g}(x) \otimes \mathbf{h}(x) = \int \mathbf{g}(w)\mathbf{h}(x-w)\,dw. \qquad (2\text{-}36)$$

One physical interpretation is that one distribution, h, is reversed and slid along the other, g, and the area of the product of the functions gives the value of the correlation as a function of the displacement x. Figure 2-10 illustrates the determination of one point on the convolution curve. From the process of acquiring the curve for the convolution we see why the convolution is a smoothing process.

TABLE 2.1   THE FIRST FEW MAXIMA AND MINIMA OF

THE NORMALIZED FUNCTION $\left[\dfrac{2J_1(\pi\rho d)}{\pi\rho d}\right]^2$

ρd	$\left[\dfrac{2J_1(\pi\rho d)}{\pi\rho d}\right]^2$
0	1.000
1.226	0
1.635	0.0175
2.233	0
2.679	0.0042
3.238	0
3.699	0.0016

Convolution of a function with a delta function simply causes the function to be moved to the location of the delta function. The operation of convolution has many of the same properties of multiplication. Convolution is commutative,

$$\mathbf{g}\otimes\mathbf{h}=\mathbf{h}\otimes\mathbf{g};\qquad(2\text{-}37)$$

associative,

$$\mathbf{f}\otimes(\mathbf{g}\otimes\mathbf{h})=(\mathbf{f}\otimes\mathbf{g})\otimes\mathbf{h};\qquad(2\text{-}38)$$

and distributive,

$$\mathbf{f}\otimes(\mathbf{g}+\mathbf{h})=\mathbf{f}\otimes\mathbf{g}+\mathbf{f}\otimes\mathbf{h}.\qquad(2\text{-}39)$$

These properties make easier the manipulation of relations involving convolutions.

The Fourier transform relations that result in a convolution can be written as

$$\mathcal{F}[\mathbf{g}(x)\mathbf{h}(x)]=\mathbf{G}(u)\otimes\mathbf{H}(u).\qquad(2\text{-}40)$$

Similarly,

$$\mathcal{F}[\mathbf{g}(x)\otimes\mathbf{h}(x)]=\mathbf{G}(u)\mathbf{H}(u).\qquad(2\text{-}41)$$

This is shown by changing the order of integration and using the shift theorem, that is,

$$\mathcal{F}[\mathbf{g}(x)\otimes\mathbf{h}(x)]=\int\mathbf{g}(w)\,\mathcal{F}[\mathbf{h}(x-w)]\,dw$$
$$=\int\mathbf{g}(w)\mathbf{H}(u)e^{-i2\pi wu}\,dw=\mathbf{G}(u)\mathbf{H}(u).\qquad(2\text{-}42)$$

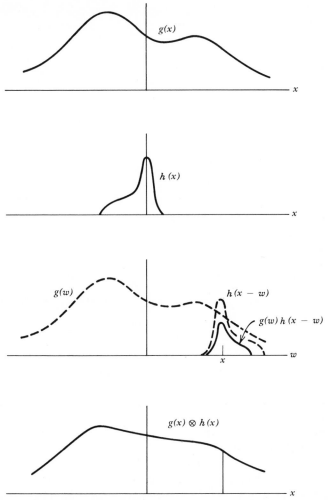

Figure 2-10 The value of $p(x)$ at any x is given by the area under the product curve $g(u)h(x-u)$.

The inverse transform of (2-40) and (2-41) yields

$$\mathbf{g}(x)\mathbf{h}(x) = \mathfrak{F}^{-1}[\mathbf{G}(u)\otimes\mathbf{H}(u)] \qquad (2\text{-}43)$$

and

$$\mathbf{g}(x)\otimes\mathbf{h}(x) = \mathfrak{F}^{-1}[\mathbf{G}(u)\mathbf{H}(u)]. \qquad (2\text{-}44)$$

Notice, however, that the only difference between the transform (2-1) and the inverse transform (2-2) is the sign in the exponent. This means that the only difference between the direct and inverse transform is a reversal of coordinates, that is, the exponent of the inverse transform can be written as exp $(i2\pi xu)$ or as exp $[-i2\pi(-x)u]$. For this reason, some authors do not distinguish between the two.

Correlation. The terms correlation and convolution are often used interchangeably. They are not, however, the same. The correlation of $g(x)$ with $h(x)$ is denoted by

$$q(x) = g(x) \star h(x) = \int g(w)h(x+w)\,dw, \qquad (2\text{-}45)$$

that is, the sliding function is not reversed. Therefore, only if h is an even function is the correlation the same as the convolution. Figure 2-10 could serve to illustrate correlation if the $g(u)$ function were not reversed. If complex functions are involved, the complex conjugate, g^*, must be used. If $g(x) = h(x)$, the term autocorrelation is applicable. When $g(x) \neq h(x)$, the term cross-correlation applies. If a substitution of variables is employed, (2-45) can be written as

$$q(x) = g(x) \star h(x) = \int g(w-x)h(w)\,dw. \qquad (2\text{-}46)$$

Note, from (2-45) and (2-46), that correlation is not a commutative operation but that if

$$q(x) = g(x) \star h(x),$$

$$q(-x) = h(x) \star g(x). \qquad (2\text{-}47)$$

The same conclusion can be reached by contemplation of Fig. 2-10. The distributive and associative laws do hold, however.

Cross-correlation, rather than being a smoothing operation, often results in a peaked distribution. This is especially true when one function is identical with a portion of the other. Figure 2-11 illustrates the correlation peak obtained when the identical parts of the functions are superimposed.

The Fourier transform of the product of two functions can be written as the convolution of the transforms of the two functions. In one case, however, the convolution can also be written as a correlation. Consider

$$\mathcal{F}[g(x)h^*(x)] = G(u) \otimes H^*(-u) \qquad (2\text{-}48)$$

where (2-15) has been used. Consideration of the integral definition of the convolution shows that (2-48) can also be written as

$$\mathcal{F}[g(x)h^*(x)] = G(u) \star H^*(u), \qquad (2\text{-}49)$$

so that we have a correlation rather than a convolution. The results are

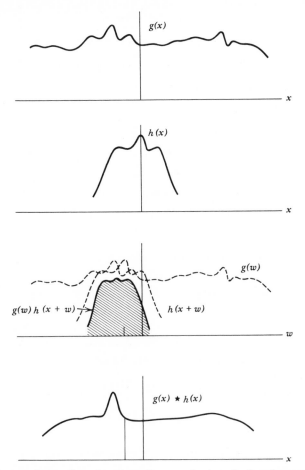

Figure 2-11 The correlation of two functions gives a peak where the functions are similar.

consistent because when the convolution reverses an already reversed function, we have a correlation.

Table of relations. Table 2-2 assembles some of the more useful relations governing convolution and correlations.

2.3   The Sampling Function

The same sampling function can be used in finding the Fourier transform of a sampled function and in obtaining the transform of a repeated distribution.

TABLE 2-2　BASIC RELATIONS GOVERNING CONVOLUTIONS
AND CORRELATIONS

$$g(x) \otimes h(x) = \int g(w)h(x-w)\,dw = \int g(x-w)h(w)\,dw$$

$$g(x) \star h(x) = \int g(w)h(x+w)\,dw = \int g(w-x)h(w)\,dw$$

$$g(x) \otimes h(x) = h(x) \otimes g(x)$$

$$f(x) \otimes [g(x) \otimes h(x)] = [f(x) \otimes g(x)] \otimes h(x)$$

$$f(x) \otimes [g(x) + h(x)] = f(x) \otimes g(x) + f(x) \otimes h(x)$$

$$q(x) = g(x) \star h(x)$$

$$q(-x) = h(x) \star g(x)$$

$$g(x) \otimes h(x) = g(-x) \star h(x)$$

$$[g(x) \otimes h(x)]^* = g^*(x) \otimes h^*(x)$$

Sampling function.　The sampling function is defined by

$$\mathrm{III}\left(\frac{x}{\sigma}\right) = |\sigma| \sum_m \delta(x - m\sigma) \qquad (2\text{-}50)$$

where σ is the spacing between samples and the integer m runs over all values. The symbol III is called the shah, comb, or sampling function. Some of the properties of the sampling function are

$$\mathrm{III}(-x) = \mathrm{III}(x) \qquad (2\text{-}51)$$

$$\mathrm{III}(x - \tfrac{1}{2}) = \mathrm{III}(x + \tfrac{1}{2}) \qquad (2\text{-}52)$$

$$\frac{1}{\sigma}\mathrm{III}\left(\frac{x}{\sigma}\right)g(x) = \sum g(m\sigma)\delta(x - m\sigma). \qquad (2\text{-}53)$$

When the shah function is used as a sampling function, the value of the sample exists only at the coordinates of the delta functions. Figure 2-12 illustrates the sampling of the function $g(x)$.

Convolution of $\mathrm{III}(x)$ with another function yields a replication of that function,

$$\frac{1}{\sigma}\mathrm{III}\left(\frac{x}{\sigma}\right) \otimes g(x) = \sum g(x - m\sigma). \qquad (2\text{-}54)$$

This property is easily proved by writing (2-54) using the integral representation of the convolution, the definition of $\mathrm{III}(x)$, and the sifting property of the delta function. The operation is illustrated in Fig. 2-13. Notice that if σ is not sufficiently large, the replicated functions overlap.

It can be shown that the Fourier transform of the III function is another III function. (See Appendix 1.) In particular,

$$\mathcal{F}\left[\text{III}\left(\frac{x}{\sigma}\right)\right] = \sigma\text{III}(\sigma u). \qquad (2\text{-}55)$$

This is a very useful relationship when sampled or replicated functions are to be Fourier transformed.

Transforms of sampled and replicated functions.   When a sampled function is Fourier transformed, the result is a replication of the spectrum of the nonsampled function. For example,

$$\mathcal{F}\left\{\mathbf{g}(x)\text{III}\left(\frac{x}{\sigma}\right)\right\} = \sigma\mathbf{G}(u)\otimes\text{III}(\sigma u). \qquad (2\text{-}56)$$

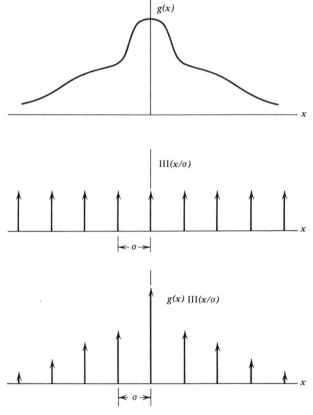

Figure 2-12   The sampling of $g(x)$ with $\text{III}(x/\sigma)$.

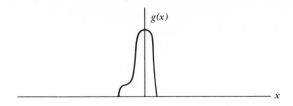

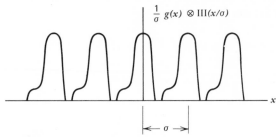

Figure 2-13   The replicating property, $g(x) \otimes (1/\sigma) \text{III}(x/\sigma)$.

The relationship described in (2-56) is illustrated in Fig. 2-14. Note that the closer the sampling in one domain, the greater the distance between replicated distributions.

A number of conclusions can be reached by study of Fig. 2-14. One is that if one section of the replicated spectrum were isolated, we could, by inverse transformation, recover the entire original distribution, $g(x)$. If the samples of $g(x)$ are not sufficiently close, the replications of the spectrum overlap. In this case, attempts to isolate one section result in inclusion of the tails of the other sections. Inverse transformation then yields the original function $g(x)$ with distortion. We can see that to avoid overlap of the spectra, the spacing $1/\sigma$ must be greater than $2u_c$ where u_c is the cutoff frequency. That means that the sample spacing, σ, must obey

$$\sigma < \frac{1}{2u_c}, \tag{2-57}$$

or there must be at least two samples per cycle of the highest frequency component of $g(x)$. If $\sigma = 1/2u_c$, the spectra just touch and $g(x)$ can be retrieved. In practice, a bit more space between the spectra is helpful! Also, of course, the function must be limited in its spectral content; it must be band limited.

If Fig. 2-14 is viewed in reverse, that is, if we consider $G(u)$ as being the

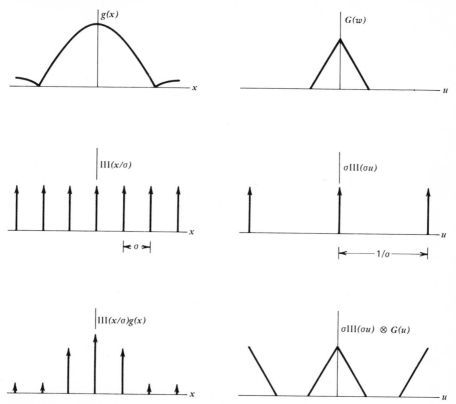

Figure 2-14 Sampling in one domain gives replication in the other.

original function and $g(x)$ its spectrum, we can see some of the properties of the transform of a periodic function. The periodic function is represented by $\sigma\text{III}(\sigma u)\otimes G(u)$ where the period of the function is $1/\sigma$. The spectrum is given by the distribution $\text{III}(x/\sigma)g(x)$, which shows the amplitudes and phases of the Fourier components. If the period $1/\sigma$ is increased, the spacing σ between the Fourier components decreases. In the limit, as $1/\sigma\to\infty$, $\sigma\to0$ and the Fourier spectrum becomes the continuous function $g(x)$.

Examples using Fraunhofer diffraction. We have seen that the Fraunhofer diffraction pattern of an aperture is related to the Fourier transform of the aperture distribution. In the examples that follow, the constants and phase terms multiplying the Fourier transform are neglected and only the transforms are given as representing the diffraction patterns.

First, consider an array of uniformly illuminated apertures represented by

$$\left[\text{II}\left(\frac{x}{a}\right) \otimes \text{III}\left(\frac{x}{\sigma}\right) \right] \text{II}\left(\frac{x}{b}\right), \tag{2-58}$$

and illustrated in Fig. 2-15a. The Fourier transform given by

$$ab\sigma \, \text{sinc} \, bu \otimes \left[\text{III}(\sigma u) \, \text{sinc} \, au \right] \tag{2-59}$$

is illustrated in Fig. 2-15b. Notice that the width of the individual lobes of the pattern is determined by the size of the array, the size of the envelope by the size of an individual element, and the distance between lobes by the

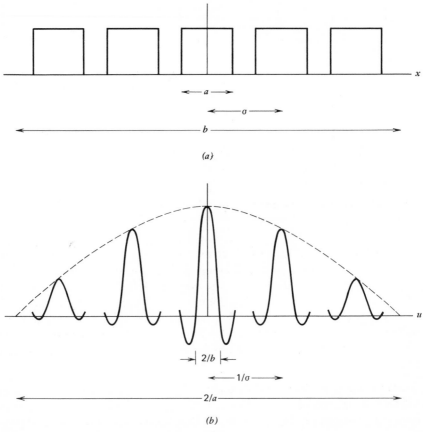

Figure 2-15 An aperture described by $[\text{II}(x/a) \otimes \text{III}(x/\sigma)]\text{II}(x/b)$ and its diffraction pattern described by $ab\sigma \, \text{sinc} \, bu \otimes [\text{III}(\sigma u)\text{sinc} \, au]$.

spacing of the elements. As the element size, a, increases, the envelope decreases in width until the zeros of the envelope fall on the secondary lobes, resulting in a single-lobed diffraction pattern whose width is determined by the size of the array.

The aperture described above can be considered as a rectangular grating as well as an array. The terminology applied to the diffraction pattern is then in terms of orders, not lobes. The central lobe is zero order, the lobes going out from the center of the pattern are positive and negative first order, and so on. The orders for a grating having transparent strips alternated with opaque strips can be described by (2-58) where $\sigma = 2a$. Equation (2-59) then describes the orders of a square wave grating or, alternatively, the spectrum of a square wave.

If the grating has a sinusoidal transmittance, it can be described by

$$t(x_1, y_1) = \left[\frac{1}{2} + \frac{M}{2} \cos\left(\frac{2\pi x_1}{\sigma} \right) \right] \Pi\left(\frac{x_1}{a} \right) \Pi\left(\frac{y_1}{b} \right) \qquad (2\text{-}60)$$

where its size is $a x b$, its spatial period is σ, and x_1 and y_1 are the coordinates in the plane of the grating. The factors of $\frac{1}{2}$ were inserted to limit the range of $t(x, y)$ from 0 to 1 when the modulation index M has its maximum value of 1. If the grating is illuminated by a uniform plane wave of unit amplitude, the wave, after passing through the grating, is described by $U(x_1, y_1) = t(x_1, y_1)$. The diffraction pattern is found by the use of $\mathcal{F}[U(x_1, y_1)] \equiv A(u, v)$.

$$A(u, v) = \left[\frac{1}{2}\delta(u) + \frac{M}{4}\delta\left(u - \frac{1}{\sigma} \right) + \frac{M}{4}\delta\left(u + \frac{1}{\sigma} \right) \right]$$

$$\otimes ab \, \mathrm{sinc}(au) \, \mathrm{sinc}(bv). \qquad (2\text{-}61)$$

By carrying out the operation of convolution, we obtain

$$A(u, v) = \frac{ab}{2} \mathrm{sinc}(bv)$$

$$\times \left\{ \mathrm{sinc}(au) + \frac{M}{2} \mathrm{sinc}\left[a\left(u - \frac{1}{\sigma} \right) \right] + \frac{M}{2} \mathrm{sinc}\left[a\left(u + \frac{1}{\sigma} \right) \right] \right\}. \qquad (2\text{-}62)$$

Notice that the spectrum has three components: one centered at $u = 0$, one centered at the spatial frequency $u = 1/\sigma$, and the other at $u = -1/\sigma$. The spatial frequency of the grating is $1/\sigma$. We can replace u and v by $x_2/\lambda z$

and $y_2/\lambda z$, respectively, where x_2 and y_2 are the coordinates of the plane of observation. The amplitude distribution in the Fraunhofer region is obtained by inserting the phase terms,

$$U(x_2,y_2) = \frac{ab}{i2\lambda z} \exp(ikz) \exp\left[\frac{ik(x_2^2+y_2^2)}{2z}\right] \operatorname{sinc}\left(\frac{by_2}{\lambda z}\right)$$

$$\cdot \left\{ \operatorname{sinc}\left(\frac{ax_2}{\lambda z}\right) + \frac{M}{2}\operatorname{sinc}\left[\frac{a(x_2-\lambda z/\sigma)}{\lambda z}\right] + \frac{M}{2}\operatorname{sinc}\left[\frac{a(x_2+\lambda z/\sigma)}{\lambda z}\right]\right\}.$$

$$(2\text{-}63)$$

The intensity is found by multiplying (2-63) by its complex conjugate. If the distributions described by the three sinc terms inside the braces are sufficiently separated, the cross terms are negligible, and

$$I(x_2,y_2) = \left[\frac{ab}{2\lambda z}\right]^2 \operatorname{sinc}^2\left(\frac{by_2}{\lambda z}\right)\left\{\operatorname{sinc}^2\left(\frac{ax_2}{\lambda z}\right)\right.$$

$$\left. + \frac{M^2}{4}\operatorname{sinc}^2\left[\frac{a(x_2-\lambda z/\sigma)}{\lambda z}\right] + \frac{M^2}{4}\operatorname{sinc}^2\left[\frac{a(x_2+\lambda z/\sigma)}{\lambda z}\right]\right\}. \quad (2\text{-}64)$$

A sketch of the intensity pattern is given in Fig. 2-16. Note that the

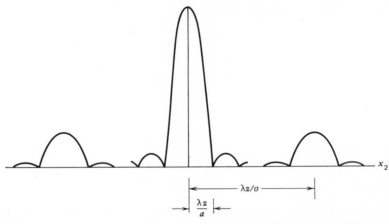

Figure 2-16 Cross section of the intensity pattern in the Fraunhofer region of a sinusoidal grating.

proportion of the intensity in the side lobes (or first orders) is a function of the square of the modulation index of the grating, and is a maximum when $M = 1$.

A comparison of the diffraction pattern of the sinusoidal grating with that of the square grating indicates that one could think of the square grating as being made up of a number of sinusoidal gratings having amplitudes as described by the coefficients obtained by a Fourier analysis of a square wave. The effect of the square grating is to produce higher orders in the diffraction pattern, each associated with a higher harmonic in the Fourier representation.

Phase apertures.   A wave can be diffracted by phase as well as amplitude or intensity perturbations, or, of course, a combination of these. For example, an aperture with a π radian phase difference between the two halves can be represented by

$$\Pi\left(\frac{x - a/2}{a}\right) + \Pi\left(\frac{x + a/2}{a}\right) e^{i\pi}. \tag{2-65}$$

The amplitude in the Fraunhofer region is

$$\frac{1}{i\lambda z} e^{ikz} \exp\left[i\pi\lambda z (u^2 + v^2)\right]\left[ae^{-i\pi au} \operatorname{sinc} au - ae^{i\pi au} \operatorname{sinc} au\right]$$

$$= -\frac{2a}{\lambda z} \exp\left[i\pi\lambda z (u^2 + v^2)\right] \sin \pi au \operatorname{sinc} au$$

$$= -\frac{2\pi a^2}{\lambda z} u \exp\left[i\pi\lambda z (u^2 + v^2)\right] \operatorname{sinc}^2 au. \tag{2-66}$$

Figure 2-17 is a plot of the magnitude of (2-66).

An example employing only phase perturbations is that of a phase grating. Consider a grating having a variation in thickness such that its transmittance is

$$\mathbf{t}(x,y) = \exp\left\{ik\left[\frac{M}{2} + \frac{M}{2} \cos\left(\frac{2\pi x_1}{\sigma}\right)\right]\right\} \Pi\left(\frac{x}{a}\right)\Pi\left(\frac{y}{b}\right) \tag{2-67}$$

where M describes the index of thickness variation. The product, kM, is the index of phase modulation, p. In this case, the analysis is easiest if the exponential function is written using the identity

$$\exp\left\{i\left[\frac{p}{2} + \frac{p}{2} \cos\left(\frac{2\pi x_1}{\sigma}\right)\right]\right\} = \sum_{m=-\infty}^{\infty} J_m\left(\frac{p}{2}\right) \exp\left[i\left(\frac{2\pi m x_1}{\sigma} + \frac{m\pi + p}{2}\right)\right]$$

$$\tag{2-68}$$

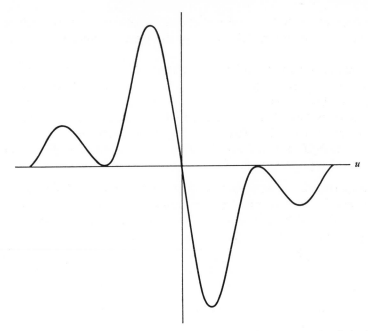

Figure 2-17 Magnitude of the amplitude pattern in the Fraunhofer region of a slit having a π radian shift in phase at the center.

where J_m is a Bessel function of the first kind, order m. If we assume that the grating is illuminated uniformly by a normally incident wave, $\mathbf{U}(x_1, y_1) = \mathbf{t}(x, y)$. The Fourier transform is given by

$$A(u,v) = [ab \operatorname{sinc} au \operatorname{sinc} bv] \otimes \sum_{m=-\infty}^{\infty} J_m\left(\frac{p}{2}\right) \exp\left[\frac{i(m\pi + p)}{2}\right] \delta\left(u - \frac{m}{\sigma}, v\right)$$

$$= \sum_{m=-\infty}^{\infty} ab \exp\left[\frac{i(m\pi + p)}{2}\right] J_m\left(\frac{p}{2}\right) \operatorname{sinc}\left[a\left(u - \frac{m}{\sigma}\right)\right] \operatorname{sinc} bv. \quad (2\text{-}69)$$

The amplitude of the field can be found by replacing u and v by $x_2/\lambda z$ and $y_2/\lambda z$ and by adding the modifying phase terms. (The $e^{ip/2}$ term is dropped as being due to the average phase delay through the grating.)

$$\mathbf{U}(x_2, y_2) = \frac{ab}{i\lambda z} \exp(ikz) \exp\left[\frac{i\pi(x_2^2 + y_2^2)}{\lambda z}\right]$$

$$\times \sum_{m=-\infty}^{\infty} \exp\left(\frac{im\pi}{2}\right) J_m\left(\frac{p}{2}\right) \operatorname{sinc}\left[\frac{a}{\lambda z}\left(x_2 - \frac{m\lambda z}{\sigma}\right)\right] \operatorname{sinc}\left(\frac{by_2}{\lambda z}\right) \quad (2\text{-}70)$$

If the overlap of the sinc terms is assumed to be negligible so that the cross terms can be dropped, the intensity pattern is described by

$$I(x_2, y_2) = \left(\frac{ab}{\lambda z}\right)^2 \sum_{m=-\infty}^{\infty} J_m^2\left(\frac{p}{2}\right) \text{sinc}^2\left[\frac{a}{\lambda z}\left(x_2 - \frac{m\lambda z}{\sigma}\right)\right] \text{sinc}^2\left(\frac{by_2}{\lambda z}\right).$$

$$(2\text{-}71)$$

The phase grating, therefore, has many more orders than the amplitude grating. For small values of p, the higher orders are negligible. Figure 2-18 shows the cross section of the amplitude pattern assuming an infinitely

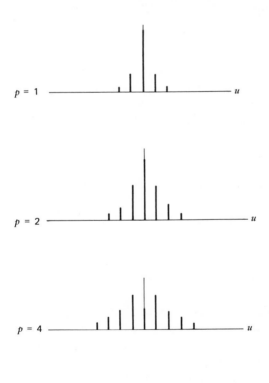

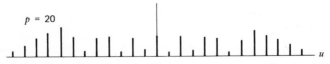

Figure 2-18 Intensity patterns in the Fraunhofer region of a very large consinusoidal phase grating for four values of p.

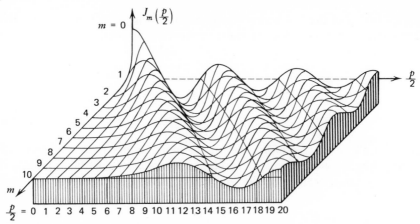

Figure 2-19 The values of $J_m(p/2)$ as a function of m and p. [After Jahnke and Emde, *Tables of Functions* (Dover Publications, 1945, p. 126.]

large grating (the sinc functions become δ functions) for four values of p. To better visualize the manner in which the different orders come into being refer to Fig. 2-19, which shows the values of $J_m(p/2)$ for various values of m and p. Notice that for some values the central order vanishes.

2.4   Table of Transforms

A summary of useful transforms is given in Table 2-3.

PROBLEMS

2-1. Prove that (2-4) holds for a separable function.

2-2. Use a change in coordinates to prove (*a*) the similarity theorem as expressed by (2-13) and (*b*) the shift theorem as expressed by (2-14).

2-3. Derive a shift theorem for the inverse transform.

2-4. Show that $\mathcal{F}[g^*(x)] = G^*(-u)$. Write $g(x)$ as $g(x) = RE(x) + RO(x) + iIE(x) + iIO(x)$ where $RE(x)$ denotes the real even part of $g(x)$, $RO(x)$ denotes the real odd part of $g(x)$, and $IE(x)$ and $IO(x)$ represent the imaginary even and odd parts. Transform both $g(x)$ and $g^*(x)$ using this representation and compare the results.

2-5. Show that if $g(x)$ is (*a*) real, (*b*) imaginary, (*c*) even, and (*d*) odd, the transform of $g^*(x)$ is (*a*) $G(u)$, (*b*) $-G(u)$, (*c*) $G^*(u)$, and (*d*) $-G^*(u)$.

2-6. Prove by change of variables that $\mathcal{F}[\delta(x-a)] = e^{-i2\pi a u}$.

2-7. Prove by change of variables that $\mathcal{F}\{\text{II}[(x-a)/b]\} = b e^{-i2\pi a u} \operatorname{sinc} bu$.

TABLE 2-3

Functions	Transforms	Functions	Transforms
$g(x)$	$G(u)$	$g(x)\otimes h(-x)$	$G(u)H(-u)$
$g(-x)$	$G(-u)$	$g(-x)\otimes h(-x)$	$G(-u)H(-u)$
$g^*(x)$	$G^*(-u)$	$g(x)\otimes h^*(-x)$	$G(u)H^*(u)$
$g^*(-x)$	$G^*(u)$	$g(x)\otimes h^*(x)$	$G(u)H^*(-u)$
$g(ax)$	$\dfrac{1}{|a|}G\left(\dfrac{u}{a}\right)$	$g(-x)\otimes h^*(-x)$	$G(-u)H^*(u)$
$g\left(\dfrac{x-a}{b}\right)$	$|b|e^{-i2\pi au}G(bu)$	$g(-x)\otimes h^*(x)$	$G(-u)H^*(-u)$
$g\left(\dfrac{x-a}{b}\right)e^{icx}$	$|b|e^{iac}e^{-i2\pi au}G\left[b\left(\dfrac{u-c}{2\pi}\right)\right]$	$g^*(-x)\otimes h^*(-x)$	$G^*(u)H^*(u)$
		$g^*(-x)\otimes h^*(x)$	$G^*(u)H^*(-u)$
$\dfrac{d}{dx}[g(x)]$	$i2\pi uG(u)$	$g^*(x)\otimes h^*(x)$	$G^*(-u)H^*(-u)$
$g(x)h(x)$	$G(u)\otimes H(u)$	$\delta(ax)$	$\dfrac{1}{|a|}\delta(x)$
$g(x)\otimes h(x)$	$G(u)H(u)$	$\text{II}\left(\dfrac{x}{a}\right)$	$|a|\operatorname{sinc}au$
$f(x)\otimes[h(x)g(x)]$	$F(u)[G(u)\otimes H(u)]$	$\text{III}\left(\dfrac{x}{\sigma}\right)$	$|\sigma|\,\text{III}(\sigma u)$
$g(x)h^*(x)$	$G(u)\otimes H^*(-u)$	$\text{circ}\left(\dfrac{r}{a}\right)$	$\dfrac{|a|J_1(2\pi a\rho)}{\rho}$
$g(x)h^*(x)$	$G(u)\star H^*(u)$	$g(x)\text{III}\left(\dfrac{x}{\sigma}\right)$	$|\sigma|G(u)\otimes\text{III}(\sigma u)$
		$g(x)\otimes\text{III}\left(\dfrac{x}{\sigma}\right)$	$|\sigma|G(u)\text{III}(\sigma u)$

2-8. Show, by substitution of variables in the integral definition of convolution, that convolution is a commutative operation.

2-9. Prove, using the integral representations, that

$$\mathcal{F}[g(x)h^*(x)]=G(u)\star H^*(u).$$

2-10. Evaluate the integral and show that $g(x)\otimes\delta(x-a)=g(x-a)$.

2-11. Discuss the differences between $\sum_m g(am)$ and $\text{III}(x/a)g(x)=\sum_m g(am)\delta(x-am)$.

2-12. The spectrum of $\text{III}(x/\sigma)g(x)$ is $\sigma G(u)\otimes\text{III}(\sigma u)$. The function $g(x)$ can be obtained by inverse transformation of only one of the replicated spectra. One of the spectra can be selected by use of a II function. Show that

$$\mathcal{F}^{-1}[\sigma G(u)\otimes\text{III}(\sigma u)]\frac{1}{\sigma}\text{II}\left(\frac{u}{\sigma}\right)=g(x).$$

2-13. Refer to (2-58) and (2-59) and to Fig. 2-15. What must be the relationship between a and σ to minimize the third order of a grating? What must the relationship be to minimize all even orders?

2-14. Find, using $\text{II}(x)$ and $x\text{II}(x)$ functions, the far field diffraction pattern of a one-dimensional aperture of unit width having the illumination linearly tapered across the aperture. The amplitude of the illumination is one-half at one end and unity at the other. How does this pattern differ from that of a uniformally illuminated aperture?

2-15. Examine the differences between $t(x,y)=[\text{II}(x/a)\otimes\text{III}(x/\sigma)]g(x)$ and $t(x,y)=\text{II}(x/a)\otimes[\text{III}(x/\sigma)g(x)]$ where $\sigma>a$. Sketch both functions and their Fourier transforms. Comment on the differences between the diffraction patterns produced if the transparencies above are illuminated with normally incident, uniform plane waves.

2-16. What are the differences between the four arrays described by
   (a) $\text{II}(x/a)\otimes[\text{III}(x/\sigma)\text{II}(x/b)e^{i2\pi\alpha x}]$
   (b) $[\text{II}(x/a)e^{i2\pi\alpha x}]\otimes[\text{III}(x/\sigma)\text{II}(x/b)]$
   (c) $[\text{II}(x/a)e^{i2\pi\alpha x}]\otimes[\text{III}(x/\sigma)\text{II}(x/b)e^{i2\pi\alpha x}]$
   (d) $[\text{II}(x/a)\otimes\text{III}(x/\sigma)]\text{II}(x/b)e^{i2\pi\alpha x}$
   where $b=m\sigma$, $m=$integer, and $\sigma>a$. Sketch the amplitudes and phases of the waves at the aperture and the Fraunhofer patterns of each. The $e^{i2\pi\alpha x}$ represents a linear phase shift.

2-17. What type of sampling process is represented by $\{\,[g(x)\otimes\text{II}(x/a)]\text{III}(x/\sigma)\}\otimes\text{II}(x/a)$?

2-18. Find the expression describing the energy distribution in the Fraunhofer region of the uniformly illuminated aperture shown in Fig. P2-18. Assume that $b=4a$ and $\sigma=1.5a$, and sketch the energy distribution along the $x_2/\lambda a$ and $y_2/\lambda z$ axes.

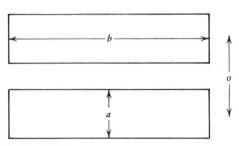

Figure P 2-18

2-19.   Assume that the square aperture in Fig. P2-19 has a square block as shown. Assume uniform illumination and find the expression describing the energy distribution in the far field. Assume $b = 2a$ and plot a cross section along the u axis. Compare with the energy distribution for the same aperture without the block. How does the diffraction pattern change as $b \to a$? Normalize the sketches to unity at the center of the pattern.

2-20.   A sinusoidal grating of width a and spatial frequency $1/\sigma$ cycles per unit length is illuminated over its entire area by waves of wavelength λ_1 and λ_2. How large must $1/\sigma$ be for the peak of the first diffraction order at one wavelength to fall at the null of the first order at the second wavelength? What determines how close λ_1 and λ_2 can be and the two first orders not overlap—the number of cycles per unit length of the grating or the total number of cycles?

2-21.   The resolution of a grating is defined as the mean wavelength $\bar{\lambda}$ divided by the minimum resolvable wavelength difference $\Delta\lambda$. The wavelengths are said to be barely resolved if the peak of one order for λ_1 falls on the first null of the corresponding order for λ_2. Use a sinusoidal grating as in problem 2-20 and show that the resolving power is proportional to the diffraction order used.

2-22.   Show that $\sigma\sum_m g(m\sigma) = \sum_m G(m/\sigma)$. Hint: Write $\mathcal{F}\left[\mathrm{III}(x/\sigma)g(x)\right] = \sigma\mathrm{III}(\sigma u) \otimes G(u)$ in the integral form and consider the point $u = 0$.

2-23.   Find the far field diffraction pattern of the distribution

$$\mathrm{III}(x/a)\mathrm{III}(y/a) + \mathrm{III}\left(\frac{x - 0.1a}{a}\right)\mathrm{III}(y/a).$$

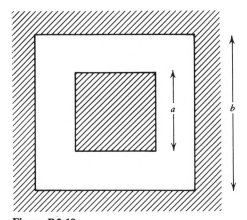

Figure P 2-19

2-24.   Use a computer or make by hand a random array of *pairs* of points. Illuminate the transparency and explain the order in the far field energy distribution [2-3].

REFERENCES

2-1.   R. Bracewell, *The Fourier Transform and its Applications* (McGraw-Hill Book Co., New York, 1965).

2-2.   A. Papoulis, *The Fourier Integral and Its Applications* (McGraw-Hill Book Co., New York, 1962).

2-3.   H. Lipson, *Optical Transforms* (Academic Press, New York, 1972), Ch. 10.

3

Wavefront Recording, Reconstruction, and Modulation

Surfaces of constant phase of a propagating wave are called phasefronts or wavefronts. These wavefronts can be modified in amplitude or phase (shape), and it is these modifications of the wavefront that carry information. If the modifications vary across the plane normal to the direction of propagation, rather than simply along the axis of propagation, spatial information can be recorded, rather than just temporal information. In order to process spatial information, various operations must be performed on the wavefront carrying the information. The most basic operations are the recording of the information on the wave, the reconstruction of a wave so that the information may be recalled, and the modulation of a wavefront so that new information can be impressed onto the wave. This chapter deals with these concepts and various techniques for implementing them. Further operations on the spatial information are discussed in the remaining chapters.

3.1 Spatial Information on a Wave

The impression of information onto an electromagnetic wave by modulating some parameter, such as amplitude, frequency, or phase, as a function of time is a familiar concept. Spatial information can also be impressed on the wave by amplitude, frequency, or phase modulation, but we must now be concerned with the whole wavefront, not just a small portion. That is, we must have an array of modulators and detectors. Figure 3-1 illustrates

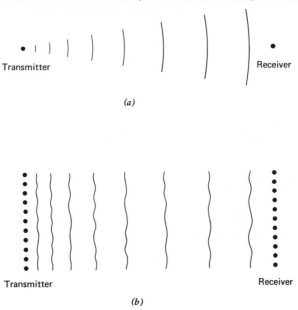

(a)

(b)

Figure 3-1 Generation and reception of waves. (a) Propagation of a wave with temporal information. The wave is described by $U(t) \cos[\omega t - \gamma r + \Psi(t)]$. (b) Propagation of a wave with no temporal variations (except ωt) but with spatial information. The wave is described by $U(x,y) \cos[\omega t - \gamma z + \Psi(x,y)]$.

the difference between the generation and detection of temporal and spatial information. Figure 3-1a represents, for example, a transmitter and receiver with a wave that is propagating from the former to the latter. The scalar representation of the wave, $U(t)\cos[\omega t - \gamma z + \Psi(t)]$, allows both amplitude and phase modulation. Gamma is simply the propagation constant, which will be dropped later. Because the entire wave has the same information, only one detector, or receiver, is necessary. Consider, however, the wave described by $U(x,y)\cos[\omega t - \gamma z + \Psi(x,y)]$ where $U(x,y)$ describes the amplitude variations in the x-y plane and $\Psi(x,y)$ represents the phase deviations from the same plane. The function U is proportional to the electric (or magnetic) field such that U^2 represents the intensity or power per unit area of the wave. Now there is no temporal modulation, but the amplitude and relative phase of the wavefront has been modulated. To transmit and detect this wave, an array of transmitters and detectors is needed. In this case, a single detector could be scanned because there is no time variation of the wave. Often, however, temporal modulation will also be present. Because of diffraction, the form of the wave changes as it propagates, but the spatial information is still preserved; it simply is represented differently.

As an example of spatial wavefront modulation, consider what happens when light strikes an object. (Assume, for the present discussion, a monochromatic wave. That is, ω has only one value.) The portions of the wave reflected from different regions of the object have different amplitudes because of variations in the color of the object, and the phasefront of the reflected wave is modulated as a function of x and y because of the surface roughness, contour, and refractive index variations. Therefore, a wave described by $U(x,y)\cos[\omega t - \gamma z + \Psi(x,y)]$ is generated. The wave propagates to the eye or camera and is acted upon by the lens so that an image is formed on the retina or film. If the object is moving, $U(x,y)$ and $\Psi(x,y)$ are functions of time as well as space. The conveyance of images by spatial modulation is but one example. We shall consider wavefront reconstruction, spatial information processing, and two-dimensional signal detection or pattern recognition in which other types of spatial information are involved.

The concept of wavefront reconstruction, or holography, is closely related to imaging. In fact, it was while considering the process of image formation that Dennis Gabor thought of wavefront reconstruction [3-1]. The line of reasoning used is best illustrated by reference to Fig. 3-2. Information concerning the object is obviously at the object, and, because an image is formed, the wave at the image must contain the necessary information. If the wave propagating from the object to the image carries the required information at those locations, the information must also be present at an intermediate plane, or indeed at every plane between the object and image. Admittedly, the information is in no readily recognizable form, but it is there. If, by some means, the wave containing this information could be recorded, then perhaps the same wave could be reconstructed at some other time and place and resume its propagation to form an image. This is the concept of holography; we now consider the means of recording the wave.

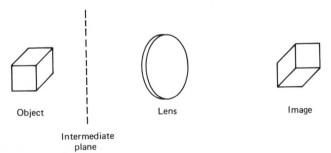

Object

Intermediate
plane

Lens

Image

Figure 3-2 Information concerning the object exists in every plane between the object and the image.

3.2 Wavefront Recording

If we simply place photographic film in the path of a wave, a piece of fogged film results with almost no information recorded. No reconstruction can be made. The reason for this failure is that film is an intensity, or energy, detector with a response time much longer than a temporal period of the wave. Consequently, the phase information is lost. The phase information can, however, be retained and recorded by heterodyne detection. Let the wave from the object be represented by

$$U(x,y)\cos[\omega_1 t - \alpha x - \gamma z + \Psi(x,y)], \tag{3-1}$$

where $U(x,y)$ and $\Psi(x,y)$ are the amplitude and phase variations in the x-y plane, and ω_1 is the temporal frequency in radians/sec. The propagation constants α and γ are related to χ, the angle between the direction of propagation of the wave and the z axis, by

$$\alpha = k \sin\chi$$

$$\gamma = k\sqrt{1-\alpha^2} = k \cos\chi \tag{3.2}$$

where $k = 2\pi/\lambda$ and λ represents the wavelength.

Now consider the introduction of a local oscillator wave (LO) described by

$$V(x,y)\cos[\omega_2 t - \zeta x - \mu z + \Phi(x,y)], \tag{3-3}$$

where $V(x,y)$ and $\Phi(x,y)$ are the amplitude and phase variations, ω_2 is the angular frequency of the local oscillator,

$$\zeta = k \sin\phi$$

$$\mu = k\sqrt{1-\zeta^2} = k \cos\phi, \tag{3-4}$$

and ϕ is the angle between the direction of propagation of the LO and the z axis (see Fig. 3-3). The addition of these waves and square law detection results in a signal described by

$$U^2(x,y)\cos^2[\omega_1 t - \alpha x - \gamma z + \Psi(x,y)] + V^2(x,y)\cos^2[\omega_2 t - \zeta x - \mu z$$

$$+ \Phi(x,y)] + U(x,y)V(x,y)\cos[(\omega_1 + \omega_2)t - (\zeta + \alpha)x - (\gamma + \mu)z$$

$$+ \Psi(x,y) + \Phi(x,y)] + U(x,y)V(x,y)\cos[(\omega_1 - \omega_2)t + (\zeta - \alpha)x$$

$$+ (\mu - \gamma)z + \Psi(x,y) - \Phi(x,y)], \tag{3-5}$$

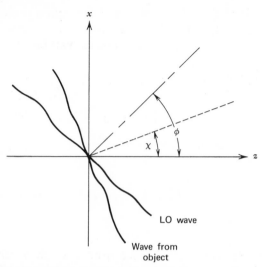

Figure 3-3 Introduction of a local oscillator wave to interfere with the wave from the object.

where it has been assumed that the polarization of the two waves is normal to the plane of incidence (the plane containing the vectors representing the directions of propagation of the two waves). The last two terms, obtained by the decomposition of the cosine product, are simply the sum and difference frequency terms. The cosine squared terms can be reduced to zero-frequency and double-frequency terms. For example,

$$U^2(x,y)\cos^2[\omega_1 t - \alpha x - \gamma z + \Psi(x,y)]$$

$$= \frac{U^2(x,y)}{2} + \frac{U^2(x,y)}{2}\cos[2\omega_1 t - 2\alpha x - 2\gamma z + 2\Psi(x,y)]. \quad (3\text{-}6)$$

Now, if the sum-frequency and double-frequency terms are dropped as varying too rapidly for the detector to respond, the resulting signal is described by

$$\frac{U^2(x,y) + V^2(x,y)}{2} + U(x,y)V(x,y)\cos[(\omega_1 - \omega_2)t + (\zeta - \alpha)x$$

$$+ (\mu - \gamma)z + \Psi(x,y) - \Phi(x,y)]. \quad (3\text{-}7)$$

That is, two zero-frequency terms and a difference-frequency term remain. Notice that both the amplitudes and phases of the original waves are preserved. If the amplitude and phase of one wave is known, the amplitude and phase of the other can be found. If electronic detectors are used, the difference or intermediate frequency (IF) can be selected, by selection of the local oscillator frequency, so that the information preserved on the IF can be processed easily. First, however, consider the case where $\omega_2 = \omega_1$, that is, the IF is zero. The result is a non-time-varying signal given by

$$\frac{U^2(x,y) + V^2(x,y)}{2} + U(x,y)V(x,y)\cos[\,(\zeta - \alpha)x$$

$$+(\mu - \gamma)z + \Psi(x,y) - \Phi(x,y)]. \qquad (3\text{-}8)$$

In this discussion, assume that $\chi = -\phi$, giving $\alpha = -\zeta$, $\gamma = \mu$, so that the expression becomes

$$\frac{U^2(x,y) + V^2(x,y)}{2} + U(x,y)V(x,y)\cos[2\zeta x + \Psi(x,y) - \Phi(x,y)].$$

$$(3\text{-}9)$$

The cosine term describes interference fringes which occur when the two waves are detected. A further simplification results if $V(x,y) = V_0$, and $\Phi(x,y) = 0$, that is, the reference beam is a uniform plane wave. The resulting expression

$$\frac{U^2(x,y)}{2} + \frac{V_0^2}{2} + V_0 U(x,y)\cos[2\zeta x + \Psi(x,y)] \qquad (3\text{-}10)$$

shows that the amplitude information is recorded as an amplitude modulation of the spatial carrier of frequency 2ζ radians per unit length and the phase information appears as a phase modulation of the same carrier. Actually, the spatial frequency should probably be referred to as a subcarrier, the carrier frequency being that of the optical wave. Figure 3-4 is a photograph of the interference fringes that occur when two plane waves of the same frequency interfere at an angle. Figure 3-5 shows the fringes that occur when a plane-wave reference interferes with an information-bearing wave.

Note that even if $V(x,y)$ of (3-10) is not a constant and $\Phi(x,y) \neq 0$, the desired information can still be extracted if V and Φ are known. Note also that if $\omega_1 \neq \omega_2$, the fringes move according to the relation

$$\cos[\,(\omega_1 - \omega_2)t + (\zeta - \alpha)x]. \qquad (3\text{-}11)$$

It is interesting to consider the moving fringes as seen by a single detector at some point on the x axis. There is no way to detect the presence of fringes, but the detector simply sees a wave varying in amplitude as $\cos[(\omega_1 - \omega_2)t]$. There is no reference from which the value of $(\zeta - \alpha)$ can be determined. On the other hand, an array of detectors sees, at any instant of

Figure 3-4 Photograph of the interference fringes formed when two uniform plane waves interfere.

time, fringes. The fringes are always there. The detectors may not record them, or may record them with reduced efficiency because of their motion, but they are always present.

Equation (3-10) would seem to indicate that two independent parameters $U(x,y)$ and $\Psi(x,y)$ are preserved even though only one parameter, intensity, is recorded. This of course, cannot be done without introducing other effects. As we shall see when we consider wavefront reconstruction, one additional effect is that a phase conjugate or secondary wave is reconstructed in addition to the original or primary wave. The discussion of resolution in Chapter 9 shows that the resolution required of the detector or recorder is double that needed to record only the original

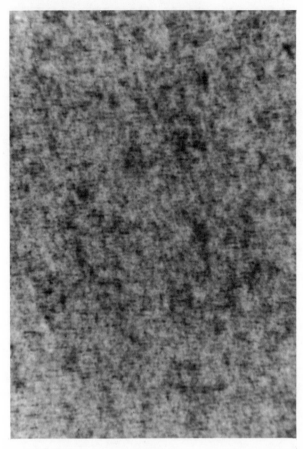

Figure 3-5 Photograph of interference fringes formed when a uniform plane reference wave intereferes with an information-bearing wave.

wavefront. If a two-parameter recorder is used (amplitude and phase separately recorded as by an array of detectors, for example) the secondary wave is not formed and a doubling of the resolution is not required.

If a volume detector is employed, such as a thick photographic emulsion or alkali halides [3-2], the fringes form in a volume. For example, if $\alpha \neq -\zeta$ and $\gamma \neq \mu$ in (3-8), variations with z occur. The angles of propagation of the waves in the medium of the detector are different from the angles outside the medium because of refraction. To indicate the difference in the angles, primes are placed upon the variables in the argument of the cosine. A plot of $\cos[(\zeta' - \alpha')x - (\gamma' - \mu')z] = $ constant is given in Fig. 3-6, showing the fringes in depth due to two interfering plane waves. In this case, because of

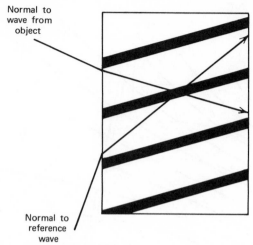

Normal to
wave from
object

Normal to
reference
wave

Figure 3-6 When two waves interfere in a volume recorder, interference fringes with depth are formed.

the variation with z, they no longer lie in a direction normal to the outer surface of the recording medium. It is easy to see that the interference planes lie along the bisector of the angle between the directions of propagation of the waves. If one of the interfering waves propagate in the negative z direction, the interference fringes can swing around until they are parallel to the outer surface of the recording medium. This occurs when $\chi = -\phi + \pi$, giving $\gamma = -\mu$ and $\alpha = \zeta$. Expression (3-8) becomes

$$\frac{U^2(x,y) + V^2(x,y)}{2} + U(x,y)V(x,y)\cos[2\mu z + \Psi(x,y) - \Phi(x,y)].$$

$$(3\text{-}12)$$

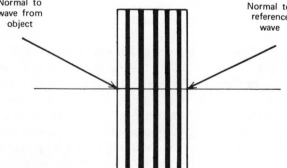

Normal to
wave from
object

Normal to
reference
wave

Figure 3-7 If the direction of propagation of one wave is reversed, the interference fringes in the volume can lie parallel to the surface of the volume.

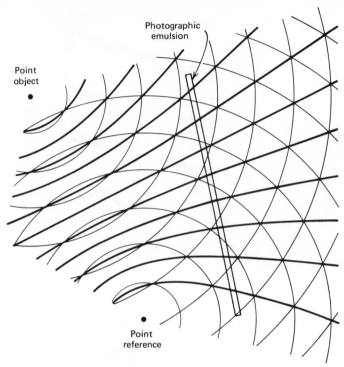

Figure 3-8 The standing wave maxima produced by the interference of waves from two point sources expose a photographic plate.

Figure 3-7 shows the fringes for two uniform plane waves traveling in accordance with the restrictions on α, ζ, and γ. If a photographic emulsion is used as the detector, planes of silver will appear after development. The resulting photographic plate will act as an interference filter causing certain wavelengths to be predominately reflected. This phenomenon will be discussed in more detail in the section on wavefront reconstruction (Section 3.3).

It is often helpful to consider waves coming from a point object. The interference fringes can be found by tracing the loci where the phases of the two waves are the same to within an integral number of wavelengths. If the reference wave also comes from a point, and is of the same frequency, a standing wave is formed with maxima falling on hyperbolas. Figure 3-8 shows the maxima intersecting a photographic plate. The exposure and development of the plate results in a transparency with regions of high transmittance as determined by the locations of the minima in the interference pattern. The manner in which this recording is used to reconstruct the wave is described in the next section.

3.3 Wavefront Reconstruction

We have seen how spatial amplitude and phase information can be recorded. Now the problem of retrieving the information is to be considered. If the recording is done with an array of detectors where the local oscillator wave is of a frequency different from that of the signal wave, the intermediate frequency is processed and its information is stored electronically. To reconstruct the same wave, an array of transmitters is required with the amplitude and relative phase of each transmitter being controlled appropriately. Electronically controlled wavefront reconstruction is discussed further in Chapter 10. This section is restricted to discussions involving photographic and other energy-sensitive recorders.

Holograms described in terms of conventional optical elements. The phenomenological description of holography can be continued with the hologram* described in terms of various types of optical elements. For example, the hologram has the properties of a lens except that the wave is intercepted and stored before being allowed to continue its propagation to form an image. The similarities between lenses and holograms have been examined in great detail [3-3], and some aspects of the analogy will be considered in Chapter 9.

Many of the earlier papers in holography considered the hologram to be a collection of Fresnel zone plates, each reconstructing a point of the image. That is, the object is assumed to be made up of point sources. Each illuminated point gives rise to a spherical wave which, when recorded in conjunction with a plane reference wave, produces a distribution similar to the far field amplitude pattern due to a pinhole. This recording is equivalent to the formation of a zone plate on the hologram. The hologram, then, is a collection of zone plates, one for each point. This interpretation has some merit and is discussed further in Chapter 9. For readers who are not familiar with the operation of a zone plate, reference to one of the introductory books on optics is suggested (e.g., [3-4]).

We are in a position to discuss holograms as being modulated gratings or modulated interference filters. In general, when a grating is illuminated by a collimated wave, numerous waves exit from the grating at different angles (again, see [3-4]). A cosine grating having a transmittance

$$\mathbf{t}(x) = C + \cos\frac{x}{\sigma} \qquad (3\text{-}13)$$

produces only three waves. Notice the similarity of (3-13) to (3-10). This similarity leads to the description of a hologram as a generalized or modulated diffraction grating. The amplitude of the grating is influenced

*The word *holography* is often used to describe the process of recording and reconstruction of a wavefront and the word *hologram* refers to the record.

by the amplitude of the wave to be reconstructed, and the position of the rulings of the grating is influenced by the phase of the wave. Consequently, the waves from the grating give rise to waves that form images, rather than plane waves. Diffraction of a wave by a grating gives rise to waves on either side of the axis of the grating that are 180° out of phase. (This is shown in Chapter 2.)

The concept of a hologram as a grating leads to the conclusion that the illumination of the hologram must be performed with a single wavelength or, in practice, a narrow band of wavelengths. Otherwise the grating effect causes not one but multiple images to be formed, with the result that a colored blur appears where the image should be. The tolerance on $\Delta \nu$ can be determined by the desired resolution in the image and the distance of the image from the hologram (grating). The angular spread of the waves due to $\Delta \nu$ multiplied by the image distance gives the amount of smear in the image. The illuminating wave need not be of the same wavelength as the original recorded wave. The effect of using a read-out wave with a different wavelength is discussed further in Chapter 9.

When a wavefront is recorded in depth, fringes such as are shown in Fig. 3-6 are obtained. Figure 3-7 illustrates the recording of two plane waves traveling in opposite directions. This figure can also represent an interference filter for a wave. Briefly, the operation of an interference filter is as follows. The incident wave is partially reflected from each surface of the filter and the spacing between surfaces is such that the reflected waves add in phase for a desired wavelength. Consequently, one wavelength is predominantly reflected while the other wavelengths are transmitted. The photographic recording of the interference of two waves forms such a filter. Different wavelengths cause interference planes of different spacings to be recorded. This effect allowed Lippmann to use black-and-white photographic emulsions in color photography in 1891 [3-5]. The same effect was described by Denisyuk [3-6] as being useful in holography at about the time that Leith and Upatnieks [3-7] were describing holography in communication theory terms.

The interference filter of Fig. 3-7 reflects only a uniform plane wave. But if the reflectivity of the filter varies across the face of the filter, spatial amplitude modulation of the wave is performed. The phase information is preserved by changes in the positions of the reflecting layers of the interference filter. See (3-12). If more than one wavelength is used to form this type of hologram, the same wavelengths are reflected when the recording is illuminated by white light. Because this technique can be used to record multiple wavelengths, it is used to reconstruct color images. Color holograms are discussed further in Chapter 9. A black and white photograph of the image from such a hologram illuminated by white light is shown in Fig. 3-9.

Figure 3.9 Black and white photograph of an image from a color hologram. The appropriate colors in the image were selected from white-light illumination by the modulated interference filter effects of the hologram. [From L. H. Lin, K. S. Pennington, G. W. Stroke, and A. E. Labeyrie, *Bell Syst. Tech. J.* **45**, 659–660 (1966).]

Secondary, or conjugate, waves.   Consider what happens when a wave described by $W_0 \cos \omega t$ illuminates a transparency described by (3-10). The resulting wave is given by

$$\frac{W_0}{2} [U^2(x,y) + V_0^2] \cos \omega t$$

$$+ W_0 V_0 U(x,y) \cos[\omega t + 2\zeta x + \Psi(x,y)]$$

$$+ W_0 V_0 U(x,y) \cos[\omega t - 2\zeta x - \Psi(x,y)]. \qquad (3\text{-}14)$$

The plane wave incident upon the wavefront recording, or hologram, is called the *read-out wave* and gives rise to the three waves described in (3-14). The first wave is modulated by the square of the amplitudes of both the plane reference wave and the object wave. This wave propagates along the same axis as the read-out wave and is usually ignored. The other two waves, because of the $\pm 2\zeta x$ terms, propagate along axes at angles $\pm 2\phi$ to the axis of the read-out wave. One wave is exactly the wave used in making the hologram (except for the new amplitude factor $V_0 W_0$ and the propagation in a different direction). This is the *original* or *primary* wave. The other wave is the complex conjugate of the recorded wave and is referred to as the *conjugate* or *secondary* wave. That is, we have reconstructed not only the original wave but also its complex conjugate. We will consider later the effect of using a read-out wave other than a uniform plane wave.

Holograms can be recorded without the reference wave being introduced at an angle to the wave from the object. This, in fact, was the manner in which Gabor [3-1] made the first hologram. In making this hologram, a transparent object was used (a transparency with dark lettering), allowing the nondiffracted portion of the wave to serve as a reference wave for the information-bearing portion diffracted by the lettering. Gabor mentioned that, in optics, a beam splitter could be used so that the reference could be altered, but because he was interested in electron optics, he considered only the case where the illuminating wave also serves as a reference wave [3-8]. An off-axis reference, a reference wave at an angle to the wave from the object, was not practical in optical holography until the development of the laser [3-7]. The difficulty with the hologram recorded with an on-axis reference wave is that both images lie on the same axis as the illuminating wave.

Real and virtual images.   Both of the off-axis reconstructed waves described by (3-14) have physical meanings. The wavefront that is reconstructed in the same form as recorded produces the primary image, and the conjugate wave forms the secondary image. The primary image is *normally* a virtual image and the conjugate wave usually forms a real image.

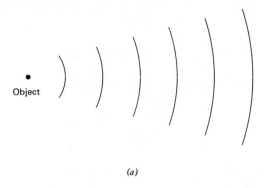

Object

(a)

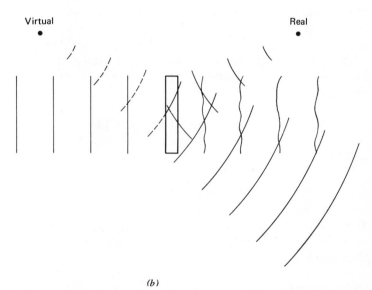

Virtual

Real

(b)

Figure 3-10 (a) The wave from a point object. (b) The reconstructed waves forming real and virtual images.

However, the type of image formed depends on the curvature of the reference and read-out waves. An example using a simple object, a point, easily illustrates the formation of real and virtual images. The wave from a point is represented in Fig. 3-10a. This wave falls on the hologram and is recorded with a uniform plane reference wave. The reconstructed wave shown in Fig. 3-10b is described by (3-14) if $\Psi(x,y)$ takes the form descriptive of a spherical wave. Two off-axis waves are reconstructed, in addition to the wave on the same axis as the read-out wave. (If the recording is not linear with intensity, more waves are formed.) One of these waves is, aside from a constant multiplier and a tilt, identical to the recorded wave and, if this wave enters the eye of a viewer, the wave appears to be coming from a point in the same location as the original point source.* This is the virtual image that can easily be seen by an observer. The conjugate wave focuses to an image on the other side of the hologram an equal distance from the hologram. This image can be used to illuminate a screen or expose photographic film. It is the real image.

Complex notation.   The expressions describing the waves and the type of modulation have thus far been written in terms of cosines. This was done because the temporal variations are more clearly shown. When the reference wave and the wave from the object are of the same frequency, it is more convenient to use the complex notation and to drop the time-varying portion. That is, $Re[e^{-i(\omega t - \gamma z)}]$ or, dropping the time variation and the indication "real part of," $e^{i\gamma z}$ will be written rather than $\cos(\omega t - \gamma z)$. One advantage of this notation is that we may very easily drop the double-frequency term. For example, consider the wave from the object described by

$$U(x,y)\exp\{i[\alpha x + \gamma z - \Psi(x,y)]\} \qquad (3\text{-}15)$$

and the reference wave

$$V(x,y)\exp\{i[\zeta x + \mu z - \Phi(x,y)]\}. \qquad (3\text{-}16)$$

The square of the absolute value of the sum of (3-15) and (3-16) is recorded by the hologram (aside from the terms describing the photographic process) as

$$U^2(x,y) + V^2(x,y)$$
$$+ U(x,y)V(x,y)\exp\{i[(\zeta-\alpha)x + (\mu-\gamma)z + \Psi(x,y) - \Phi(x,y)]\}$$
$$+ U(x,y)V(x,y)\exp\{-i[(\zeta-\alpha)x + (\mu-\gamma)z + \Psi(x,y) - \Phi(x,y)]\}$$

$$(3\text{-}17)$$

---

*In this case the point appears to be in the same location as the original point. Chapter 9 deals with cases where the read-out wave differs from the reference wave, causing, among other effects, magnification of the apparent distance to the point.

or

$$U^2(x,y) + V^2(x,y)$$
$$+ 2U(x,y)V(x,y)\cos[(\zeta - \alpha)x + (\mu - \gamma)z + \Psi(x,y) - \Phi(x,y)]. \quad (3\text{-}18)$$

Compare this expression with (3-8). The only difference is the factor of 2, which we will ignore. The 2 arises because the average value of $\cos^2$ is $\frac{1}{2}$.

In describing the process of wavefront reconstruction using complex notation, the expression for the read-out wave is multiplied by (3-17). The resulting expressions describe the on-axis wave and the two off-axis images. While (3-17) rather compactly shows the gist of the three reconstructed waves, (3-18) is purely real and better shows the means by which the amplitude and relative phase information are preserved by the hologram.

General read-out waves. Consider the case where the read-out wave has some general amplitude variation and a general phase front described by $W(x,y)\exp(i\Xi x)\exp[i\Omega(x,y)]$, where Ξ is the propagation constant in the x direction. Let the wave from the object be described by $U(x,y)\exp(i\alpha x)\exp[i\Psi(x,y)]$ and the reference wave by $V(x,y)\exp(i\zeta x)\exp[i\Phi(x,y)]$. The two reconstructed waves of interest are described by

$$U(x,y)V(x,y)W(x,y)\exp[i(\alpha - \zeta + \Xi)x]$$
$$\times \exp\{i[\Omega(x,y) + \Psi(x,y) - \Phi(x,y)]\} \quad (3\text{-}19)$$

and

$$U(x,y)V(x,y)W(x,y)\exp[i(\zeta - \alpha + \Xi)x]$$
$$\times \exp\{i[\Omega(x,y) - \Psi(x,y) + \Phi(x,y)]\}. \quad (3\text{-}20)$$

The expression (3-19) represents the primary reconstructed wave related to the original wave and (3-20) represents the secondary wave related to the conjugate of the original wave. Note that the resulting wave described by (3-20) is not the conjugate of the wave described by (3-19). A number of conclusions can be drawn from (3-19) and (3-20). We find, for example, that a general reference wave can be used and the phase structure of the primary wave can be reconstructed accurately if the read-out wave is identical to the reference wave. The amplitude structure of the wave can also be reconstructed accurately if $W(x,y) = V(x,y)^{-1}$. Errors in the amplitude structure are not so important as errors in phase, however, and in practice, $W(x,y)$ need not equal $V(x,y)^{-1}$. If a wave from another object is used as a general reference wave, the same wave can be used as a read-out wave if the hologram is carefully placed back onto its original

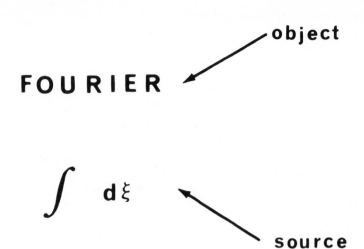

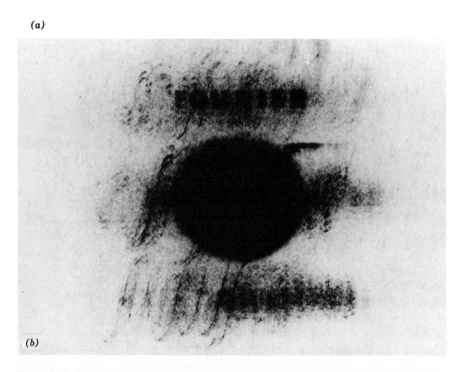

Figure 3-11 Image reconstructed when the read-out wave is the same generalized wave that was used as a reference. (*a*) Object and reference or source. (*b*) Reconstructed image formed when the hologram is illuminated by the same source. [From G. W. Stroke, R. Restrick, A. Funkhouser, and D. Brumm, *Phys. Lett.* **18**, 274 (1965).]

position after development. Any misalignment is equivalent to changing the phase distribution $\Omega(x,y)$. Figure 3-11 shows an image reconstructed by use of the same general read-out wave as was used as a reference. To retrieve the conjugate wave, $\Omega(x,y)$ must be known (see problem 3-4).

Wave reconstruction using volume recordings. An important difference between surface holograms (no depth information in the detector) and volume holograms (recorded with a detector capable of recording depth information) is that the conjugate or secondary image can be eliminated. If the wavefront recording is such that it is described by (3-12), then we are no longer attempting to record two parameters, wave amplitude and phase, by detecting only one, intensity, for a given point x_i,y_i. We now have additional information stored in the z direction. As a consequence, only one wave is reconstructed.

Figure 3-8 is helpful in understanding the effect of a volume or thick recording medium. Let the recording medium be placed between the reference and object point sources. Assume that a photographic plate, for example, is normal to the line connecting the sources, and that the emulsion is thick enough to record a maximum of the standing wave pattern. Development of this recording results in the formation of a partial reflector (deposited silver in this case) along the surface of the standing wave maxima. Figure 3-12 shows the record of one of the maxima of Fig. 3-8 and its illumination by the point reference. The wave reflected from the hologram is identical to the object wave. *Only one wave, the reflected object*

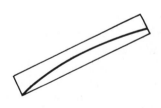

Figure 3-12 Illumination of the reflection surface of a hologram by one point source produces an image of the other point.

wave, comes from the hologram. This is true whether or not the recording is linear. Illumination of the hologram by the wave from the point object causes a reconstruction of the reference wave. In the case of a more complex object, the amplitude distribution of the wave will be nonuniform, producing regions in the hologram where the surfaces of silver are less dense than in others. This produces amplitude modulation of the read-out wave so that the read-out wave takes on the same spatial amplitude distribution as the original object wave. The phase of the read-out wave is modified by the shape of the reflecting surface, imparting the correct phase to the wave. The result is one reflected wave, the primary wave.

Equation (3-20) can be used to see that if the read-out wave is the conjugate of the reference wave, the conjugate of the object wave is reconstructed. From Fig. 3-8 we see that the conjugation of the reference wave consists of a simple reversal of the direction of propagation of the wave. The phase $\Phi(x,y)$ is then measured with respect to a reversed direction of propagation so that, even though the shape of the wavefront is the same in the figure, the phase is now described by $-\Phi(x,y)$. It is interesting to note that use of the optimum illumination angles for the primary and the conjugate images results in the illumination of the opposite sides of the surfaces formed in the recording medium. A region of the reflecting surfaces that causes a phase advance in that region of the hologram for one image causes a phase retardation for the other.

Even if the reference wave is not introduced in such a manner as to cause the fringes to be parallel to the surface of the detecting surface, similar results are seen. If $\alpha = -\zeta$, for example, the fringes are normal to the surface of an emulsion and both waves can be reconstructed. If the hologram is illuminated by a wave propagating normal to the surface of the hologram, both the original and conjugate waves are reconstructed with equal intensity. If the read-out wave is introduced at an angle, however, one wave, and consequently one image, can be made brighter at the expense of the other. The primary image has maximum intensity when the read-out wave comes from the same angle as did the reference wave. We can see in either Fig. 3-6 or Fig. 3-8 that at this angle the light is *reflected* by the surfaces in the recorder, along the axis of propagation of the image wave. The angle at which the image appears with respect to the direction of illumination is determined by the fringe spacing on the hologram. The angle of maximum wave amplitude and the angle at which the image is located coincide only under the conditions described above. The thicker the recording medium, the larger the reflecting surfaces, and the more critical the alignment conditions become. The effect of the thickness of the recorder is discussed further in Chapter 6.

In the case of plane or spherical reference and read-out waves, the primary image could also be reconstructed by turning the hologram so that the read-out wave arrives at the angle of the wave from the *object* plus 180°. Two problems arise if this is done. Normally the hologram is recorded so that the reconstructed image appears centered in the hologram aperture (framed) when the hologram is rotated to provide maximum image brightness. If the hologram is turned as described above, the image is no longer framed properly. The other problem is that the entire image cannot be brightly formed at the same time. This is because object waves from different angles interfere with the reference wave to produce surfaces within the hologram at different angles. Figure 3-13 illustrates the recording and reconstruction of two plane waves from an object. A plane reference wave is assumed. Figure 3-13*a* shows the geometry for the recording of the hologram and the surfaces of deposited silver (for the case of a photographic emulsion) resulting from the interference between the two object waves and the reference wave. There is also interference between the two object waves, but it is neglected here. The planes within the hologram bisect the angle between the interfering waves. The effects of refraction upon the waves are not shown. Figure 3-13*b* illustrates the optimum illumination of the hologram producing maximum brightness of both waves. If the read-out wave is introduced 180° from the *average* direction of the object waves, a compromise is necessary. Figure 3-13*c* shows this type of illumination and the impossibility of illuminating both sets of surfaces within the hologram such that both waves are reconstructed with maximum amplitude. Remember that the angles at which the diffracted wave must propagate are fixed by the spacing of the surfaces within the hologram.

Advantages of holograms for imaging.   Some advantages of holograms are readily apparent. Because the reconstructed wave can be identical to the recorded wave, all of the original information is preserved. There is a requirement to refocus the eyes or camera to properly view objects at different depths, and all parallax is preserved. Figure 3-14 shows these effects as seen by a camera in photographing the virtual image. When an image formed by wavefront reconstruction is viewed, the effect is the same as looking through a window the size of the hologram at the object illuminated with light from the laser.

In addition to the three-dimensional aspects of the image, the large range of illumination intensity or reflectivity of the object is maintained. In a conventional image recording system, the film recording the image has a limited dynamic range. This restricts the range of intensities seen in viewing the negative, for example. In the case of a holographic recording, the bright parts of the object illuminate the entire hologram. Upon reconstruction of the wave, the entire hologram can direct energy to the bright

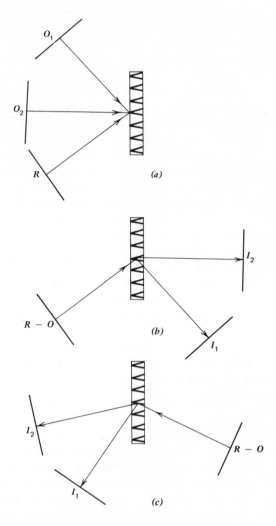

Figure 3-13 The recording and reconstruction of object waves O_1 and O_2 with the reference wave R and read-out wave R-O. (a) Recording planes in the hologram. (b) Reconstruction of both object waves with maximum brightness. (c) Reconstruction of object waves, but not with brightness of both maximized.

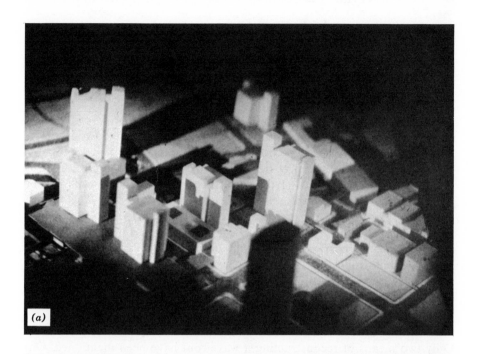

Figure 3-14 Photographs of reconstructed virtual image showing the effects of changes in camera focus and position.

portion of the image. In this manner, the same photographic emulsion can form an image with a much greater range of brightness when it is used to make a hologram than when it is used to make a conventional photograph. When the image is formed by wavefront reconstruction, the range of brightness in the image is related more to the size of the hologram than to the dynamic range of the emulsion.

Another property of holographic imaging which is well known is that only a portion of the hologram is necessary to form an image. Any surprise at discovering this is due to comparing the hologram to a photograph rather than to a lens which has a memory. Tear off a portion of a photograph, and a portion of the image is lost; break off a portion of a hologram and only the resolution and brightness are affected. The latter is analogous to reducing the aperture size of a lens. This will become more obvious after study of Chapter 9. The property that the entire image remains even if a portion of the hologram is damaged makes holographic image storage useful in information storage. This topic is considered in Chapter 10.

3.4   Wavefront Modulation

Only two means of reconstructing a wavefront have been mentioned. One is the possibility of controlling amplitude and phase in an array of transmitters to directly form a wavefront described by $U(x,y) \exp[i\Psi(x,y)]$. The other is to cause the transmission of a photographic negative (or positive—see problem 3-2) to vary in accordance with a distribution described by the modulated spatial cosine or carrier frequency as described by (3-9). (Neglect any effects of recorder depth.) The amplitude variations in the multiplier of the cosine govern the amplitude variations of the reconstructed wave and the spacing or spatial phase variations of the cosine govern the relative phase of the reconstructed wave.

Techniques that employ transmission-variable materials modulate only the amplitude of the wavefront. That is, immediately behind the wavefront modulator only the relative amplitude of the wave is affected. This may, at first, seem to contradict the statement that the original wavefront, with both amplitude and phase information, is reconstructed. What is reconstructed, however, is the original wave and its conjugate, the sum of which is real. Immediately behind the hologram, therefore, any phase variation is due to the read-out wave only. The resulting wave propagates and can be separated into three parts—the wave associated with the original wave, its phase conjugate, and the wave described by the U^2 and V^2 terms of (3-14). Wavefront modulation of this type is spatial amplitude modulation.

In the case of temporal modulation of a wave, either amplitude or phase modulation can be used to impress the signal upon the wave. Similarly, either spatial amplitude modulation or spatial phase modulation can be employed. Spatial phase modulation does not refer to using only the phase information $\Psi(x,y)$ to reconstruct a wave but, instead, refers to the manner in which information concerning both the amplitude and phase are impressed on the wave by varying the phase front. In some cases, phase modulation is preferable to amplitude modulation in image formation by holography. The image is brighter, a greater portion of the light incident upon the hologram goes into image formation, and a number of materials are suitable for use as phase modulators. There are disadvantages to phase modulation (PM) holograms, however. As many communication texts point out, phase modulation can produce intermodulation noise [3-9]. If the phase index of modulation is small, the intermodulation noise is small [3-10, 3-11].

Phase modulation of a wave can be achieved by varying the thickness of a material or the index of refraction or both. Means of obtaining phase modulation are discussed in Chapter 6. Any of the means can be used to provide a transparency with a transmittance

$$t(x,y) = D_0 \exp\left(ip \left\{ U^2(x,y) + V_0^2 + 2U(x,y) V_0 \cos[2\zeta x + \Psi(x,y)] \right\} \right)$$

$$(3\text{-}21)$$

where p is the modulation index. If the index of modulation is small, the exponential can be approximated by the first few terms of the power series expansion. The result is

$$D_0 \exp(ip \{ \cdots \}) = D_0 \left(1 + ip \{ \cdots \} - \frac{p^2 \{ \cdots \}^2}{2} + \cdots \right) \quad (3\text{-}22)$$

where, to conserve space, the terms in braces have not been written, but are the same as in (3-21). If the value of p is sufficiently small, the higher-order terms can be neglected and (3-22) becomes

$$D_0 \exp(ip \{ \cdots \})$$

$$\approx D_0 \left(1 + ip \left\{ U^2(x,y) + V_0^2 + 2U(x,y) V_0 \cos[2\zeta x + \Psi(x,y)] \right\} \right)$$

$$= D_0 + ipD_0 [U^2(x,y) + V_0^2] + i2pD_0 V_0 U(x,y) \cos[2\zeta x + \Psi(x,y)].$$

$$(3\text{-}23)$$

The only differences between the cosine term of (3-23) and that of (3-10), an expression assuming amplitude modulation, are the multipliers i, p, and

D_0. The constant D_0 is of no consequence; p, the index of modulation, influences the amplitude of the reconstructed wave; and i shows that the reconstructed wave is 90° out-of-phase with the read-out wave. When an intensity detector is employed, the phase difference has no effect because of the physical separation of the reconstructed waves.

If the modulation index is not small, the higher-order terms of (3-22) must be included. It is easily seen by multiplying out the p-squared term, for example, that in addition to waves being reconstructed at twice the angle of the first reconstructed wave, there are terms that describe additional waves propagating at the same angle as the desired wave. This gives rise to noise inherent to the phase modulation process. For example, one wave appearing at the angle of the desired wave is described by $V_0 U^3(x,y)$ $\exp[i\Psi(x,y)]$. The phase is correct for this wave contributed by the p^2 term but the amplitude is not. Other waves related to higher-order terms in the expansion of (3-22) have incorrect phases. In general, amplitude errors are not so serious as phase errors. In the case of the recording of two plane waves, there is no problem. The phases may be such that the extra waves subtract from the one produced by the linear terms but no distortion is introduced. This is the situation with a sinusoidal phase grating where a large modulation index is permissible. (See Figs. 2-18 and 2-19.) All of the higher-order terms cause contributions at the same angle as the desired wave. A total phase deviation of from 0.2 to 0.6 radians has been found to be near the optimum for minimizing noise and maximizing the amount of energy forming the image [3-11, 3-12]. The exact value of p permissible obviously depends on the form of $U(x,y)$. If the amplitude is almost uniform, little distortion is introduced into the image. Often the amplitude information is discarded and only the phase information is used in making a hologram. This is particularly effective if the wave is from a diffuse reflector. Such phase information holograms are discussed in Chapter 9. Because a larger value of p results in more energy being directed to the image, p should be as large as possible within the limits set by noise.

A different analysis is used for holograms operating as reflectors. In that case, the phase information is impressed onto the read-out wave by the variations in shape of the reflecting surface. The surface in the case of a phase hologram reflects because of the different indices of refraction of the surface and the surrounding medium. The amplitude information is impressed by the variation in index of refraction of the surface, thereby changing the reflectivity in selected locations.

Modulators. The most common modulating medium for wavefront reconstruction is photographic plate. This, of course, gives amplitude modulation but also causes phase modulation (PM) due to variations in refractive index and thickness of the emulsion. If the plate is bleached, the amplitude modulation is removed and a PM hologram is obtained [3-12]. The phase modulation is due to both index of refraction changes within the emulsion and changes in the emulsion thickness. Numerous studies of the phase structure of bleached photographs have been made in which these effects have been separately measured ([3-13] through [3-17]). The effects of variations in emulsion thickness can be eliminated by immersing the photographic plate into a fluid that matches the index of refraction of the emulsion. In addition to photographic materials, phase holograms have been constructed using transparent gelatine [3-18], dichromated gelatin [3-19], and lithium niobate [3-20]. Any material that can modify the relative amplitude or phase of a wave can serve as a holographic material. These include photochromics, thermoplastics, oil films, and numerous crystals. Other materials are mentioned in Chapters 9 and 10. Emphasis will not be on any particular material, however, because the development of new materials is too rapid. Rather, *types* of modulators and their effects will be the primary interest in all of the following discussions of wavefront modulators. Further discussions of the types of holograms and recording techniques and materials appear in Chapters 6 and 9.

PROBLEMS

3-1. Equation (3-8) describes the detected interference fringes when both waves are polarized normal to the plane containing the normals to the reference wave and the wave from the object. Use vector notation to find the appropriate representation when the waves are polarized parallel to this plane. How are the fringes described when the polarization of one wave has an arbitrary relation to the polarization of the other?

3-2. Assume that the transmittance of a hologram is described by $t(x,y)$ and that a negative of the hologram can be described as having a transmittance of $A_0 - t(x,y)$. Let $t(x,y) = K\{U^2(x,y) + V_0^2 + 2V_0 U(x,y)\cos[2\zeta x + \Psi(x,y)]\}$. Compare the images formed by positive and negative holograms when, during recording, the reference wave is introduced (a) along the same axis as the object wave and (b) at an angle sufficiently different from the angle of the object wave to cause separation of the images. In particular, in what case, if any, is a negative image formed?

3-3. Assume that in recording a hologram, an off-axis plane wave is introduced in the y-z plane as shown in Fig. P3-3 as a reference wave. During the reconstruction process, the read-out wave is directed along the same axis as was the reference wave. What is the effect on the images of rotating the hologram 180° about the (a) x axis, (b) y axis, (c) z axis while viewing? Neglect the effects of the thickness of the hologram.

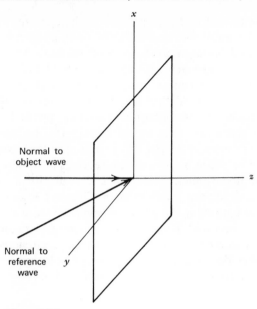

Figure P3-3

3-4. Assume that a curved reference wave described by $\exp[ik(x^2+y^2)/\rho]$ is used in forming a hologram. What read-out wave is necessary to reconstruct, as accurately as possible, (a) the virtual image and (b) the real image?

3-5. If the hologram is analyzed as a grating, find the spread of wavelengths, $\Delta\lambda$, which could appear in the read-out wave if the image is formed 10 cm from the hologram, at an angle of 30° from the direction of propagation of the read-out wave, and must resolve a spot 100 μm in diameter. Assume that the mean wavelength is 0.5 μm.

3-6. Refer to Fig. 3-6. Show that $\zeta=\zeta'$ and $\alpha=\alpha'$ and that consequently, the separation of the fringes in the x direction can be calculated using the external angles.

3-7. Obtain a general expression for the normal separation of the surfaces shown in Fig. 3-6. Write the answer in terms of χ, ϕ, λ, and the index of refraction, n, of the recording material.

3-8. Assume that two plane waves of wavelength 0.5 μm are incident upon a recording medium of index $n=1.5$. The bisector of the angle between the normals of the waves is normal to the medium and $\chi=-\phi=15°$. Find the fringe spacing in cycles per millimeter.

3-9. Describe, using equations of the form of (3-17), the geometry used in illuminating this type hologram so that the conjugate image is obtained. Give a phenomenological description of the manner in which the recorded planes physically cause either a primary or a conjugate image. Assume that the illumination is with the same wavelength as used in the recording of the hologram.

REFERENCES

3-1. D. Gabor, "A New Microscopic Principle," *Nature* **161**, 777 (May 15, 1948). This paper and Gabor's two fundamental papers on holography are reproduced as appendices in the book by G. W. Stroke, *Introduction to Coherent Optics and Holography* (Academic Press, New York, 1966).

3-2. P. J. Van Heerden, "Theory of Optical Information Storage in Solids," *Appl. Opt.* **2**, 393–400 (1963).

3-3. R. W. Meier, "Cardinal Points and the Novel Imaging Properties of a Holographic System," *J. Opt. Soc. Am.* **56**, 219–223 (1966).

3-4. F. A. Jenkins and H. E. White, *Fundamentals of Optics* (McGraw-Hill Book Co., New York, 1957).

3-5. M. Born and E. Wolf, *Principles of Optics* (MacMillan Co., New York, 1964).

3-6. Y. U. Denisyuk, "Photographic Reproduction of the Optical Properties of an Object in Its Own Scattered Radiation Field," *Sov. Phys. Dok.* **7**, 543–545 (1962).

3-7. E. N. Leith and J. Upatnieks, "Reconstructed Wavefronts and Communication Theory," *J. Opt. Soc. Am.* **52**, 1123–1130 (1962).

3-8. D. Gabor, "Microscopy by Reconstructed Wavefronts," *Proc. Roy. Soc.* **A 197**, 454–487 (1949).

3-9. H. S. Black, *Modulation Theory* (D. Van Nostrand Co., Inc., Princeton, N.J., 1960), Ch. 12.

3-10. W. T. Cathey, Jr., "Spatial Phase Modulation of Wavefronts in Spatial Filtering and Holography," *J. Opt. Soc. Am.* **56**, 1167–1171 (1966).

3-11. J. C. Urbach and R. W. Meier, "Properties and Limitations of Hologram Recording Materials," *Appl. Opt.* **8**, 2269–2281 (1969).

3-12. W. T. Cathey, Jr., "Three-Dimensional Wavefront Reconstruction Using a Phase Hologram," *J. Opt. Soc. Am.* **55**, 457 (1965).

3-13. J. H. Altman, "Microdensitometry of High Resolution Plates by Measurement of the Relief Image," *Photogr. Sci. Eng.* **10**, 156–159 (1966).

3-14. J. H. Altman, "Pure Relief Images on Type 649-F Plates," *Appl. Opt.* **5**, 1689–1690 (1966).

3-15. H. Hannes, "Interferometrische Messungen an Phasenstrukturen für die Holographie," *Optik* **26**, 363–380 (1967).

3-16. H. Hannes, "Interferometric Measurements of Phase Structures in Photographs," *J. Opt. Soc. Am.* **58**, 140–141 (1968).

3-17. H. M. Smith, "Photographic Relief Images," *J. Opt. Soc. Am.* **58**, 533–539 (1968).

3-18. G. L. Rogers, "Experiments in Diffraction Microscopy," *Proc. Roy. Soc., Edinburgh* **63A**, 193–221 (1952).

3-19. T. A. Shankoff, "Phase Holograms in Dichromated Gelatin," *Appl. Opt.* **7**, 2101–2105 (1968).

3-20. F. S. Chen, J. T. LaMacchia, and D. B. Fraser, "Holographic Storage in Lithium Niobate," *Appl. Phys. Lett.* **13**, 223–225 (1968).

4

Coherence

The behavior of a wave when diffracted, recorded, or used for information storage is dependent on its degree of *coherence*. A perfectly coherent wave has a phase structure that is fixed with respect to time. This is the sort of wave considered in most of the discussions so far. We shall be concerned primarily with the aspects of coherence relating to the formation of holograms or the processing of spatial information. The goal of this chapter is to describe the manner in which the degree of coherence of a wave influences the recording of interference fringes and, consequently, the formation of a hologram. This introduction to the subject should, however, serve to prepare the reader for the discussions of coherence and image formation in Chapter 8 and incoherent holography in Chapter 9.

In recording a wavefront, the sensitivity or time constant of the detector determines the integration time necessary for the recording of the fringes (either visible or electronic). If the relative phase of the wave changes during the recording time, the fringes move and the quality of the recording is degraded. If the source is a nonmonochromatic source, such as a thermal source, the phase varies randomly. Or, if a random phase disturbance exists in one path, the fringes move during recording. If such a random process is present, the recording of the fringes is best described in terms of the mutual coherence function. Certain other cases occur where the phase function varies, not in a random manner, but in a way that can be described deterministically. The effect of a moving object is one example. First, we note the effect of shifts in the frequency of a wave upon the recording of that wave. Next, we consider spatial and temporal coherence and means of measuring each. Finally, the problems of laser modes and the coherence properties of solid and gaseous lasers are treated.

For a more thorough treatment of the subject of coherence and partially coherent waves, reference is made to a number of books and review articles [4-1 through 4-6]. We shall not delve into such aspects as the quantum-mechanical description of coherence [4-7] or the intriguing subject of fourth-order correlation as related to intensity interferometry [4-8], but shall be concerned with the aspects of coherence more closely related to the formation of an image or the recording of a hologram.

4.1 Frequency Shifts

If the frequency of one wave changes with respect to the frequency of another wave, the interference fringes move as described by (3-7). This expression, restated below, shows how the interference term of the intensity distribution varies with time as a function of frequency difference:

$$I(x,y,t) = U^2(x,y) + V^2(x,y) + 2U(x,y)V(x,y)\cos[(\omega_1 - \omega_2)t$$

$$+ (\zeta - \alpha)x + (\mu - \gamma)z + \Psi(x,y) - \phi(x,y)]. \tag{4-1}$$

The recording of the pattern described by (4-1) takes place during the finite integration time of the detector. If the frequency difference is fixed at $\Delta\omega$, integration (exposure) for time τ yields

$$\mathcal{E}(x,y) = \tau\left\{ U^2(x,y) + V^2(x,y) + 2\,\text{sinc}\left(\frac{\Delta\omega\tau}{2}\right)U(x,y)V(x,y)\right.$$

$$\left. \times \cos[(\zeta - \alpha)x + (\mu - \gamma)z + \Psi(x,y) - \phi(x,y)]\right\}, \tag{4-2}$$

where $\mathcal{E}$ is the energy per unit area on the detector.

Changes in source frequency. A frequency difference between the object and the reference waves can be caused by a drift in the frequency of the source if the path length from the source to the recorder is not the same for the two waves. If the two paths are equal, a shift of frequency of the source has no effect. Notice, however, that if one wave is tilted with respect to the other, the paths can never be equal except at one point on the recorder.

Object motion. If, while recording a hologram, the object or some element in the optical system is vibrating or moving in some other fashion, a phase fluctuation of the wave may be produced. This can be written in terms of a frequency shift. The only real concern is the phase or frequency fluctuation during the fringe recording time; if the recording time is short enough, the object motion can be large and acceptable fringes can still be recorded.

Object motion can be considered in two parts—motion transverse to the direction of wave propagation and motion along the propagation path. Transverse motion has the effect of changing the form of the diffraction pattern resulting in changes in $U(x,y)$ and $\Psi(x,y)$ but the effect is small.

The more critical effects are due to the component of motion along the path of propagation. This component introduces a Doppler shift of the frequency of the wave from the object, causing the fringes to move. If the recording time is such that the fringes move only a fraction of a fringe width during the recording, the motion causes no adverse effects. (This is often the case when a pulsed laser is used. The pulse times are often shorter than a nanosecond, with many megawatts of illuminating power in a pulse.) Because the fringes move one position for a change of one wavelength in the path of the object wave, this restricts longitudinal object motion to less than $\lambda/4$. For a good recording, it should be less than $\lambda/8$. Techniques of controlling the fringe position exist and use of such control systems allows the recording of a hologram under many circumstances that would normally prevent fringes from being recorded.

4.2 Mutual Coherence Function

If the phase fluctuations of the wave are random, the *mutual coherence function* is a convenient means of describing the wave. The mutual coherence function is defined in terms of the cross-correlation of the complex functions describing the wave at two points in space at different times. The description will be simplified by assuming that the waves are all polarized in the same sense and are stationary in time. The assumption of identical polarization allows us to neglect the vectorial aspects of the waves. If a function is stationary in time, the time average is independent of the choice of temporal origin, that is, it does not matter when the average is taken. The assumption of stationarity allows the simplification of many integrals. The mutual coherence function can be written as

$$\Gamma(x_1, x_2, \tau) \equiv \Gamma_{12}(\tau) = \langle U(x_1, t+\tau) U^*(x_2, t) \rangle, \qquad (4\text{-}3)$$

where U is the *complex* analytic signal associated with the scalar field, the subscript $_{12}$ is used in place of the arguments x_1 and x_2, and the angular brackets denote the time average

$$\Gamma \langle \cdots \rangle \equiv \lim_{T \to \infty} \frac{1}{T} \int_0^T (\cdots) \, dt. \qquad (4\text{-}4)$$

Notice that the assumption of stationarity allows Γ to be a function of the time difference, not the absolute time.

To illustrate the effect of the degree of coherence on interference and fringe formation, let us consider the double slit experiment illustrated in Fig. 4-1. An extended source S illuminates both slits 1 and 2 at positions x_1 and x_2. Waves from these slits propagate to point P where they interfere.

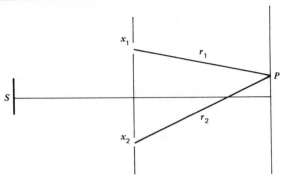

Figure 4-1 The wave from source S illuminates both slits at x_1 and x_2. The waves passing through the slits interfere at point P.

The wave at P is the superposition of waves from x_1 and x_2:

$$\mathbf{U}(P,t) = \mathbf{K}_1 \mathbf{U}(x_1, t - t_1) + \mathbf{K}_2 \mathbf{U}(x_2, t - t_2), \qquad (4\text{-}5)$$

where t_1 and t_2 are the periods of time required for the waves to propagate from x_1 and x_2 to P:

$$t_1 = \frac{r_1}{c}, \qquad t_2 = \frac{r_2}{c}. \qquad (4\text{-}6)$$

The factors $\mathbf{K}_1$ and $\mathbf{K}_2$ are proportional to the aperture size, the angle of incidence of the primary wave, and the values of r_1 and r_2. The intensity at any point P of the observation plane is described by

$$I(P) = \langle |\mathbf{U}(P,t)|^2 \rangle = \mathbf{K}_1 \mathbf{K}_1^* \langle \mathbf{U}_1(t - t_1) \mathbf{U}_1^*(t - t_1) \rangle + \mathbf{K}_2 \mathbf{K}_2^*$$

$$\times \langle \mathbf{U}_2(t - t_2) \mathbf{U}_2^*(t - t_2) \rangle + \mathbf{K}_1 \mathbf{K}_2^* \langle \mathbf{U}_1(t - t_1) \mathbf{U}_2^*(t - t_2) \rangle$$

$$+ \mathbf{K}_2 \mathbf{K}_1^* \langle \mathbf{U}_2(t - t_2) \mathbf{U}_1^*(t - t_1) \rangle. \qquad (4\text{-}7)$$

The time average is used because, in the case of a noncoherent wave, the phases are fluctuating, thus making the average intensity the most meaningful quantity. The assumption of stationarity allows (4-7) to be written as

$$I(P) = |\mathbf{K}_1|^2 \langle \mathbf{U}_1(t) \mathbf{U}_1(t) \rangle + |\mathbf{K}_2|^2 \langle \mathbf{U}_2(t) \mathbf{U}_2(t) \rangle$$

$$+ |\mathbf{K}_1 \mathbf{K}_2| \langle \mathbf{U}_1(t + \tau) \mathbf{U}_2^*(t) \rangle + |\mathbf{K}_1 \mathbf{K}_2| \langle \mathbf{U}_1^*(t + \tau) \mathbf{U}_2(t) \rangle, \qquad (4\text{-}8)$$

where $\tau = t_2 - t_1$. The form $|\mathbf{K}_1 \mathbf{K}_2|$ can be employed because $\mathbf{K}_1$ and $\mathbf{K}_2$ are purely imaginary factors. (This is shown in Chapter 1. See also [4-1].)

The intensity at point P is related to $\Gamma_{11}(0)$ at point x_1 by

$$|\mathbf{K}_1|^2 \Gamma_{11}(0) = I_1 \tag{4-9}$$

where $\mathbf{K}_1$ is the propagation constant. Equation (4-9) can be verified by blocking one slit and evaluating (4-8). This gives, for example, with slit 2 blocked,

$$I(P) = |\mathbf{K}_1|^2 \langle \mathbf{U}_1(t)\mathbf{U}_1^*(t) \rangle. \tag{4-10}$$

From (4-3) we see that

$$\langle \mathbf{U}_1(t)\mathbf{U}_1^*(t) \rangle = \Gamma_{11}(0). \tag{4-11}$$

Consequently, (4-8) can be reduced to

$$I(P) = I_1 + I_2 + 2|\mathbf{K}_1\mathbf{K}_2|\text{Re }\Gamma_{12}(\tau) \tag{4-12}$$

by use of (4-3) and the stationarity requirements, and noting that the imaginary parts of Γ_{12} and Γ_{21} cancel. The result given in (4-12) can also be written as

$$I(P) = I_1 + I_2 + 2|\mathbf{K}_1\mathbf{K}_2||\Gamma_{12}(\tau)|\cos[\alpha_{12}(\tau) - \delta] \tag{4-13}$$

where

$$\alpha_{12}(\tau) - \delta \equiv \arg \Gamma_{12}(\tau) \tag{4-14}$$

and

$$\delta = \frac{2(r_2 - r_1)}{\bar{\lambda}}, \tag{4-15}$$

where $\bar{\lambda}$ is the mean wavelength. The function δ arises because of the geometrical relations between the locations of the slits and the point P, that is, δ is the same regardless of the degree of coherence of the wave. The form of the equation in (4-13) more closely resembles the form previously considered in recording wavefronts, but the form in (4-12) is more commonly used.

The mutual coherence function is usually normalized so that

$$\gamma_{12}(\tau) = \frac{\Gamma_{12}(\tau)}{\sqrt{\Gamma_{11}(0)}\ \sqrt{\Gamma_{22}(0)}}, \qquad 0 < |\gamma_{12}| < 1. \tag{4-16}$$

A general interference law for stationary fields can then be written as

$$I(P) = I_1(P) + I_2(P)$$
$$+ 2\sqrt{I_1(P)}\ \sqrt{I_2(P)}\ |\gamma_{12}(\tau)|\cos[\alpha_{12}(\tau) - \delta]. \tag{4-17}$$

If the waves are mutually coherent, $|\gamma_{12}| = 1$ and

$$I(P) = I_1(P) + I_2(P) + 2\sqrt{I_1(P)}\ \sqrt{I_2(P)}\ \cos\left[\alpha_{12}(\tau) - \delta\right], \quad (4\text{-}18)$$

showing that the amplitudes add, forming interference fringes in the intensity distribution. If the waves are mutually incoherent, $\gamma_{12} = 0$ and

$$I(P) = I_1(P) + I_2(P), \quad (4\text{-}19)$$

indicating that the intensities add. For values of $|\gamma|$ between zero and one, the fields are partially coherent.

An alternate way of writing the expression for the degree of mutual coherence is to separate the terms as follows:

$$I(P) = |\gamma_{12}(\tau)| \left\{ I_1(P) + I_2(P) + 2\sqrt{I_1(P)}\ \sqrt{I_2(P)} \right.$$
$$\left. \times \cos\left[\alpha_{12}(\tau) - \delta\right] \right\} + \left\{ 1 - |\gamma_{12}(\tau)| \right\} \left\{ I_1(P) + I_2(P) \right\}. \quad (4\text{-}20)$$

The first brace describes a coherent superposition of waves of intensity $|\gamma_{12}(\tau)|I_1(P)$ and $|\gamma_{12}(\tau)|I_2(P)$ and relative phase difference of $\alpha_{12}(\tau) - \delta$. The second brace represents an incoherent superposition of two beams of intensities $[1 - |\gamma_{12}(\tau)|]I_1(P)$ and $[1 - |\gamma_{12}(\tau)|]I_2(P)$. The field distribution at P can be thought of as a mixture of coherent and incoherent waves with intensities of the ratio

$$\frac{I_{\text{coh}}}{I_{\text{incoh}}} = \frac{|\gamma_{12}(\tau)|}{1 - |\gamma_{12}(\tau)|} \quad (4\text{-}21)$$

$$\frac{I_{\text{incoh}}}{I_{\text{total}}} = |\gamma_{12}|. \quad (4\text{-}22)$$

The combination of a coherent and an incoherent field is sometimes a convenient way of thinking of a partially coherent wave, but neither completely coherent nor completely incoherent waves exist.

4.3 Relation of Fringe Visibility to Coherence

The value of $|\gamma|$ can be determined by measuring the visibility of fringes formed. The visibility of the fringes is defined by

$$\mathcal{V} = \frac{I_{\max} - I_{\min}}{I_{\max} + I_{\min}} \quad (4\text{-}23)$$

If the expression for $I(P)$ is written as

$$I(P) = I_1(P) + I_2(P)$$
$$+ 2\sqrt{I_1(P)} \ \sqrt{I_2(P)} \ |\gamma_{12}(\tau)| \cos[\alpha_{12}(\tau) - \delta], \qquad (4\text{-}24)$$

the maximum and minimum values of $I(P)$ are seen to be

$$I_{max} = I_1(P) + I_2(P) + 2\sqrt{I_1(P)} \ \sqrt{I_2(P)} \ |\gamma_{12}(\tau)| \qquad (4\text{-}25)$$

and

$$I_{min} = I_1(P) + I_2(P) - 2\sqrt{I_1(P)} \ \sqrt{I_2(P)} \ |\gamma_{12}(\tau)|. \qquad (4\text{-}26)$$

The locations of the maxima and minima are determined by $\alpha_{12}(\tau)$ and δ. Substitution of (4-25) and (4-26) into (4-23) results in

$$\upsilon = \frac{2\sqrt{I_1} \ \sqrt{I_2} \ |\gamma_{12}(\tau)|}{I_1 + I_2}. \qquad (4\text{-}27)$$

From this we find that

$$|\gamma_{12}(\tau)| = \frac{(I_1 + I_2) \ \upsilon}{2\sqrt{I_1} \ \sqrt{I_2}}, \qquad (4\text{-}28)$$

or, if $I_1 = I_2$,

$$|\gamma_{12}(\tau)| = \upsilon. \qquad (4\text{-}29)$$

Note that

$$\upsilon \sim |\langle \mathbf{U}_1(t + \tau) \mathbf{U}_2^*(t) \rangle| \qquad (4\text{-}30)$$

so that the duration of the time average affects the visibility. We can have *short-term coherence* and not have *long-term coherence*. The result (4-29) means, of course, that if the degree of coherence between points x_1 and x_2 of a wave is unknown, the magnitude of the value can be found by measuring the visibility of the fringes formed. For our application of recording spatial information on waves, however, it is more meaningful to read (4-29) backward. The quality of the recording or hologram is a direct function of the degree of coherence between the two waves. A low degree of coherence means a poor fringe recording and a poor hologram. Figure 4-2 shows the effect of the degree of coherence on the fringe visibility. The graphs show the increase in the background or incoherent portion de-

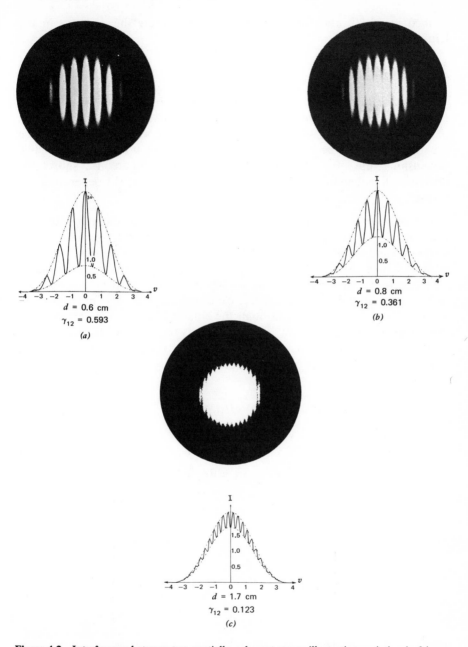

Figure 4-2 Interference between two partially coherent waves illustrating variation in fringe visibility. (*a*) Observed patterns. (*b*) Theoretical intensity curves. The slit separation is given by *d*. (*c*) Note the phase reversal for *d* = 1.7 cm. [After B. J. Thompson and E. Wolf, *J. Opt. Soc. Am.* **47**, 895 (1957).]

scribed by (4-20) and the photographs clearly show the decrease in the fringe visibility.

4.4   Spatial Coherence

If $\tau = 0$, that is, $r_1 = r_2$, on-axis fringes in the double slit experiment give a measure of $\gamma_{12}(0)$. By varying the slit separation, a measure of the degree of *spatial coherence* is obtained. The region over which the slits can be separated and fringes still be observed is the *coherence width*, or, in two dimensions, the *area of coherence* of the wave. Figure 4-2 shows the decrease of coherence, hence fringe visibility, as the slit separation increases.

Effect of source size.   It is possible to determine the angular size of a thermal or spatially incoherent source by using two slits to measure the spatial coherence of the wave produced by the source. One description of why the fringe visibility, hence the spatial coherence depends on the size of an incoherent source can be found by superimposing the fringes formed by light from each point on the source. Consider the path of light from point x on the source of extent S through the slits of Fig. 4-3. We can describe the fringes formed near point P on the x'' axis by (4-18) if the function δ is written as $-\bar{k}x''d/r$ where $\bar{k} = 2\pi/\bar{\lambda}$ and $|x'' - d| \ll r$, and if $\alpha_{12}(\tau)$ is written as $\bar{k}xd/R$, for $R \gg |x - d|$. Consequently, (4-18) can be written as

$$I(P) = I_1(P) + I_2(P) + 2\sqrt{I_1(P)}\ \sqrt{I_2(P)}\ \cos\left[\bar{k}\left(\frac{d}{R}\right)x + \bar{k}\left(\frac{d}{r}\right)x''\right].$$

$$(4\text{-}31)$$

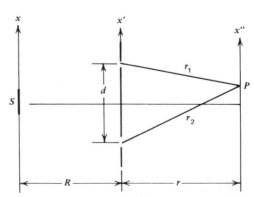

Figure 4-3   The size of the thermal source can be determined by measurement of the visibility of the fringes along the x'' axis.

This describes the fringes due to the light from a single point on the object. The other points on the source cause a superposition of the fringes due to the light from each point. For a uniformly illuminated one-dimensional source extending from $-S/2$ to $S/2$, the resulting fringe pattern is found by

$$I_1(P) + I_2(P) + 2\sqrt{I_1(P)}\ \sqrt{I_2(P)}\ \int_{-S/2}^{S/2} \cos\left[\bar{k}\left(\frac{d}{R}\right)x + \bar{k}\left(\frac{d}{r}\right)x''\right]dx$$

$$= I_1(P) + I_2(P) + 2\sqrt{I_1(P)}\ \sqrt{I_2(P)}\ S$$

$$\times \text{sinc}\left[\left(\frac{d}{\lambda R}\right)S\right]\cos\left[\bar{k}\left(\frac{d}{r}\right)x''\right]. \qquad (4\text{-}32)$$

Therefore, the fringes vanish if

$$d = \frac{\bar{\lambda}R}{S}. \qquad (4\text{-}33)$$

Expressed in terms of the angular size of the source θ_s,

$$d = \frac{\bar{\lambda}}{\theta_s}. \qquad (4\text{-}34)$$

Notice that (4-32) has a sinc function of the same form as found for the diffraction pattern of a slit illuminated by coherent light. This is an example where the concept of the sum of a coherent and an incoherent wave is useful. The coherent wave causes interference; the incoherent portion does not. The expression (4-32) indicates that the visibility of the fringes, hence γ_{12}, is proportional to sinc $(Sd/\lambda R)$. That is, if γ_{12} were plotted as a function of d, S, or $1/R$, a sinc function would be traced out. The visibility of the fringes goes to zero and then goes negative. The negative value indicates that the fringes shift position by 180° in phase (see Fig. 4-2c). If the incoherent source had been circular, the visibility of the fringes would have been proportional to

$$\frac{J_1(\pi Sd/\lambda R)}{\pi Sd/\lambda R}$$

rather than sinc $(Sd/\lambda R)$, and the slit separation at which the fringes disappear would be

$$d = 1.22 \frac{\bar{\lambda}}{\theta_s}.$$ (4-35)

Rearranging (4-35) gives

$$\theta_s = 1.22 \frac{\bar{\lambda}}{d},$$ (4-36)

which is a familiar equation in another context; it gives the angular resolution of an aperture used as a lens or antenna when the illumination is coherent.

Note that the slit separation at which the fringes vanish, hence the region of spatial coherence, increases as the distance to the source increases or as the source size decreases. In other words, the degree of spatial coherence of the wave increases with propagation. An incoherent source produces a partially coherent wave. Note also that the value of d at which the fringe visibility is a minimum will depend on the intensity distribution across the source.

The van Cittert-Zernike theorem. The relation between the source distribution and the degree of coherence between two points is known as the van Cittert-Zernike theorem [4-1]. This theorem states that the normalized mutual coherence function for $\tau = 0$ in one plane produced by an incoherent source having an intensity distribution $I(x_s,y_s)$ in another plane is described by the same mathematics as the diffraction pattern produced by a coherent wave converging on point P_1 after passing through a transparency having an amplitude transmittance proportional to the intensity distribution of the incoherent source. That is, the mutual coherence between points $P_1(x_1,y_1)$ and $P_2(x_2,y_2)$, at distances R_1 and R_2 from the source respectively, is given by

$$\gamma_{12}(0) = \frac{\int_S I(x_s,y_s) e^{i\bar{k}(R_1 - R_2)} dS}{\int I(x_s,y_s) dS}$$ (4-37)

If points P_1 and P_2 are closely spaced and far from a small source, (4-37) becomes a normalized Fourier transform. The function $\gamma_{12}(0)$ for a source having a uniform intensity distribution would then be of the form of the Bessel function if the source is circular or of the form of the sinc function if the source is a slit.

4.5 Temporal Coherence

If, in some manner, we can take a wave at some point in space, split it into two parts, delay one portion, and then recombine the two portions to form an interference pattern, we can measure the *temporal coherence* of a wave. The degree of temporal coherence is a measure of the correlation of a wave at one time with a wave at a later time. It may also be referred to as *longitudinal coherence* as opposed to transverse or spatial coherence. The time during which the wave is coherent, multiplied by the velocity of propagation of the wave, gives the *coherence length* of the wave. Spatial coherence, measured in two transverse directions determines the coherence *area* of the wave. Consequently, the two measures, spatial and temporal coherence, describe the degree of coherence of a wave within a volume of space.

The most common technique used to measure temporal coherence is the Michelson interferometer (not to be confused with the stellar interferometer) shown in Fig. 4-4. The beamsplitter B divides the waves, sending one portion to mirror M_1 and allowing the other to fall on mirror M_2. One of the paths, $B-M_2$ for example, is variable in length. The resulting interference fringes are used as a measure of $\gamma_{11}(\tau)$. The subscripts on γ are the same, either 11 or 22, because the interfering waves come from the same point in space. When the path difference is zero ($\tau=0$), the same wave is split and recombined to interfere with itself. In the case of the other extreme, $\tau \to \infty$, the phases of the two waves are completely random with respect to each other, and no fringes are visible. Between the two extremes, there is partial correlation or coherence, and fringes can be seen with visibility varying as a function of τ.

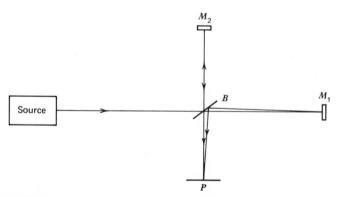

Figure 4-4 Michelson interferometer for measurement of temporal coherence. The path from beamsplitter B to mirror M_2 is variable.

The characteristics of the source of the wave obviously affect the temporal coherence of the wave. If a purely monochromatic wave is phase modulated, the bandwidth or spectral width of the wave must increase. If the modulating function is known, then in certain applications the temporal coherence can be considered to be preserved. If the phase modulation is random, the phase at one instant can be predicted for only a short time after the instant at which the phase is known. The time interval Δt over which the phase can be predicted is related to the bandwidth of the signal because the bandwidth, or spectral width, imposes a limit on how rapidly the phase can vary. The *minimum* time interval over which the wave is coherent is given approximately by

$$\Delta t = \frac{1}{\Delta \nu} \qquad (4\text{-}38)$$

where $\Delta \nu$ is the bandwidth of the wave. The time difference Δt is the coherence time of the wave and the corresponding length $c\,\Delta t$ is the coherence length, where c is the wave velocity.

4.6 Laser Sources

Lasers are commonly thought of as single-frequency, highly coherent optical sources. However, they are often multiple-frequency sources with relatively low coherence. This section treats some of the factors contributing to the coherence characteristics of laser outputs, and describes some techniques for increasing coherence. The greatest problems are related to multiple frequency operation and the effect on the temporal coherence of the source. Lack of spatial coherence is now rarely a problem.

Lasers have resonators or cavities that provide discrete frequency operation. One implementation is shown schematically in Fig. 4-5. The mirrors reflect the wave back and forth to produce standing waves within the resonator. This determines the axial mode of the resonator. Normally, the laser resonator is adjusted so that the phase is uniform across each reflector, producing what is called a *transverse electromagnetic mode* of zero order, or TEM_{00}. This is the lowest-order mode and is illustrated in Fig. 4-6a. The amplitude of the wave varies smoothly across the laser exit

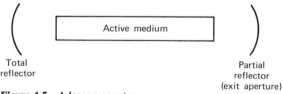

Total
reflector

Active medium

Partial
reflector
(exit aperture)

Figure 4-5 A laser resonator.

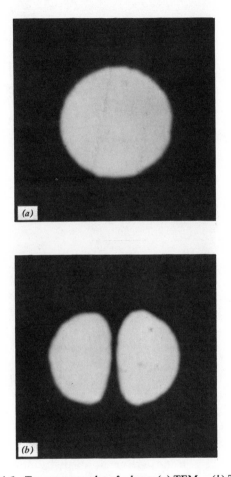

Figure 4-6 Transverse modes of a laser. (*a*) TEM_{00}, (*b*) TEM_{10}.

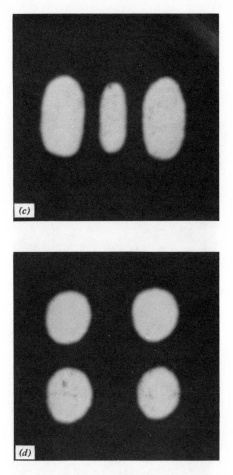

Figure 4-6   Transverse modes of a laser. (c) TEM_{20}, (d) TEM_{11}.

aperture, resembling a truncated gaussian distribution. The phase is uniform. If the reflectors are misaligned, higher-order modes can appear, as illustrated in Fig. 4-6b to d. These modes have phase reversals and the subscripts indicate the number of reversals in each direction. The modes within laser resonators have been studied extensively from the time of the earlier work [4-9 through 4-12] until recently [4-13]. We shall be primarily interested in the mode structure along the axis of the resonator because of the possibility of multiple modes and, consequently, multiple frequencies in the output. We shall assume that the laser is adjusted to operate in the lowest-order transverse mode, TEM_{00}.

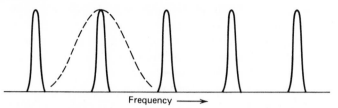

Figure 4-7 Resonant frequencies of a typical laser cavity. The dashed line shows the gain curve of a narrow-line medium.

The temporal coherence of a laser can vary over a large range depending on the type of laser and the number of axial modes allowed. Semiconductor lasers are normally considered to have the shortest coherence time because of the wide spectral line emitted. The coherence time is being increased, however, with newer semiconductor techniques. The output of solid-state pulsed lasers, such as ruby lasers, have variable frequencies due to transient heating effects. In addition, solid-state lasers tend to have many more modes, although mode control techniques can be used to obtain coherence lengths approaching the pulse length. The gas lasers have, in general, a longer coherence length than do other types. Consequently, much wavefront recording is performed with gas lasers operating continuously. Uses of pulsed lasers in holography are discussed in Chapter 10. The discussion that follows treats the mode structures of lasers in general and are applicable to both types, gaseous and solid-state, but examples are given in terms of gas lasers.

The reciprocal relationship between linewidth and coherence time or length has been mentioned. The active medium of a laser has a gain curve determined by the emission linewidth, but the linewidth of the radiated energy is narrowed considerably by the effect of the laser resonator, or cavity. The structure forming the laser resonator allows many resonant frequencies, as shown in Fig. 4-7. If the gain curve of the active medium is narrow enough for only one resonant frequency to be enveloped, the degree of temporal coherence of the laser is very high. This type of operation is common with CO_2 lasers where the width of the gain curve is approximately 5×10^7 Hz at a wavelength of 10.6 μ. The width of the gain curve is determined primarily by the collisions occurring between molecules and the Doppler effect caused by their motion. For a discussion of the line broadening factors, see, for example, [4-14]. The spacing between resonator lines is

$$\delta\nu = \frac{c}{2L} \qquad (4\text{-}39)$$

where c is the velocity of light in the medium and L is the resonator length. For a resonator length of 1 meter, $\delta\nu = 1.5 \times 10^8$ Hz so that, with length adjustment, one resonance can be aligned with the peak of the gain curve. The result is a single line, with its width determined by the quality of the resonator. With a high-quality resonator and good frequency stability, the linewidth of a CO_2 laser can be on the order of 200 Hz.

Most other lasers have wider gain curves, allowing oscillation at many frequencies at the same time. The helium-neon laser, for example, can normally run at four or more cavity resonances within one spectral linewidth. Techniques are available to force the laser to run in only one axial mode (single resonance), but before examining such techniques let us consider the effect of more than one axial mode.

A laser operating with only one mode has a coherence length related to the reciprocal of the spectral width of the emitted radiation, and interference fringes can be formed with path differences up to the coherence length. If more than one mode is present, the effect on the visibility of the fringes is as shown in Fig. 4-8. The repeating of the maxima under the envelope at intervals of twice the cavity length and the reduction in visibility between the maxima can be explained in terms of the different frequency waves adding in and out of phase to give an output with frequency *beats*. These beats describe an envelope of the energy transmitted by the laser and can be found by Fourier transforming the spectral output of the laser. When the laser beam is split and recombined with a

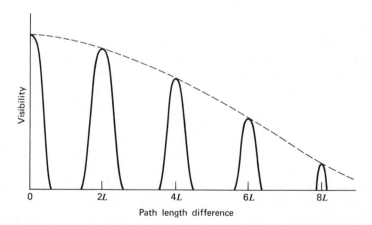

Figure 4-8 Visibility of fringes formed as a function of path length difference, as in a Michelson interferometer. The dashed line shows the variation of the visibility of the fringes with path difference when a laser with a single mode is used. The solid line shows the effect of multiple axial modes in the laser, producing a multiple-frequency output. The peaks in visibility occur at intervals of $2L$ where L is the laser resonator length.

delay in one arm, one of the beams may be at a null. If the delay is such that the peaks coincide, fringes of high visibility are formed. An alternate method of describing the variation in visibility is in terms of the superposition of the fringes formed by each axial mode [4-15].

If a laser is operating in more than one axial mode, care must be taken to maintain path length differences to within a few centimeters, or the quality of the hologram recording will decrease. Figure 4-8 illustrates that the path difference could be $2nL$ plus or minus a few centimeters.

The effects of multiple modes on the visibility of the fringes are shown in more detail in Fig. 4-9. As the number of modes increase, the sharpness of the peaks increases. The visibility drops to zero between peaks only when the mode frequencies are symmetrical with respect to the gain curve of the active medium. If the mode energies are not symmetrically distributed, the fringe visibility does not drop to zero [4-15, 4-16].

Various forms of mode control can be used to cause operation of a laser in a single axial mode [4-14, Ch. 5]. The simplest technique is to use two resonators. One resonator is the regular laser cavity and the other is formed either by placing the entire laser in a second resonant structure or by placing a second structure in the original laser cavity. The latter is the more common, with the second resonator being a plate of polished quartz or other material, depending on the laser operating wavelength, with plane parallel sides. The reflection from the two surfaces, due to differences in the index of refraction, form the second resonator. This approach leads to single-mode operation by adjusting the laser resonator length so that the two resonators have a common resonant frequency. Because the spacings of the resonant frequencies of the laser cavity of length L are $c/2L$ and the spacings of the other resonator of length L' are $c/2L'$, superposition of resonant frequencies at one frequency normally eliminates resonances at other frequencies under the natural emission line of the medium. The laser would then operate at only one wavelength. Figure 4-10 shows that some care is necessary to position properly the resonances, or there may be multiple superpositions of the resonance frequencies. (There may also be other superpositions under another emission line of the material.)

Another technique for obtaining all of the laser energy in one axial mode employs two phase modulators to cause the entire laser output to be in one axial mode [4-17, 4-18]. One phase modulator inside the cavity causes the modes to have the phases and amplitudes comparable to the side bands of a frequency-modulated wave. The output of the laser then passes through the second phase modulator, which is driven at the same frequency but 180° out of phase from the first modulator. The result is to demodulate the output of the laser, producing a single-mode output from the entire system.

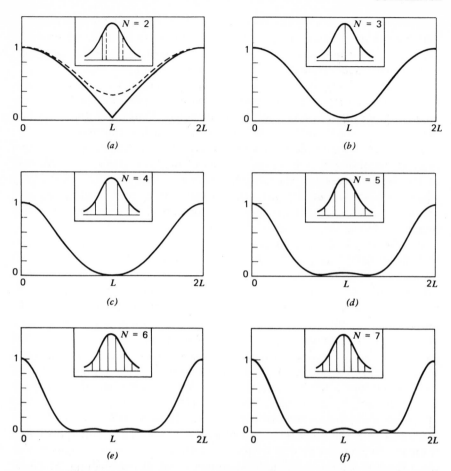

Figure 4-9 Plot of the magnitude of the fringe visibility for multiple axial modes as a function of path difference. The inset in each figure shows the number of modes N, the relative amplitude, and the position of the operating frequencies with respect to the gain curve of the active medium. Note the effects of asymmetry in (a). [Adapted from J. W. Foreman, *Appl. Opt.* **6**, 821–826 (1967).]

It is evident that, by using the proper precautions or controls, a laser with both spatial and temporal coherence properties can be obtained, and little attention need be given to path differences if operation in a single axial mode is achieved.

Holograms are formed and wavefronts recorded at wavelengths other than optical. Microwave and acoustical holograms are discussed in Chapter 10. Sources for these wavelengths are normally very coherent, eliminating any concern of limited coherence length.

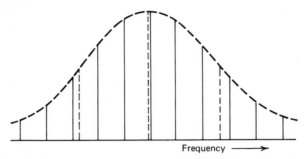

Figure 4-10 Resonant lines, caused by two cavities, superimposed under the gain curve of the active material. The solid vertical lines represent resonances of the larger cavity and the dotted lines represent the resonances of the small cavity.

PROBLEMS

4-1. Obtain two expressions in terms of $\Delta\omega$ and τ for the visibility of the fringes of the distribution described in (4-2). First, assume that $U^2 \neq V^2$ and, second, assume that the intensities are equal.

4-2. Use (4-2) to show how the visibility or contrast of the fringes affects the fraction of the energy of the read-out wave that appears in the image. Find the ratio of energy in one image to the energy going straight through a hologram. Assume no z variations in the fringes and a zero-thickness amplitude-modulation hologram.

4-3. The angular diameter of the star Betelgeuse was found by Michelson and Pease to be 0.047 sec of arc. At what separation of the mirrors M_1 and M_4 of the stellar interferometer would the interference fringes caused by the starlight disappear? (Figure P4-3.)

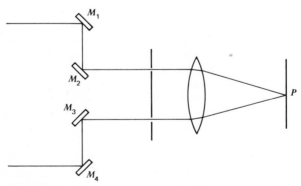

Figure P4-3

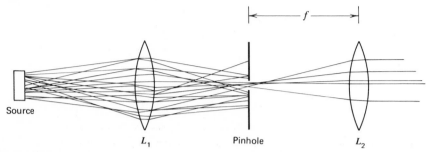

Figure P4-7

4-4. Assume that a double-slit interferometer is illuminated by sunlight. At what slit separation would fringes no longer be visible? Assume $\bar{\lambda} = 5000$ Å.

4-5. What would be the area of coherence of a uniform plane wave of diameter d?

4-6. If a wave with randomly varying phase is passed through a narrow band filter such that its bandwidth is reduced by a factor of 5, what can be said about the effect on the temporal coherence of the wave?

4-7. A source produces incoherent light which is to be collimated by the lens L_2 as shown in Fig. P4-7. If a coherence area of $\pi d^2 / 4$ is needed, where d is the separation of slits at which fringes vanish, what diameter pinhole must be used?

4-8. Assume that a laser produces two waves of equal amplitude cos $(\omega_1 t - k_1 z)$ and cos $(\omega_2 t - k_2 z)$ where $\omega_2 = \omega_1 + 2\pi(c/2L)$, $k_1 = 2\pi/\lambda_1$, and $k_2 = 2\pi/\lambda_2$. Allow the output of the laser to be split and recombined using a Michelson interferometer. Time average to obtain an expression, in terms of the difference in the two path lengths, for the visibility of the fringes formed.

REFERENCES

4-1. M. Born and E. Wolf, *Principles of Optics* (Macmillan Co., New York, 1964), Ch. 10.

4-2. L. Mandel, "Fluctuations of Light Beams," in *Progress in Optics*, Vol. II, E. Wolf, ed. (North-Holland Publishing Co., Amsterdam, 1963), Ch. 5.

4-3. H. Gamo, "Matrix Treatment of Partial Coherence," in *Progress in Optics*, Vol. III, E. Wolf, ed. (North-Holland Publishing Co., Amsterdam, 1964), Ch. 3.

4-4. M. J. Beran and G. B. Parrent, Jr., *Theory of Partial Coherence* (Prentice-Hall, Inc., Englewood Cliffs, N. J., 1964).

4-5. L. Mandel and E. Wolf, "Coherence Properties of Optical Fields," *Rev. Mod. Phys.* 37, 231–287 (1965).

4-6. C. L. Mehta, "Coherence and Statistics of Radiation," in *Lectures in Theoretical Physics*, Vol. VII C (University of Colorado Press, Boulder, Colo., 1965). Also available as AD 632 419.

4-7. R. J. Glauber, "The Quantum Theory of Optical Coherence," *Phys. Rev.* **130**, 2529–2539 (1963); "Coherent and Incoherent States of the Radiation Field," *Phys. Rev.* **131**, 2766–2788 (1963).

4-8. R. Hanbury Brown and R. Q. Twiss, "A New Type of Interferometer for Use in Radio Astronomy," *Phil. Mag.* **45**, 663–682 (1954).

4-9. G. D. Boyd and J. P. Gordon, "Confocal Multimode Resonator for Millimeter through Optical Wavelength Masers," *Bell Syst. Tech. J.* **40**, 489–508 (1961).

4-10. A. G. Fox and T. Li, "Resonant Modes in a Maser Interferometer," *Bell Syst. Tech. J.* **40**, 453–488 (1961).

4-11. G. D. Boyd and H. Kogelnik, "Generalized Confocal Resonator Theory," *Bell Syst. Tech. J.* **41**, 1347–1369 (1962).

4-12. A. G. Fox and T. Li, "Modes in a Maser Interferometer with Curved and Tilted Mirrors," *Proc. IEEE* **51**, 80–89 (1963).

4-13. H. K. V. Lotsch, "The Scalar Theory for Optical Resonators and Beam Waveguides," *Optik* **26**, 112–130 (1967); "The Fabry-Perot Resonator," *Optik* **28**, 65–75, 328–345, and 555–574 (1968/69); **29**, 130–145 and 622–623 (1969); "The Confocal Resonator System," Part I, *Optik* **30**, 1–14 (1969); Part II, *Optik* **30**, 181–201 (1969); Part III, *Optik* **30**, 217–233 (1969); Part IV, *Optik* **30**, 563–576 (1970).

4-14. D. C. Sinclair and W. E. Bell, *Gas Laser Technology* (Holt, Rinehart and Winston, Inc., New York, 1969), pp. 26–35.

4-15. W. T. Cathey, "Multiple Frequency Wavefront Recording," *Opt. Acta* **15**, 35–45 (1968).

4-16. E. F. Erickson and R. M. Brown, "Calculation of Fringe Visibility in a Laser-Illuminated Interferometer," *J. Opt. Soc. Am.* **57**, 367–371 (1967).

4-17. G. A. Massey, M. K. Oshman, and R. Targ, "Generation of Single-Frequency Light Using the FM Laser," *Appl. Phys. Lett.* **6**, 10–11 (1965).

4-18. S. E. Harris and O. P. McDuff, "Theory of FM Laser Oscillation," *IEEE J. Quant. Electron.* **QE−1**, 245–262 (1965).

5

Imaging and Fourier Transforming Elements

In this chapter it is pointed out that many optical operations can best be described in terms of Fourier transforms. We find that the lens acts as a Fourier transformer of spatial distributions, providing a Fourier transform relationship between the optical distributions in its front and back focal planes. We shall also see that a lens can, under different conditions, produce an image of an object, replicating the original wave distribution, rather than Fourier-transforming it.

The diffraction equations are used to derive the conditions under which an image is formed and those that are necessary for a Fourier transform of the object to be obtained. First, however, the focusing elements themselves are examined to find their effect on an incident wavefront.

5.1 Effects of Focusing Elements on a Wavefront

Any element focusing a beam causes, in the geometrical optics approximation, the beam to converge to a point. If an element causes a wave to be focused to a point at a distance d_2 from the element, the wave must have, in the region before it focuses, a curvature described by $\exp(-ikr)$, where r is the distance to the focus. This result can be inferred from the fact that a wave from a point source has a phasefront described by $\exp(ikr)$. Figure 5-1 shows a wave propagating to a focal point P'. We can see from the geometry that

$$r^2 = d^2 + x^2 + y^2 \tag{5-1}$$

where x and y are the rectangular coordinates in a plane before the focus.

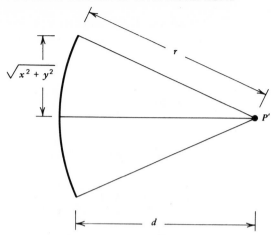

Figure 5-1 Spherical wave converging to point P'.

If the range of x and y is small with respect to d, the approximation

$$r = d + \frac{x^2 + y^2}{2d} \tag{5-2}$$

is valid. Substitution of (5-2) shows that from the lens to the focus there is an average phase shift kd radians and another phase shift that is a function of x and y. The average phase shift is of no concern; we are only interested in the relative phase shift terms that describe the phasefront of the wave. The phase of the wave in the plane just past the lens can be represented by

$$\exp\left[\frac{-ik(x^2 + y^2)}{2d} \right] \tag{5-3}$$

if $x^2 + y^2 \ll d^2$.

Figure 5-2 shows a diverging wave propagating to the right and being focused by a lens. If the wave before focusing has a phasefront described

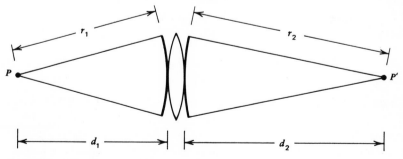

Figure 5-2 Diverging wave focused by a lens.

by $\exp[ik(x^2+y^2)/2d_1]$, it appears to have come from a point d_1 in front of the lens. The phase of the wave just to the right of the lens (neglecting lens thickness) can be written as

$$\exp\left[\frac{ik(x^2+y^2)}{2d_1}\right]\mathbf{p}_l=\exp\left[\frac{-ik(x^2+y^2)}{2d_2}\right] \tag{5-4}$$

where $\mathbf{p}_l$ is the effect of the lens on the phase. Solving for $\mathbf{p}_l$ yields

$$\mathbf{p}_l=\exp\left[-ik(x^2+y^2)\left(\frac{1}{2}\right)\left(\frac{1}{d_1}+\frac{1}{d_2}\right)\right], \tag{5-5}$$

which means that if the wave coming from point P is to converge at P', the lens must change the phase by the amount given in (5-5). The focal length of a lens is defined in terms of the effect of the lens on a plane wave. A plane wave can be described as being produced by a point at infinity; that is, let $d_1\rightarrow\infty$. For this case we find that

$$\mathbf{p}_l=\exp\left[\frac{-ik(x^2+y^2)}{2d_2'}\right] \tag{5-6}$$

where d_2' is the value of d_2 when $d_1=\infty$. This distance d_2' is called the focal length, f, of the lens, so that the phase effect of the lens is normally written as

$$\mathbf{p}_l=\exp\left[\frac{-ik(x^2+y^2)}{2f}\right]. \tag{5-7}$$

A comparison of (5-5) and (5-7) shows that for $d_1\neq\infty$, a lens of focal length

$$\frac{1}{f}=\frac{1}{d_1}+\frac{1}{d_2} \tag{5-8}$$

is needed to cause the wave to converge to a point at a distance d_2 from the lens. If it is desired to change the phasefront such that the wave appears to have come from a point nearer the lens, a similar analysis can be performed to show that the lens must have a negative focal length. Figure 5-3 shows the effect on a wave from point P a distance d_1 before

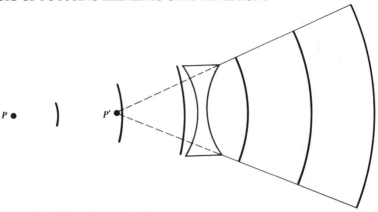

Figure 5-3 Effect of a negative lens.

the lens. The resulting wave appears to have come from point P' a distance d_2 from the lens. The phase effect of the lens is then

$$\mathbf{p}_l = \exp\left[\frac{ik(x^2+y^2)}{2} \left(\frac{1}{d_2} - \frac{1}{d_1} \right) \right].$$
(5-9)

If $d_1 = \infty$, we see, by comparison with (5-7), that $f = -d_2$. A plane wave passing through such a lens will be changed into a diverging wave that appears to have originated from a point a distance f in front of the lens. This is shown in Fig. 5-4.

We have seen what a lens must do. Now let us see *how* it does it.

Geometry of a lens. To determine how a lens can produce the required phase effect, it is helpful to consider the differences in path lengths through the lens at different points on the lens. Again, assume a thin lens so that the mechanism is simplified. The problem is to determine the phase

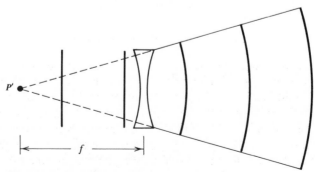

Figure 5-4 Effect of a negative lens on a plane wave.

P_1 P_2

T_0

Figure 5-5   Lens and the chosen reference planes.

shift encountered on passing through the lens. Figure 5-5 illustrates the lens and the planes P_1 and P_2 at which the phase is to be determined. If the lens is assumed to be sufficiently thin that a ray entering at coordinates x,y also leaves at those coordinates, the phase difference between the planes P_1 and P_2 is given by

$$k\mathbf{n}T_l(x,y) + k[T_0 - T_l(x,y)] \qquad (5\text{-}10)$$

where $k = 2\pi/\lambda$, $\mathbf{n}$ is the index of refraction, T_0 is the distance between planes P_1 and P_2, and $T_l(x,y)$ is the thickness of the lens. The expression (5-10) gives the phase shift due to the lens and the free space between the surfaces of the lens and the reference planes. It can be rewritten in the form

$$kT_0 + (\mathbf{n} - 1)kT_l(x,y). \qquad (5\text{-}11)$$

If we assume that the lens has spherical surfaces (which is the most common type), $T(x,y)$ can be found in terms of the radii of curvature of the surfaces. Refer to Fig. 5-6, which shows the left half of the lens. It is assumed that the wave comes from the left, and it is conventional to assign a positive number to the radius of a surface with its center of curvature to the right of the surface. From Fig. 5-6 we see that $T_{l1}(x,y)$, the thickness of the left half of the lens is related to the distance between the left reference

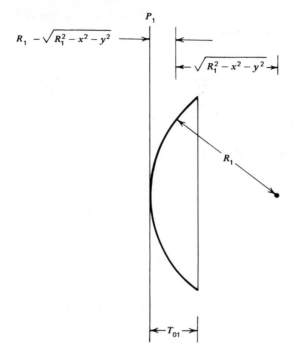

Figure 5-6   Surface of the left portion of a lens.

plane and the center of the lens, T_{01}, by

$$T_{l1}(x,y) = T_{01} - \left(R_1 - \sqrt{R_1^2 - x^2 - y^2} \right)$$

$$= T_{01} - R_1 \left(1 - \sqrt{1 - \frac{x^2 + y^2}{R_1^2}} \right). \tag{5-12}$$

When the center of curvature is to the left of the surface, the radius of curvature is negative. Consequently, for the expression associated with the right section of the lens R_1 is replaced by $-R_2$, giving

$$T_{l2}(x,y) = T_{02} + R_2 \left(1 - \sqrt{1 - \frac{x^2 + y^2}{R_2^2}} \right), \tag{5-13}$$

so that when the negative value of R_2 is inserted, the correct operation results. The total thickness function is

$$T_l(x,y) = T_0 - R_1\left(1 - \sqrt{1 - \frac{x^2+y^2}{R_1^2}}\right) + R_2\left(1 - \sqrt{1 - \frac{x^2+y^2}{R_2^2}}\right).$$

(5-14)

If the maximum values of x^2 and y^2 are small compared with R_1 or R_2, the approximation

$$\sqrt{1 - \frac{x^2+y^2}{R^2}} \cong 1 - \frac{x^2+y^2}{2R^2}$$

(5-15)

is valid. This allows (5-14) to be reduced to

$$T_l(x,y) = T_0 - \left(\frac{1}{R_1} - \frac{1}{R_2}\right)\frac{x^2+y^2}{2}.$$

(5-16)

Substitution of (5-16) into (5-11) gives a total phase shift due to the lens, $\mathbf{p}_l$, of

$$\mathbf{p}_l = \exp\left(i\left\{kT_0 + (\mathbf{n}-1)k\left[T_0 - \left(\frac{1}{R_1} - \frac{1}{R_2}\right)\frac{x^2+y^2}{2}\right]\right\}\right)$$

$$= \exp(ik\mathbf{n}T_0)\exp\left[-ik(\mathbf{n}-1)\left(\frac{1}{R_1} - \frac{1}{R_2}\right)\frac{x^2+y^2}{2}\right].$$

(5-17)

The focal length of the lens is identified as

$$\frac{1}{f} = (\mathbf{n}-1)\left(\frac{1}{R_1} - \frac{1}{R_2}\right)$$

(5-18)

and

$$\mathbf{p}_l = \exp(ik\mathbf{n}T_0)\exp\left[\frac{-ik(x^2+y^2)}{2f}\right].$$

(5-19)

Normally, the $\exp(ik\mathbf{n}T_0)$ term is ignored because it is simply a constant.

Reference to (5-18) and the condition allowing (5-15) indicates that the restriction is equivalent to requiring that the diameter of the lens be small in comparison with the focal length. That is, the F-number, defined as

$$F^{\#} = \frac{f}{D},\tag{5-20}$$

where D is the diameter of the lens, must be large for the preceding assumptions to be valid. The mathematics does, however, allow for different types of lenses. For example, the planoconvex lens, with $R_1 = \infty$ and R_2 a finite negative number, results in a positive focal length.

We shall normally ignore deviations of the lens from the perfect shape required to provide the spherical phasefront. If the F-number is large, simple spherical lenses are adequate. For small F-numbers the lens is normally made of a number of elements that may deviate from a spherical shape. Such compound lenses are illustrated in introductory optics texts [5-1].

In the figures showing thin lenses, the focal lengths have been measured from the center of the lens to the focal point. In the case of a thick or a compound lens, the focal length is measured from what are called principal planes, or unit planes, associated with the lens [5-2]. We shall not be concerned with these planes except to realize that the exact focal length of a compound system is not easy to specify by a simple inspection. Indeed, the focal length may be a factor of 2 or more greater than the distance from the back surface of the lens to the focal plane.

Parabolic elements. Any focusing element that produces a wavefront with constant phase over a converging spherical surface will be acceptable as a Fourier transforming element. The paraboloid will produce a converging spherical wave when a plane wave is reflected from its surface. The parabola, shown in Fig. 5-7, is a surface such that the distance from a plane to a point on the surface P and then to F is a constant. Consequently, the plane wave propagating into a paraboloid is focused, as shown, at point F. The phase shift due to the parabola, $\mathbf{p}_p$, is

$$\mathbf{p}_p = \exp\left(\frac{-ikr^2}{f}\right).\tag{5-21}$$

5.2 Fourier Transformation

Many examples of a lens producing a Fourier transform can be seen. A uniform plane wave incident upon a lens produces a point in the focal plane if the lens is infinitely large or the wavelength is zero. This is

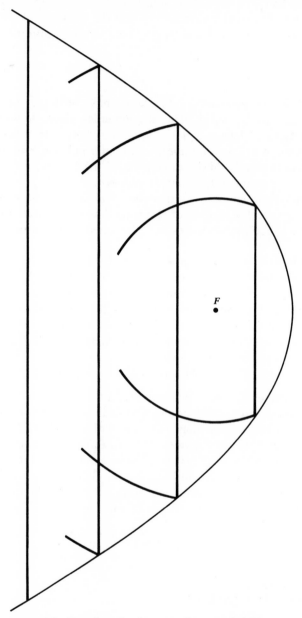

Figure 5-7   Focusing of a plane wave by a paraboloid.

equivalent to saying that the Fourier transform of unity is a delta function. If the lens has a finite size, we find the Fourier transform of a circle function, a $J_1(r)/r$ distribution, in the focal plane.

There are many analytical methods to show that an element that produces a quadratic phase factor will perform a Fourier transformation. We use the diffraction formula because it allows a general approach to the problem, and the exercise demonstrates the technique of using the equations. We first calculate the distribution U_2, in the x_2,y_2 plane shown in Fig. 5-8, assuming that the distribution U_1 in the x_1,y_1 plane represents $t(x,y)e(x,y)$ where $e(x,y)$ describes the complex amplitude of the illuminating wave and $t(x,y)$ is the amplitude transmission factor for the transparency. The effect of the lens on U_2 is added and the new distribution used to calculate the distribution U_3 in the x_3,y_3 plane. An examination of the resulting equation will show the conditions under which U_3 is a Fourier transform of U_1. The effect of illuminating $t(x,y)$ with a wave having a spherical phasefront is then examined.

Plane wave illumination. Assume that a transparency, $t(x,y)$, in the x_1,y_1 plane is illuminated by a uniform plane wave of unit amplitude and that the wave immediately past the transparency, $U_1(x_1,y_1)$ is the same as $t(x,y)$. By use of the diffraction formula, we find that the distribution just before the lens is described by

$$U_2(x_2,y_2) = \frac{\exp(ikd_1)}{i\lambda d_1}$$

$$\times \int\int U_1(x_1,y_1)\exp\left\{\frac{ik\left[(x_2-x_1)^2+(y_2-y_1)^2\right]}{2d_1}\right\} dx_1\,dy_1, \quad (5\text{-}22)$$

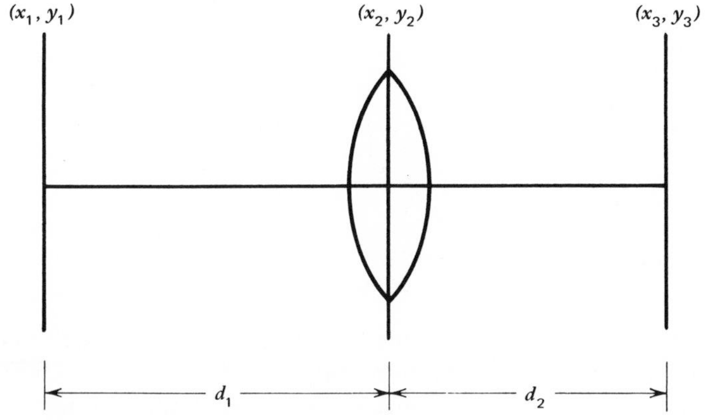

Figure 5-8 Fourier-transforming system.

where d_1 is the distance between the x_1, y_1 and the x_2, y_2 planes. The expression (5-22) can be written as

$$U_2(x_2, y_2) = \frac{\exp(ikd_1)}{i\lambda d_1} \exp\left[\frac{ik(x_2^2 + y_2^2)}{2d_1} \right]$$

$$\times \int \int U_1(x_1, y_1) \exp\left[\frac{ik(x_1^2 + y_1^2)}{2d_1} \right] \exp\left[\frac{-ik(x_1 x_2 + y_1 y_2)}{d_1} \right] dx_1 \, dy_1.$$

$$(5\text{-}23)$$

Or, using the definition

$$\mathbf{h}(x, y; d) \equiv \exp\left[\frac{ik(x^2 + y^2)}{2d} \right], \qquad\qquad (5\text{-}24)$$

$$U_2(x_2, y_2) = \frac{\exp(ikd_1)}{i\lambda d_1} \mathbf{h}(x_2, y_2; d_1)$$

$$\times \int \int U_1(x_1, y_1) \mathbf{h}(x_1, y_1; d_1) \exp\left[\frac{-ik(x_1 x_2 + y_1 y_2)}{d_1} \right] dx_1 \, dy_1. \quad (5\text{-}25)$$

To find the distribution in the x_3, y_3 plane, U_2 must be multiplied by $\mathbf{p}_l$, the phase effect of the lens, and $\mathbf{P}(x, y)$, the pupil function, including phase errors as well as the size of the lens aperture. The product is used in the diffraction formula to yield

$$U_3(x_3, y_3) = \frac{\exp(ikd_2)}{i\lambda d_2} \mathbf{h}(x_3, y_3; d_2) \int \int \mathbf{P}(x_2, y_2) U_2(x_2, y_2)$$

$$\times \mathbf{h}(x_2, y_2; -f) \mathbf{h}(x_2, y_2; d_2) \exp\left[\frac{-ik(x_2 x_3 + y_2 y_3)}{d_2} \right] dx_2 \, dy_2 \quad (5\text{-}26)$$

where the term $\exp(iknT_0)$ due to the lens thickness was dropped because it is constant across the plane. The pupil function $\mathbf{P}(x_2, y_2)$ is zero outside the lens and unity (or any other amplitude variation) within the lens. Assume for now that there are no phase errors to be included in $\mathbf{P}(x, y)$. The exponential terms of all of the equations could be combined into one $\exp(\cdots)$ or $h(\cdots)$ but they are separated to make a physical interpretation easier. To find the function U_3 in terms of U_1, (5-25) and (5-26) are combined and the order of integration reversed. The phase term $\exp[ik(d_1 + d_2)]$, giving the total phase shift of the path, is also dropped, yielding

$$U_3(x_3,y_3) = \frac{1}{i^2\lambda^2 d_1 d_2} h(x_3,y_3;d_2) \int\int U_1(x_1,y_1)$$

$$\times h(x_1,y_1;d_1) \int\int P(x_2,y_2) P\left[\frac{ik(x_2^2+y_2^2)}{2}\left(\frac{1}{d_1}+\frac{1}{d_2}-\frac{1}{f}\right)\right]$$

$$\times \exp\left\{-i2\pi\left[\frac{x_1x_2+y_1y_2}{\lambda d_1}+\frac{x_2x_3+y_2y_3}{\lambda d_2}\right]\right\} dx_2\,dy_2\,dx_1\,dy_1. \quad (5\text{-}27)$$

Let us first consider the evaluation of (5-27) when the lens is assumed to be sufficiently large that $P(x_2,y_2)$ can be taken as unity over the range of integration. Equation (5-27) can be written as

$$U_3(x_3,y_3) = \frac{1}{i^2\lambda^2 d_1 d_2} h(x_3,y_3;d_2) \int\int U_1(x_1,y_1) h(x_1,y_1;d_1)$$

$$\times \int\int h(x_2,y_2;w)\exp\left\{-i2\pi\left[x_2\left(\frac{x_1}{\lambda d_1}+\frac{x_3}{\lambda d_2}\right)+y_2\left(\frac{y_1}{\lambda d_1}+\frac{y_3}{\lambda d_2}\right)\right]\right\}$$

$$\times dx_2\,dy_2\,dx_1\,dy_1, \quad\quad\quad\quad (5\text{-}28)$$

where

$$\frac{1}{w} \equiv \frac{1}{d_1}+\frac{1}{d_2}-\frac{1}{f}. \quad\quad\quad\quad (5\text{-}29)$$

The inner integral describes the Fourier transform of $h(x,y;d)$, which is shown in Appendix 2 to be $i\lambda d h(u,v;-1/\lambda^2 d)$. Therefore the inner integral becomes

$$i\lambda w h\left(\frac{x_1}{\lambda d_1}+\frac{x_3}{\lambda d_2}, \frac{y_1}{\lambda d_1}+\frac{y_3}{\lambda d_2}; \frac{-1}{\lambda^2 w}\right). \quad\quad (5\text{-}30)$$

A regrouping of the exponentials in (5-28) yields

$$U_3(x_3,y_3) = \frac{w}{i\lambda d_1 d_2} \exp\left[\frac{ik}{2d_2}(x_3^2+y_3^2)\left(1-\frac{w}{d_2}\right) \right]$$

$$\times \int\int U_1(x_1,y_1) \exp\left[\frac{ik}{2d_1}(x_1^2+y_1^2)\left(1-\frac{w}{d_1}\right) \right]$$

$$\times \exp\left[-\frac{i2\pi}{\lambda}(x_1x_3+y_1y_3)\frac{w}{d_1d_2} \right] dx_1\,dy_1. \qquad (5\text{-}31)$$

This can be a Fourier transform if the exponential with $x_1^2+y_1^2$ is eliminated. This can be accomplished by requiring that

$$w = d_1 \qquad (5\text{-}32)$$

which, from (5-29), is the same as requiring that

$$d_2 = f. \qquad (5\text{-}33)$$

This causes (5-31) to become

$$U_3(x_3,y_3) = \frac{1}{i\lambda f} \exp\left[\frac{ik}{2f}(x_3^2+y_3^2)\left(1-\frac{d_1}{f}\right) \right]$$

$$\times \int\int U_1(x_1,y_1) \exp\left\{ -i2\pi\left[x_1\left(\frac{x_3}{\lambda f}\right) + y_1\left(\frac{y_3}{\lambda f}\right) \right] \right\} dx_1\,dy_1 \qquad (5\text{-}34)$$

which describes a Fourier transform of U_1. The phase error multiplying the Fourier transformation varies depending on the value of d_1. If $d_1 = f$, the transform is exact with no phase error. If $d_1 \neq f$, the phase error indicates that the Fourier transform lies on the surface of a sphere* of radius $f^2/(f-d_1)$.

The transformation to the $x_3/\lambda f$, $y_3/\lambda f$ domain means that the scale of the distribution in the transform plane is dependent on the wavelength of the illumination and the focal length of the lens. The variables $x_3/\lambda f$, $y_3/\lambda f$

---

*The error in amplitude due to variations in distance to the transform surface is negligible. Recall that in deriving the diffraction formula, the second-order terms were dropped and the approximation $r = z$ was made.

are often replaced by u and v, which is acceptable when we assume that the values of x and y are small in comparison with f. If we assume that $d_1 = f$ and make the above mentioned substitution, (5-34) can be written as

$$U_3(u,v) = \frac{1}{i\lambda f} \int \int U_1(x_1,y_1) \exp[-i2\pi(ux_1 + vy_1)] \, dx_1 \, dy_1$$

$$= \frac{1}{i\lambda f} A_1(u,v). \tag{5-35}$$

That is, the spectrum of the distribution in the front focal plane of the lens appears in the back focal plane. One convenience of this notation is that u and v are normalized coordinates.

Spherical wave illumination. We have found that a Fourier transform relationship exists between the wave distributions in the front and back focal planes of a lens. By writing $U_1(x_1,y_1)$ as

$$U_1(x_1,y_1) = e(x,y)t(x,y) \tag{5-36}$$

where $e(x,y)$ describes the complex amplitude of the illuminating wave and $t(x,y)$ the amplitude transmission factor for the transparency, we see that, unless the illumination is with a uniform plane wave, we do not have the Fourier transform of the distribution described by the transparency. The convolution theorem indicates that we have the convolution of the transform of $t(x,y)$ with the transform of $e(x,y)$. It is particularly interesting to examine the case of illumination with a uniform spherical wave. Assume, for example, that

$$e(x,y) = \frac{1}{i\lambda\rho} \exp\left[\frac{ik(x_1^2 + y_1^2)}{2\rho}\right] = \frac{1}{i\lambda\rho} h(x_1,y_1;\rho), \tag{5-37}$$

which indicates that the illumination is produced by a point source a distance ρ in front of the transparency. (A converging wave is represented by a negative ρ.) The Fourier transform of $e(x,y)$ is another quadratic distribution in u and v:

$$\mathcal{F}[e(x,y)] = E(u,v) = \exp[-i\pi\lambda\rho(u^2 + v^2)] = h\left(u,v,;\frac{-1}{\lambda^2\rho}\right). \tag{5-38}$$

Consequently,

$$\mathcal{F}[U_1(x,y)] = \left[T(u,v) \otimes h\left(u,v;\frac{-1}{\lambda^2\rho}\right)\right], \tag{5-39}$$

where $T(u,v)$ is the Fourier transform of $t(x,y)$. Comparison of (5-39), with u and v replaced by $x_2/\lambda z$ and $y_2/\lambda z$, respectively, with the diffraction formula (1-24) repeated below shows that the effect of the spherical illumination is the same as smearing the transform of $t(x,y)$ due to the diffraction effects over a distance z^2/ρ:

$$U_2(x_2,y_2) = \frac{1}{i\lambda z} \exp(ikz) U_1(x_2,y_2) \otimes \exp\left[\frac{ik(x_2^2+y_2^2)}{2z}\right]. \quad (1\text{-}24)$$

Notice that when $\rho \to \infty$, the exponential term in (5-39) resembles the wave from a point source with vanishing radius, that is, the convolution is with a delta function. [See Appendix 2, equation (A2-7).]

The precise result is easier to obtain if we modify (5-31) such that the exponential term with x_1^2 and y_1^2 reads

$$\exp\left[\frac{ik(x_1^2+y_1^2)}{2}\left(\frac{1}{d_1} - \frac{w}{d_1^2} + \frac{1}{\rho}\right)\right] \quad (5\text{-}40)$$

where the $\exp[ik(x_1^2+y_1^2)/2\rho]$ term is added because of the spherical wave illumination. Now, to obtain a Fourier transform,

$$\frac{1}{d_1} - \frac{w}{d_1^2} + \frac{1}{\rho} = 0, \quad (5\text{-}41)$$

which leads to the condition

$$\frac{1}{d_2} = \frac{1}{f} - \frac{1}{\rho+d_1}. \quad (5\text{-}42)$$

If $d_1 = f$,

$$d_2 = f + \frac{f^2}{\rho}. \quad (5\text{-}43)$$

Note that the position of the *transform of* $t(x,y)$ depends, in general, on d_1 as well as ρ and is in the plane of $d_2 = f$ only when $\rho = \infty$, the radius of curvature indicating plane wave illumination of $t(x,y)$. The *transform of* $U_1(x_1,y_1)$ is still in the plane at $d_2 = f$.

It can be shown that, for spherical wave illumination, (5.34) becomes

$$U_3(x_3,y_3) = \frac{1}{i\lambda f} \exp\left[\frac{ik}{2d_2}(x_3^2+y_3^2)\frac{(d_1+\rho)(f-d_1)}{\rho f}\right]$$

$$\times \int\int t(x_1,y_1) \exp\left[\frac{-i2\pi(\rho+d_1-f)}{\lambda\rho f}(x_1x_3+y_1y_3)\right] dx_1\, dy_1. \quad (5\text{-}44)$$

Of particular interest is the scaling of the transform as indicated by the exponential inside the integral. If $d_1 = f$, the relation becomes

$$U_3(x_3,y_3) = \frac{1}{i\lambda f} \int \int t(x_1,y_1) \exp\left[\frac{-i2\pi}{\lambda f}(x_1 x_3 + y_1 y_3)\right] dx_1\, dy_1.$$

$$(5\text{-}45)$$

Consequently, even if $\rho \neq \infty$, the scale of the transform is the same as when $\rho = \infty$, if $d_1 = f$. The plane in which the transform appears is, however, given by (5-43).

Transparency behind the lens. If the transparency $t(x_2,y_2)$ appears behind the lens as shown in Fig. 5-9, the transform appears in the back focal plane if the lens is illuminated with a uniform plane wave. The transparency cannot be close to the focal plane. If an analysis similar to the preceding one is used, we find that

$$U_3(x_3,y_3) = \frac{f}{i\lambda d_2^2} \exp(ikf)h(x_3,y_3;d_2)T\left(\frac{x_3}{\lambda d_2}, \frac{y_3}{\lambda d_2}\right), \qquad (5\text{-}46)$$

where T is the Fourier transform of t. Notice the scale factor d_2 instead of f as before. This indicates that, when the transparency is behind the lens, the scale can be changed by moving the transparency. An illuminating plane wave of unity amplitude is assumed in the derivation of (5-46).

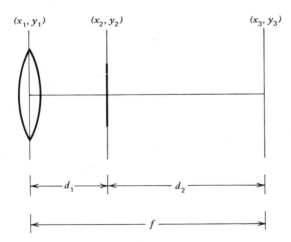

Figure 5-9 The transparency can be on the same side of the lens as the transform.

Vector distribution in the focal plane of a lens.   Scalar diffraction theory gives a Fourier transform relation between the wave distributions in the front and back focal planes of a lens. We must not forget the restrictions on this theory, however. Indeed, a longitudinal component of electric field along the axis in the focal plane has been predicted and measured.

Effect of the pupil function $P(x_2, y_2)$.   If $P(x_2, y_2)$ is retained as a multiplier of $h(x_2, y_2; w)$ in (5-28), the inner integral of (5-28) defines a Fourier transform of $P(x_2, y_2)h(x_2, y_2; w)$. The convolution theorem states that

$$\mathcal{F}\left[Ph\right] = \tilde{P} \otimes \tilde{h} \tag{5-47}$$

where the tilde ~ denotes the transform. If we use the convolution theorem and put the result into (5-28), we obtain

$$U_3(x_3, y_3) = \frac{1}{i^2 \lambda^2 d_1 d_2} h(x_3, y_3; d_2) \int\int U_1(x_1, y_1) h(x_1, y_1; d_1)$$

$$\times \left[\tilde{P}\left(\frac{x_1}{\lambda d_1} + \frac{x_3}{\lambda d_2}, \frac{y_1}{\lambda d_1} + \frac{y_3}{\lambda d_2} \right) \otimes \tilde{h}\left(\frac{x_1}{\lambda d_1} + \frac{x_3}{\lambda d_2}, \frac{y_1}{\lambda d_1} + \frac{y_3}{\lambda d_2}; \frac{-1}{w\lambda^2} \right) \right] dx_1 \, dy_1.$$

$$\tag{5-48}$$

   Appendix 2 shows that

$$\tilde{g}(x,y) \otimes \tilde{h}(x,y; -d)$$

$$= h(\lambda dx, \lambda dy; -\lambda^2 d^3)\left[g(\lambda dx, \lambda dy) \otimes h(\lambda dx, \lambda dy; \lambda^2 d^3) \right]. \tag{5-49}$$

Consequently,

$$\tilde{P}\left(\frac{x_1}{\lambda d_1} + \frac{x_3}{\lambda d_2}, \frac{y_1}{\lambda d_1} + \frac{y_3}{\lambda d_2} \right) \otimes \tilde{h}\left(\frac{x_1}{\lambda d_1} + \frac{x_3}{\lambda d_2}, \frac{y_1}{\lambda d_1} + \frac{y_3}{\lambda d_2}; \frac{-1}{\lambda^2 w} \right)$$

$$= h\left(\frac{wx_1}{d_1} + \frac{wx_3}{d_2}, \frac{wy_1}{d_1} + \frac{wy_3}{d_2}; -w \right)$$

$$\times \left[P\left(\frac{wx_1}{d_1} + \frac{wx_3}{d_2}, \frac{wy_1}{d_1} + \frac{wy_3}{d_2} \right) \otimes h\left(\frac{wx_1}{d_1} + \frac{wx_3}{d_2}, \frac{wy_1}{d_1} + \frac{wy_3}{d_2}; w \right) \right]. \tag{5-50}$$

The use of (5-50) in (5-48) and the expansion of the $\mathbf{h}(\cdots ; -w)$ term give

$$\mathbf{U}_3(x_3,y_3) = \frac{1}{i^2\lambda^2 d_1 d_2}\mathbf{h}(x_3,y_3; d_2)\mathbf{h}\left(x_3,y_3; -\frac{d_2^2}{w}\right)$$

$$\times \int\int \mathbf{U}_1(x_1,y_1)\mathbf{h}(x_1,y_1; d_1)\mathbf{h}\left(x_1,y_1; -\frac{d_1^2}{w}\right)$$

$$\times \left[\mathbf{P}\left(\frac{wx_1}{d_1} + \frac{wx_3}{d_2}, \frac{wy_1}{d_1} + \frac{wy_3}{d_2}\right)\otimes\mathbf{h}\left(\frac{wx_1}{d_1} + \frac{wx_3}{d_2}, \frac{wy_1}{d_1} + \frac{wy_3}{d_2}; w\right)\right]$$

$$\times \exp\left\{-i2\pi\left[x_1\left(\frac{wx_3}{\lambda d_1 d_2}\right) + y_1\left(\frac{wy_3}{\lambda d_1 d_2}\right)\right]\right\} dx_1\, dy_1. \qquad (5\text{-}51)$$

We can see that if $w = d_1$ two of the $\mathbf{h}$'s cancel, and (5-51) takes the form $\mathcal{F}[\mathbf{U}_1(\mathbf{P}\otimes\mathbf{h})]$. Letting $w = d_1$ means, from (5-29), that $d_2 = f$. That is, the Fourier transform appears one focal length behind the lens. Equation (5-51) becomes

$$\mathbf{U}_3 = \frac{1}{i^2\lambda^2 d_1 f}\exp\left[\frac{ik}{2f}(x_3^2+y_3^2)\left(1-\frac{d_1}{f}\right)\right]\int\int \mathbf{U}_1(x_1,y_1)$$

$$\times \left\{\mathbf{P}\left(x_1 + \frac{d_1 x_3}{f}, y_1 + \frac{d_1 y_3}{f}\right)\otimes\mathbf{h}\left[x_1 + \frac{d_1 x_3}{f}, y_1 + \frac{d_1 y_3}{f}; d_1\right]\right\}$$

$$\times \exp\left\{-i2\pi\left[x_1\left(\frac{x_3}{\lambda f}\right) + y_1\left(\frac{y_3}{\lambda f}\right)\right]\right\} dx_1\, dy_1. \qquad (5\text{-}52)$$

Because the convolution of $\mathbf{P}$ with $\mathbf{h}$ represents the distribution arising from the Fresnel diffraction of $\mathbf{P}$, $\mathbf{U}_1$ is vignetted by $\mathbf{P}$ as seen at the input to the system.

If $d_1 = 0$, $h(\cdots)$ within the integral of (5-52) becomes a delta function [see (A2-7) of Appendix 2], and (5-52) becomes the Fourier transform of $\mathbf{U}_1\mathbf{P}$, indicating that $\mathbf{U}_3$ is proportional to the convolution, $\mathbf{A}_1\otimes\tilde{\mathbf{P}}$. If d_1 is small, diffraction plays a small part, and the vignetting of $\mathbf{U}_1$ can be described geometrically. The finite size of the lens then restricts the region of $\mathbf{U}_1$ which can contribute to the transform. Consideration of the variables of $\mathbf{P}(x_1 + d_1 x_3/f, y_1 + d_1 y_3/f)$ indicates that the effect is the same as having an aperture the size of the lens placed in the x_1, y_1 plane with its center at $(-d_1 x_3/f, -d_1 y_3/f)$. Another way of interpreting the effect is to say that the lens restricts the region of $\mathbf{U}_1$ seen from any position x_3, y_3 to be that visible through the lens. See Fig. 5-10, which shows the region of $\mathbf{U}_1$ seen from x_3, y_3. This means, of course, that only a portion of $\mathbf{U}_1$ can contribute to the value of the transform at x_3, y_3, and that the contributing region is different for different values of x_3, y_3.

If the lens is of sufficient size that $\mathbf{P}$ can be considered to be unity, $1 \otimes \mathbf{h} = i\lambda d$, and (5-52) becomes (5-34).

5.3 Image Formation

We consider two different image-formation techniques—double-lens and single-lens. The double-lens system is described in terms of a double Fourier transform, and the imaging properties of a single lens are derived using the scalar diffraction formula. The effect of a lens of finite size on the imaging process is determined during the scalar diffraction analyses. We restrict ourselves here to an examination of the conditions under which an image is formed. A different approach to the analysis of an imaging system is given in Chapter 8.

Two lens imaging.   We have seen that a single lens can produce a Fourier transform of a distribution in its front focal plane. This property can be used to perform two Fourier transformations with two lenses to return to the original distribution. If a transparency $\mathbf{t}(x,y)$ is illuminated by a uniform plane wave, $\mathbf{t}(x,y)$ is the same as $\mathbf{U}_1(x_1, y_1)$, and an imaging system

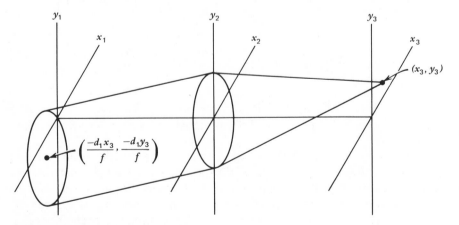

Figure 5-10   Effect of the lens aperture.

can be formed as shown in Fig. 5-11. The transform of the wave $U_1(x_1,y_1)$ appears in the x_2,y_2 plane as

$$U_2(x_2,y_2) = \frac{1}{i\lambda f_1} A_1\left(\frac{x_2}{\lambda f_1}, \frac{y_2}{\lambda f_1}\right) \qquad (5\text{-}53)$$

where A_1 is the Fourier transform of U_1, the lens is assumed to be large, and the transformation is from the x_1,y_1 domain to the $x_2/\lambda f, y_2/\lambda f$ domain. The next lens picks up $U_2(x_2,y_2)$ and transforms it to yield, in the x_3,y_3 plane,

$$U_3(x_3,y_3) = \left(\frac{1}{i\lambda f_1}\right)\left(\frac{1}{i\lambda f_2}\right) \int \int A_1\left(\frac{x_2}{\lambda f_1}, \frac{y_2}{\lambda f_1}\right)$$

$$\times \exp\left[\frac{-i2\pi(x_2 x_3 + y_2 y_3)}{\lambda f_2}\right] dx_2\, dy_2. \qquad (5\text{-}54)$$

Two modifications of (5-54) are necessary. First, an inverse transform is needed to go from A_1 to U_1 and, second, the variable is normalized by λf_1 but the exponential contains f_2, the focal length of the second lens. The inverse transformation is obtained by reversing the sign of x_3 and y_3, that is, incorporating the negative sign of the exponent into the coordinate. The

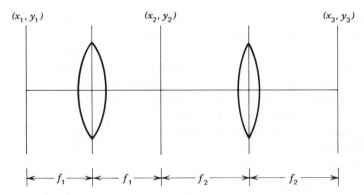

Figure 5-11   A two-lens imaging system.

variables are corrected by factoring f_1 from the exponent and the differential. That is,

$$U_3(x_3,y_3) = \frac{f_1}{f_2} \int \int A_1\left(\frac{x_2}{\lambda f_1}, \frac{y_2}{\lambda f_1}\right) \exp\left\{ i2\pi\left[\left(\frac{-f_1 x_3}{f_2}\right)\left(\frac{x_2}{\lambda f_1}\right)\right.\right.$$

$$\left.\left. + \left(\frac{-f_1 y_3}{f_2}\right)\left(\frac{y_2}{\lambda f_1}\right)\right]\right\} d\left(\frac{x_2}{\lambda f_1}\right) d\left(\frac{y_2}{\lambda f_1}\right). \qquad (5\text{-}55)$$

The multiplying factors indicate a change in amplitude of the wave describing the image, and the negative sign in the relationship

$$U_3(x_3,y_3) = \frac{f_1}{f_2} U_1\left(\frac{-f_1 x_3}{f_2}, \frac{-f_1 y_3}{f_2}\right) \qquad (5\text{-}56)$$

indicates that U_3 is an inverted image of U_1. The image is magnified by the factor $f_2/f_1 \equiv M$,

$$U_3(x_3,y_3) = \frac{1}{M} U_1\left(\frac{-x_3}{M}, \frac{-y_3}{M}\right). \qquad (5\text{-}57)$$

The negative signs in (5-57) can be interpreted as being due to the inability of a lens to perform an inverse transformation—only direct Fourier transforms.

Single-lens imaging. So far we have seen that a lens can perform a Fourier transform and two of them can form an image. If this were all we knew about optics, it might appear strange that a single lens can also form an image. Actually, many imaging systems use only one lens (the eye, e.g.). The diffraction formula can be used to show the imaging capabilities of a single lens and, at the same time, indicate the effect of the lens size.

Assume that a transparency having transmittance $t(x,y)$ is placed in the x_1,y_1 plane of Fig. 5-12 and is illuminated by a wave described by $e(x,y)$. Consequently,

$$U_1(x_1,y_1) = e(x,y)t(x,y). \qquad (5\text{-}58)$$

The distribution in the x_2,y_2 plane is found as for the Fourier-transforming

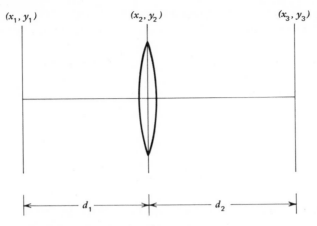

Figure 5-12   Single-lens imaging system.

case. The expression (5-28) can be written in the form

$$U_3(x_3,y_3) = \frac{1}{\lambda^2 d_1 d_2} \mathbf{h}(x_3,y_3; d_2) \int \int U_1(x_1,y_1)\mathbf{h}(x_1,y_1; d_1)$$

$$\times \int \int \mathbf{P}(x_2,y_2)\mathbf{h}(x_2,y_2; w) \exp\left\{ -i2\pi\left[x_2\left(\frac{x_1}{\lambda d_1} + \frac{x_3}{\lambda d_2} \right) \right. \right.$$

$$\left. \left. + y_2\left(\frac{y_1}{\lambda d_1} + \frac{y_3}{\lambda d_2} \right) \right] \right\} dx_2\, dy_2\, dx_1\, dy_1, \qquad (5\text{-}59)$$

where the i^2 or minus sign has been dropped.

Before considering the effect of the pupil function, **P**, let us find the conditions for an image to be formed. Assume that the lens is perfect and sufficiently large that $\mathbf{P}(x_2,y_2)$ can be replaced by unity over the region of integration. Inspection of (5-59) reveals that U_3 can be written in terms of U_1 if the integral over x_2 and y_2 produces a delta function which would take U_1 out of the integral over x_1 and y_1. If $\mathbf{h}(x_2,y_2; w)$ is unity, the integral over x_2 and y_2 becomes the Fourier transform of unity giving $\delta(x_1/\lambda d_1 + x_3/\lambda d_2,\ y_1/\lambda d_1 + y_3/\lambda d_2)$. The function $\mathbf{h}(x_2,y_2; w)$ is unity if $1/w$ is zero. Equation (5-29) shows that this means

$$\frac{1}{d_1} + \frac{1}{d_2} = \frac{1}{f}, \qquad (5\text{-}60)$$

which is the condition for image formation.

With (5-60) satisfied and $P(x_2, y_2)$ taken as unity, (5-59) becomes

$$U_3(x_3, y_3) = \frac{1}{\lambda^2 d_1 d_2} h(x_3, y_3; d_2) \int \int U_1(x_1, y_1) h(x_1, y_1; d_1)$$

$$\times \delta\left(\frac{x_1}{\lambda d_1} + \frac{x_3}{\lambda d_2}, \frac{y_1}{\lambda d_1} + \frac{y_3}{\lambda d_2}\right) dx_1 dy_1. \qquad (5\text{-}61)$$

The relation

$$\delta(ax) = \frac{1}{|a|} \delta(x) \qquad (5\text{-}62)$$

can be used to rewrite the delta function of (5-61) as

$$\delta\left[\frac{1}{\lambda d_1}\left(x_1 + \frac{d_1 x_3}{d_2}\right), \frac{1}{\lambda d_1}\left(y_1 + \frac{d_1 y_3}{d_2}\right)\right]$$

$$= |\lambda d_1|^2 \delta\left(x_1 + \frac{d_1 x_3}{d_2}, y_1 + \frac{d_1 y_3}{d_2}\right). \qquad (5\text{-}63)$$

Consequently, (5-61) becomes

$$U_3(x_3, y_3) = \frac{d_1}{d_2} h(x_3, y_3; d_2) U_1\left(-\frac{d_1 x_3}{d_2}, -\frac{d_1 y_3}{d_2}\right) h\left(\frac{d_1 x_3}{d_2}, \frac{d_1 y_3}{d_2}; d_1\right)$$

$$= \frac{1}{M} h(x_3, y_3; d_2) U_1\left(-\frac{x_3}{M}, -\frac{y_3}{M}\right) h\left(\frac{x_3}{M}, \frac{y_3}{M}; d_1\right), \qquad (5\text{-}64)$$

where

$$h(-x, -y; d) = h(x, y; d) \qquad (5\text{-}65)$$

and

$$M \equiv \frac{d_2}{d_1}. \qquad (5\text{-}66)$$

The two h's of (5-64) can be combined so that the image is described by

$$U_3(x_3, y_3) = \frac{1}{M} \exp\left[\frac{ik}{2d_2}(x_3^2 + y_3^2)\left(\frac{1}{M} + 1\right)\right] U_1\left(\frac{-x_3}{M}, \frac{-y_3}{M}\right). \qquad (5\text{-}67)$$

The difference between (5-67) and (5-57), the equation describing imaging with two lenses, is the phase term that arises when a single lens is used. If a uniform spherical wave illuminates the object, this phase term can be canceled. Let U_1 be replaced by the expression in (5-58) and $e(x,y)$ be represented by $h(x_1, y_1; \rho)$. Equation (5-67) then can be written as

$$U_3(x_3, y_3) = \frac{1}{M} \exp\left[\frac{ik}{2}(x_3^2 + y_3^2)\left(\frac{1}{Md_2} + \frac{1}{d_2} + \frac{1}{M^2\rho}\right) \right] t\left(\frac{-x_3}{M}, \frac{-y_3}{M}\right).$$

(5-68)

If

$$\frac{1}{\rho} = -M\left(\frac{1}{d_2} + \frac{M}{d_2}\right) = -M\left(\frac{1}{d_1} + \frac{1}{d_2}\right)$$

(5-69)

or

$$\rho = -\frac{f}{M},$$

(5-70)

the phase term vanishes. Physically this means that, in the absence of a transparency in plane x_1, y_1, the illuminating wave would focus one focal length in front of the imaging lens and be collimated by the lens.

It is interesting to note that, while spherical wave illumination changes the location of the *transform* of $t(x,y)$, it does not affect the location of the *image*.

Now let us consider the effect of the pupil function $P(x_2, y_2)$ in more detail. The integral over x_2 and y_2 of (5-59) can be written in terms of the convolution of the transforms of P and h. That is,

$$U_3(x_3, y_3) = \frac{1}{\lambda^2 d_1 d_2} h(x_3, y_3; d_2)$$

$$\times \int\int \left\{ \tilde{P}\left(\frac{x_1}{\lambda d_1} + \frac{x_3}{\lambda d_2}, \frac{y_1}{\lambda d_1} + \frac{y_3}{\lambda d_2}\right) \otimes \mathfrak{F}[h(x_2, y_2; w)] \Big|_{\left(\frac{x_1}{\lambda d_1} + \frac{x_3}{\lambda d_2}, \frac{x_2}{\lambda d_1} + \frac{x_3}{\lambda d_2}\right)} \right\}$$

$$\times U_1(x_1, y_1) h(x_1, y_1; d_1) \, dx_1 \, dy_1,$$

(5-71)

which has the form of the correlation of $[\tilde{P} \otimes \tilde{h}]$ with $U_1 h$.

If the $\tilde{h}$ term could be removed, (5-71) would represent the correlation of an image distribution with the transform of the pupil function. If $w = \infty$, $h(x_2, y_2; w) = 1$ and the Fourier transform yields a delta function having the form $|\lambda d_1|^2 \delta(x_1 + \frac{d_1}{d_2} x_3, y_1 + \frac{d_1}{d_2} y_3)$. The requirement $w = \infty$ means that

$$\frac{1}{d_1} + \frac{1}{d_2} = \frac{1}{f}$$

(5-72)

which is the imaging condition. If the resulting delta function is convolved with $\tilde{\mathbf{P}}$, (5-71) can be rewritten as

$$\mathbf{U}_3(x_3,y_3) = \frac{d_1}{d_2}\mathbf{h}(x_3,y_3;d_2) \int \int \tilde{\mathbf{P}}\left(x_1 + \frac{d_1}{d_2}x_3, y_1 + \frac{d_1}{d_2}y_3\right)$$

$$\times \mathbf{U}_1(x_1,y_1)h(x_1,y_1;d_1)\,dx_1\,dy_1. \qquad (5\text{-}73)$$

Notice that the correlation form of (5-73) can be changed to look like a convolution if the coordinates are reversed, that is,

$$\mathbf{g}(x) \star \mathbf{h}(x) = \mathbf{g}(x) \otimes \mathbf{h}(-x). \qquad (5\text{-}74)$$

Consequently,

$$\mathbf{U}_3(x_3,y_3) = \frac{d_1}{d_2}\mathbf{h}(x_3,y_3;d_2)\left\{\tilde{\mathbf{P}}\left(\frac{d_1 x_3}{d_2}, \frac{d_1 y_3}{d_2}\right)\right.$$

$$\left.\otimes\left[\mathbf{U}_1\left(\frac{-d_1}{d_2}x_3, \frac{-d_1}{d_2}y_3\right)\mathbf{h}\left(\frac{-d_1}{d_2}x_3, \frac{-d_1}{d_2}y_3; d_1\right)\right]\right\}, \qquad (5\text{-}75)$$

which can be written as

$$\mathbf{U}_3(x_3,y_3) = \frac{1}{M}\mathbf{h}(x_3,y_3;d_2)\left\{\tilde{\mathbf{P}}\left(\frac{x_3}{M}, \frac{y_3}{M}\right)\right.$$

$$\left.\otimes\left[\mathbf{U}_1\left(\frac{-x_3}{M}, \frac{-y_3}{M}\right)\mathbf{h}\left(\frac{x_3}{M}, \frac{y_3}{M}; d_1\right)\right]\right\}. \qquad (5\text{-}76)$$

This expression shows that the image is magnified (or demagnified), inverted, and that the resolution is reduced by the convolution with $\tilde{\mathbf{P}}$. If $\mathbf{P}$ is a circle function representing a uniformly transparent, circular lens, the convolution is with $J_1(r)/r$ where $J_1(r)$ is a Bessel function. We will examine the effects of $\mathbf{P}$ further in Chapter 8.

PROBLEMS

5-1. Examine all possible relations between R_1 and R_2, the radii of curvature of a lens, and indicate whether the lens has a positive or negative focal length. Include the special cases of R_1 or $R_2 = \infty$. Sketch typical lenses.

5-2. Derive (5-46) using the techniques employed in obtaining (5-35). Ignore the effect of a finite lens diameter.

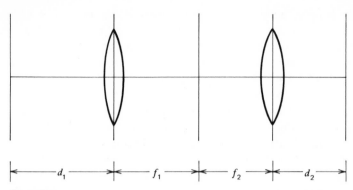

Figure P5-4

5-3. Let the input to a single lens imaging system be a point represented by $\delta(x_1,y_1)$ and find the distribution in the image plane when the lens is a perfect lens of diameter D. That is, work out the impulse response of the imaging system.

5-4. A coherent imaging system is set up to use a two-lens imaging system as shown in Fig. P5-4. Instead of the transparency of transmittance $t(x,y)$ being in the focal plane in front of the lens, however, it is a distance d_1 in front of the first lens. Where will the image of $t(x,y)$ appear?

5-5. A transparency has an amplitude transmittance described by

$$t(r) = \left(\frac{1}{2} + \frac{1}{2} \cos ar^2 \right) \mathrm{circ}\left(\frac{r}{b} \right).$$

Describe the resulting waves when the transparency is illuminated by a uniform plane wave. Compare the effect of the transparency with that of a lens. Find the equivalent focal lengths. (Note that this is the transmittance of a hologram of a point object when the reference is a uniform plane wave propagating coaxially with the wave from the point object.)

5-6. A transparency has a transmittance of

$$t(r) = \left[\frac{1}{2} + \frac{1}{2} \mathrm{sgn}(\cos ar^2) \right] \mathrm{circ}\left(\frac{r}{b} \right).$$

Describe the resulting waves when plane wave illumination is used. Write $\mathrm{sgn}(\cos ar^2)$ in terms of a Fourier series in ar^2. Find a relation giving the relative power in the resulting foci of each wave. Show that the power *density* in each focus is the same.

5-7.   A grating having a transmittance of $t(x,y) = 1 + \cos(2\pi x/\sigma)$ is moved with a velocity $\mathbf{v}_x$ in the front focal plane of a lens. Assume a normally incident, uniform plane wave and describe the distribution occurring in the back focal, or Fourier transform, plane. In particular, find the temporal frequency shift of the off-axis spots [5-3].

5-8.   Assume a single lens imaging system where a transparency $t(x_1, y_1)$ is illuminated by a converging spherical wave. Review (5-68) through (5-70) and show that the illuminating wave focuses one focal length in front of the imaging lens when the phase term in the image vanishes. Sketch the system and show the illuminating wave as it passes through.

5-9.   Under what conditions will the Fourier transform plane coincide with the image plane?

5-10.   Assume $\mathbf{P}(x_2, y_2) = 1$ in (5-59) and evaluate the integrals using the results of Appendix 2 to obtain (5-34).

5-11.   A two-dimensional signal is to be Fourier analyzed. The lowest spatial frequency component expected is 20 cycles/mm and the highest is 200 cycles/mm. Design a spatial spectral analyzing system to display the highest and lowest spectral components with a separation of 90 mm in the spectral plane. Use a collimated illumination beam at $\lambda = 500$ nm and a single Fourier transforming lens. Find the focal length of the transforming lens needed.

REFERENCES

5-1.   See, for example, F. A. Jenkins and H. E. White, *Fundamentals of Optics* (McGraw-Hill Book Co., New York, 1957).

5-2.   See, for example, W. Brouwer, *Matrix Methods in Optical Instrument Design* (W. A. Benjamin, Inc., New York, 1964).

5-3.   T. Suzuki and R. Hioki, "Translation of Light Frequency by Moving Grating," *J. Opt. Soc. Am.* **57**, 1551 (1967).

6

Properties of Recorders and Spatial Modulators

In most of the cases of recording or reconstructing a wavefront discussed so far, perfect recording materials have been assumed. That is, the materials have been assumed to be linear and to have infinite resolution. Before treating the subjects of spatial filtering and holography in detail, we must consider recording media having nonlinearity and limited resolution. Some of the effects of finite recording range and resolution are discussed more appropriately in later chapters, but general properties of many available materials are described here.

A large portion of the present chapter is devoted to a description of photographic emulsions. Film is used for most recordings of optical information, and the techniques of describing the effects of film characteristics are applicable to other recorders. The characteristics of other materials used for holograms are presented briefly. In particular, thermoplastics, photochromics, magnetic materials, and crystals are discussed.

6.1 Exposure Mechanism of Photographic Emulsions

The photographic emulsion contains the photosensitive materials that change composition when struck by a photon. The base on which the emulsion is placed is normally an acetate film, polyester film, or glass. For use in holography, glass has the advantage of preventing changes in the shape of the hologram between the wavefront recording and reconstruction stages.

Sensitivity to light. The *exposure* of an emulsion is the product of the intensity of the wave and the time for which it falls on the emulsion.

Consequently, the units of exposure are energy per unit area, normally microjoules per square centimeter. The fundamental photosensitive components of the emulsion are the silver halides, which are sensitive only to shorter wavelengths such as blue and violet. The emulsion can be made sensitive to a wide range of wavelengths by the addition of sensitizing dyes. Examples of the spectral sensitivity of some emulsions commonly used in holography are shown in Figs. 6-1 and 6-2. The Eastman Kodak emulsion, 649-F has a relatively flat spectral response, which makes it useful in multiple color holographic recordings. The four Agfa-Gavaert emulsions were specially sensitized to be sensitive near the 514-nm line of the argon laser and the 694.3-nm line of the ruby laser. The curves of Figs. 6-1 and 6-2 cannot be accurately compared at the shorter wavelengths. The curve for the Kodak material is for equal energy at all wavelengths, whereas the Agfa curves are tungsten spectrograms which have the effect of suppressing the curves at progressively shorter wavelengths. The Agfa-Gavaert series ending in 70 is no longer generally available, but was widely used and is referenced extensively in holographic literature.

The sensitivity of the emulsion is roughly inversely proportional to the *resolution*, where resolution is an indication of how small a spatial period can be resolved. This effect can be described as being a function of the size of the photographic *grain*. Each grain has one or more *development centers* which are affected by incident photons. If only one of these centers is activated by a photon, *development* of the emulsion with a photographic developer causes the entire grain to decompose into silver and a halide.

The form of the deposited silver is dependent on the type of developer [6-1]. If the silver ions are provided primarily by the silver halide crystals,

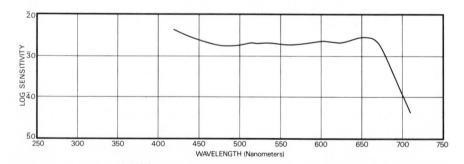

Figure 6-1 Spectral sensitivity curves of Kodak 649-F materials developed in D-19 for 5 min at 20°C. [After *Kodak Plates and Films for Science and Industry* (Eastman Kodak Co., Rochester, N. Y., 1967).] (Reproduced with permission from a copyrighted Eastman Kodak publication.)

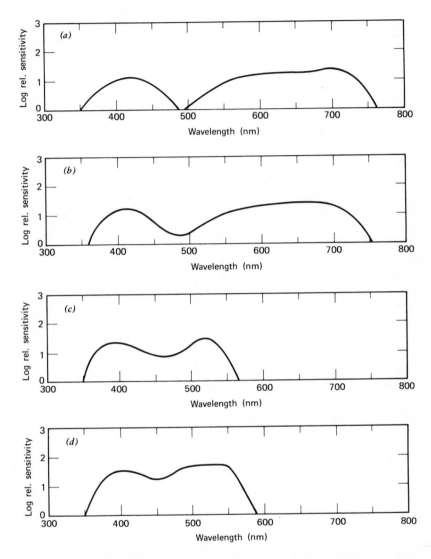

Figure 6-2 Spectral sensitivity curves of Agfa-Gevaert. (*a*) 8E75, (*b*) 10E75, (*c*) 8E56, and (*d*) 10E56. [After *Photographic Materials for Holography* (Agfa-Gevaert, Belgium, 1972).]

125

the grain can be a mass of silver filaments as shown in Fig. 6-3a. In extremely high resolution emulsions, single sections of a filament are normal. This reaction is called *chemical development*. If the silver ions are provided by the developer in the form of a soluble silver salt, for example, the grains are solid as shown in Fig. 6-3b. This type of reaction occurs in *physical development*. Obviously, greater density can be obtained with chemical development than with physical development. A combination process, referred to as *solution physical development*, occurs when there is a silver halide solvent in the developer.

The process of *fixing* the emulsion removes the unexposed silver halide so that exposure to light will not cause further decomposition of the compound and further darkening of the photographic plate. Development is not necessary to cause the deposition of silver. A large exposure will darken the plate without development. A thorough description of the photographic process is given by Mees and James [6-2]. Appendix 3 describes the developing procedures normally followed in making holograms.

If high resolution is required, the grains are smaller, needing a greater number of photons activating a greater number of development centers to produce the same density of silver in the developed emulsion. Consequently, the emulsion is less sensitive. A sensitive emulsion has large grains with many development centers, any one of which could cause the development of the entire grain. Table 6-1 gives the resolution and sensitivity of some of the common emulsions. The peak sensitivity is indicated. Note that the inverse relationship between sensitivity and resolution depends on the width of the spectral sensitivity. Another deviation is the large decrease in sensitivity when very high resolutions are required.

Photographic density. The *photographic density* $\mathfrak{D}$ of a developed transparency was defined by F. Hurter and V. C. Driffield in 1890 using the relation

$$\mathfrak{D} = \log\left(\frac{1}{\tau}\right) \qquad (6\text{-}1)$$

where τ is the intensity transmittance of the transparency. That is, the density is the logarithm of the opacity of the silver deposit. The manner in which the density increases as the exposure increases is an important characteristic of photographic film. A plot of the density as a function of the logarithm of the exposure is referred to as the Hurter-Driffield or H & D curve. A typical curve is shown in Fig. 6-4. The flat region below the toe is the density of the substrate or emulsion support plus the density produced by development of an unexposed emulsion. This density is known as *gross fog*. The region just above the gross fog is the *threshold*

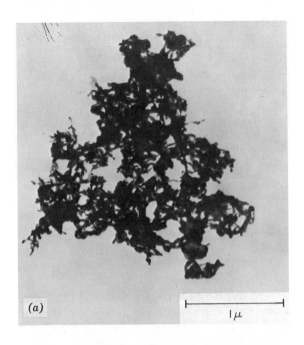

(a)

1 μ

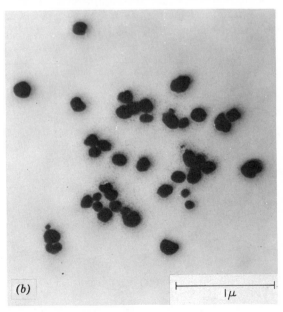

(b)

1 μ

Figure 6-3 Electron micrographs of developed silver formed by (*a*) chemical development and (*b*) physical development [6-1].

TABLE 6-1   RELATIVE SENSITIVITY AND RESOLUTION OF SOME EMULSIONS

Emulsion	Relative Sensitivity	Resolution (lines/mm)
I-F	125	69–95
II-F	56	69–95
III-F	18	96–135
IV-F	8	136–225
V-F	0.36	Above 225
649-F	0.005	Above 2000
14C75	9.5	1500
10E75	0.57	2800
8E75	0.14	3000

exposure, that is, the minimum exposure that will produce an increase in density. The linear region is of interest if the film is used to record an image with a large range of exposures. It is *not* the region of exposure that results in a linear variation in the amplitude transmittance of a transparency as a function of the exposure. The *shoulder* of the curve is the region where saturation begins.

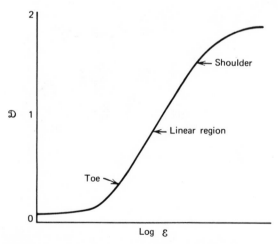

Figure 6-4   A typical Hurter-Driffield (H&D) curve.

6.2   Resolution and Modulation Transfer Function

Specifications of the *resolution* of an emulsion or any recording material cannot give a complete picture of the recording capabilities of the material. A specification is needed of *how well* fringes of a certain frequency are recorded. Once this information is available, we can determine the visibility of the fringes, hence the efficiency of the recorder when used as a hologram.

Modulation transfer function. The modulation of the density of a transparency is caused by recording a sinusoidal fringe pattern. The ratio of the visibility of the fringes of the image to the visibility of the fringes of the exposing pattern is called the MTF or *modulation transfer function* of the recording medium. If the concept is generalized, it can be applied to bleached emulsions or other phase modulators by determination of the change in index of refraction rather than density of the material. The change in index can be determined by measurement of the effectiveness of the recording as a grating. By measurement of the levels of the diffracted waves, the level of each spatial frequency component of the recorded fringes can be obtained.

Figure 6-5 shows the modulation transfer functions for a variety of emulsions and developers. The technique used in measuring the MTF of the emulsion of Fig. 6-5 is described in reference [6-3]. Figure 6-6 is a MTF curve for 649-F bleached in mercuric chloride. Notice that the resolution of a particular film cannot be specified uniquely, but the spatial frequency

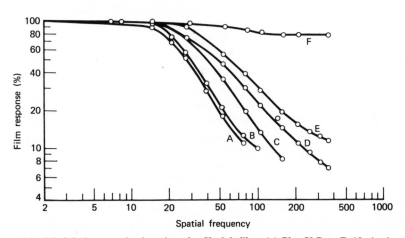

Figure 6-5   Modulation transfer functions for Kodak films (*a*) Plus-X Pan, D-19, 4 min; (*b*) Tri-X Pan, D-19, 4 min; (*c*) Pan-X, D-19, 4 min; (*d*) Panchromatic Separation, D-76, 6.5 min; (*e*) High Contrast Copy, D-19, 4 min; (*f*) 649-F, D-19, 5 min [6-3].

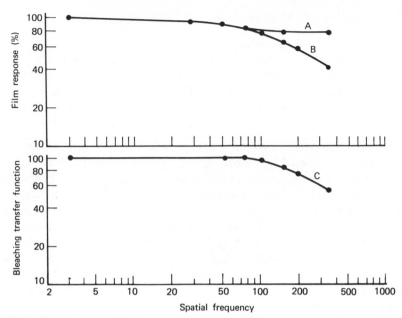

Figure 6.6 The modulation transfer function for Kodak 649-F plates bleached in mercuric chloride: (*a*) unbleached, (*b*) bleached, and (*c*) bleaching transfer function [6-3].

beyond which the MTF drops below a certain level can be given. For some applications, operation in the flat part of the curve is necessary.

The importance of the MTF is related to the diffraction efficiency of the hologram. If the fringe spacing is too close, the fringes are not recorded and the resolution of the reconstructed image is reduced, the field of view is reduced, or both, depending on the type of hologram recorded. These effects are discussed in Chapter 9.

The visibility of the recorded fringes is found by multiplying the visibility of the exposing fringes by the value of the MTF at the fringe spatial frequency. For example, the exposing fringes formed by two plane waves of amplitude U and V are described by

$$I(x) = U^2 + V^2 + 2UV \cos 2\zeta x \qquad (6\text{-}2)$$

where the direction cosines of the U and V waves are ζ and $-\zeta$, respectively. The visibility is then given by

$$\upsilon = \frac{2UV}{U^2 + V^2}. \qquad (6\text{-}3)$$

The intensity distribution (6-2) can be written in terms of the visibility

$$I(x) = (U^2 + V^2)[1 + \upsilon \cos 2\zeta x].\tag{6-4}$$

In evaluating the hologram (of a plane wave in this case), υ is multiplied by the MTF at ζ/π cycles/mm. The result determines the amplitude of the reconstructed plane wave.

6.3 Effects of Nonlinearities and Grain Noise

Distortions and multiple images are the result of a nonlinear recording medium. Scattering of unwanted light into the image is due to grain noise. A brief discussion of the effect of nonlinearities is given so that the efforts to achieve linearity described in Section 6.6 can be appreciated. Grain noise cannot be eliminated and can be reduced only by reducing the size of the grain.

Grain noise.   Let us first consider grain noise. Grain noise results from the statistical fluctuation of the number of grains in a developed emulsion. Grain noise is therefore low in regions of low exposure and increases as the exposure increases. That is, grain noise is signal dependent. In holography, the primary effect of grain noise is to scatter unwanted light into the image. The effect of this noise has been studied using various models of the photographic emulsion, and the signal-to-noise (S/N) ratio was found to be independent of the size and resolution of the hologram if the object has a diffuse surface [6-4]. On the other hand, if the object is a large number of point sources, the S/N ratio was found to be directly proportional to the product of the hologram size and resolution. This affects the density of data that can be stored holographically. The film grain models used in most studies are the checkerboard and overlapping circular-grain model discussed by O'Neill [6-5], but some workers have used models having arbitrary size and shape [6-6, 6-7].

The amount of energy scattered in unwanted directions is dependent, of course, on the grain size or resolution of the film, but it also varies with transmittance and angle. Some of the scattering caused by a hologram is the result of variations in the thickness of the emulsion after exposure and development. This effect contributes strongly to the scatter at low spatial frequencies, but has a much smaller effect at high spatial frequencies. Variations in emulsion thickness are treated in the next section.

Figure 6-7 shows the scattering of 14C70, 10E70, 8E70, and 649-F as measured by Biedermann [6-8]. The development was in Agfa-Gevaert Varitol N for 5 min at 20°C. The mean amplitude transmittance was 0.45.

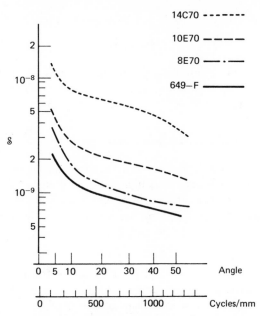

Figure 6-7 Scattering at 633 nm of Agfa-Gevaert Scientia 14C70 film, 10E70 and 8E70 plates, and Kodak Spectroscopic Plate 649-F developed in Varitol N for 5 min at 20°C. Exposure was to 633 nm. The abscissa is given in both cycles per millimeter and the angle at which the scattering appears [6-8].

The scattered power is normalized to the incident power and to a two-dimensional bandwidth. This allows the scattering to be measured in terms of noise power per unit bandwidth per unit input power. The spatial frequency at which the noise appears is determined by calculating the spatial frequency of a grating that would diffract energy at the angle of measurement.

The Agfa-Gevaert emulsions 14C70, 10E70, and 8E70 have been discontinued and the use of the series with a 75 suffix is suggested. The same decrease in scatter with increase in resolution of the emulsion would still be expected, however. The resolution of the emulsions used to gather data for Fig. 6-7 is lowest for Agfa 14C70 film and is highest for Kodak 649-F. Some of the additional scatter of the 14C70 film is due to its film base. Figure 6-8 shows the scatter produced by various bases.

The scattering as a function of the amplitude transmittance is plotted in Fig. 6-9. Note that the scatter peaks near an amplitude transmittance of 0.6. The solid curve shows the scatter predicted using a circular grain model for the emulsion [6-9]. It appears that the scattering is not a function

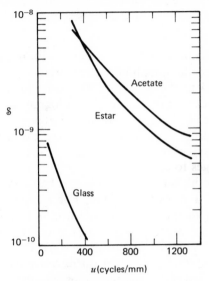

u(cycles/mm)

Figure 6-8 Scattering as a function of spatial frequency for three bases [6-9].

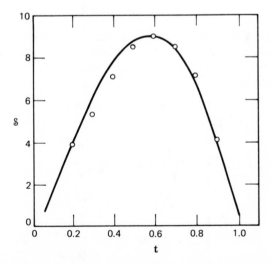

Figure 6-9 Scattering as a function of amplitude transmittance **t**. Solid curve: calculated scattering using an overlapping circular grain model for the emulsion. Experimental points: scatter from Kodak 649-F at 500 cycles/nm [6-9].

of the ratio of the reference wave intensity to the object wave intensity [6-10]. However, the effects of nonlinearities are strongly dependent on this ratio.

Distortions due to nonlinearities.   In working with transparencies, we have thus far assumed that the amplitude transmittance of the transparency is linearly proportional to the exposure. This was implicit in the equations relating the amplitude transmittance of a hologram to the intensity distribution given by the squared sum of the reference and object waves. In addition to higher-order images, a nonlinear relation between amplitude transmittance and exposure produces noise in the first-order image, and can cause false images to appear.

The production of higher-order images can be described in terms of the spatial spectrum of a distorted grating. For example, if the object and reference wave are plane, a cosine distribution in intensity is produced. If the recording is nonlinear, the fringes are not cosinusoidal fringes but are distorted. A Fourier analysis of the fringes results in higher harmonics, which obviously produce higher-order diffracted waves. To see how noise can appear in the first-order image, let us assume a nonlinear relation between the amplitude transmittance of a transparency and the exposure.

Figure 6-10 shows a typical curve of amplitude transmittance as a function of exposure for a photographic recorder. Obviously the relation between t and $\mathscr{E}$ is linear in only a small region. The nonlinear relation can be written in terms of a series

$$t = c_0 + c_1\mathscr{E} + c_2\mathscr{E}^2 + \cdots. \tag{6-5}$$

The constant c_0 is the base or zero exposure transmittance, and the $c_1\mathscr{E}$ term is the linear relation assumed previously. All higher-order terms contribute extraneous images and distortions of the first-order images [6-11]. The $c_2\mathscr{E}^2$ term is

$$c_2\left\{[\mathbf{UU^*}+\mathbf{VV^*}+\mathbf{UV^*}+\mathbf{U^*V}]\Delta t\right\}^2 \tag{6-6}$$

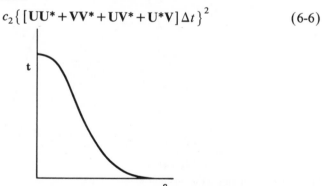

Figure 6-10 Typical curve of amplitude transmittance as a function of exposure.

where Δt is the exposure time. The phase terms describing the angles of propagation of U and V are included in the phases of the complex U and V. Evaluating (6-6) gives

$$c_2(\Delta t)^2 \left[(UU^*)^2 + (VV^*)^2 + 4UU^*VV^* + 2UV(U^*)^2 + U^*V^*U^2 \right.$$

$$\left. + 2U^*V^*V^2 + 2UV(V^*)^2 + U^2(V^*)^2 + V^2(U^*)^2 \right]. \qquad (6\text{-}7)$$

The first three terms of (6-7) represent waves that appear on the same axis as the read-out wave. The last two terms describe waves that appear as second-order waves. Notice that not only is the angle of propagation doubled, but also the rest of the phase. For example, the $U^2(V^*)^2$ term appears at twice the angle as the linear UV^* term and has a squared amplitude distribution and a *doubled* phase variation. These doubled phase variations can sometimes be used to advantage, but more often they simply represent further distortions of the image. The location of the image is also modified. Consider, for example, the wave from a point object and a plane wave reference. The squaring of the essentially uniform amplitude has little effect, but the doubling of the phase term changes the radius of curvature of the phasefront by a factor of 2. Consequently, the wave appears to be coming from a point image at only one-half the distance from the point object to the hologram.

The sixth and seventh terms of (6-7) represent waves propagating at the same angle as the first-order waves and can produce noise in the first-order images. However, if the reference wave V is uniform in amplitude, VV^* is simply a constant and no noise is introduced.

The fourth and fifth terms of (6-7) always cause noise in the first-order images. These terms can be written as $2U^*V(UU^*)$ and $2UV^*(UU^*)$. From this we see that a distribution described by UU^* multiplies the desired image distribution and appears at the same angle as the first-order images. The degree to which this degrades the image depends on the ratio c_2/c_1. Expansion of the $c_3 \mathcal{E}^3$ term will show that it, too, contributes noise to the image.

Operation can be restricted to small variations of exposure about an average value located near the center of the linear region. This is most often accomplished by increasing the ratio of reference wave to object wave intensity. The effect is easier to see when the reference intensity is factored out as

$$\mathcal{E} = \Delta t V^2 \left[1 + \frac{U^2}{V^2} + \frac{U}{V}\cos(2\zeta x) \right]. \qquad (6\text{-}8)$$

Assume that **V** represents a uniform plane wave. For a constant exposure, the variation in exposure is reduced when the magnitude of **V** is increased. The diffraction efficiency is, however, reduced. An alternative method of biasing the exposure is to preexpose with a uniform illumination. The effect on the diffraction efficiency is the same—the visibility of the recorded fringes is reduced.

An indication of the effect of nonlinearity on the images formed holographically is best given for two types of objects. A diffusely reflecting or diffusely illuminated object is best described with one technique and point objects or transparent objects by another.

Diffuse objects. The term *diffuse objects* is used to refer to diffusely reflecting or diffusely illuminated objects. In essence, the phase of the amplitude of the object wave is assumed to be random. Goodman and Knight have shown (using what is called a Fourier transform hologram) that the nonlinearities cause distributions to appear in the image plane that are autoconvolutions of the image [6-12]. For example, second-order nonlinearities cause third-order autoconvolutions. Higher-order terms of (6-5) produce higher-order correlations.

Predicted effects of the nonlinearities of 649-F on the image of a uniformly bright diffuse square are shown in Fig. 6-11. Experimental results are shown in Fig. 6-12. Figure 6-12a shows the image plane with a reference-to-object intensity ratio K of 1.4 to 1 and an average amplitude transmittance of 0.5. Part b shows the additional, second-order, images and the increased distortion of the first-order images when the average amplitude transmittance is 0.24. The reference-to-object intensity ratio is the same, but the average exposure places the operating point in a region of greater nonlinearities. These figures illustrate the general result that nonlinearities of the recording medium cause a diffuse halo about the image. The effects are different when the object is not diffuse. The correlation function can cause discrete additions to the image rather than a smearing.

Point objects. We have seen how a nonlinear recorder can produce second-order point images at double the angle of the first-order image and in an image plane only half as far from the hologram. The fourth and fifth terms of (6-7) also cause extra point images to appear, but these terms produce images at the first-order and in the same image plane. As an example, consider two-point objects and a uniform plane reference wave.

The wave from a point at $x = a$, $y = 0$ and a distance d from the hologram can be described by $\mathbf{h}(x - a, y; d)$ and a wave from the second point at a distance d but located at $x = -a$, $y = 0$ is described by $\mathbf{h}(x + a, y; d)$. The fifth term of (6-7), $2UV^*U^*U$ describes a wave traveling in the

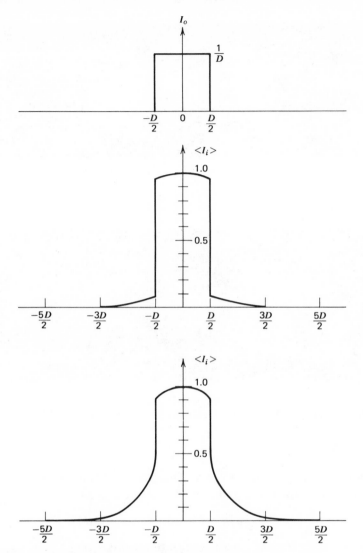

Figure 6-11 Predicted holographic images using 649-F plate. The object is a uniformly bright diffuse square. (*a*) Object, (*b*) first-order image with the reference-to-object wave intensity ratio K of 2.5 to 1 and (*c*) first-order image when K is 1.25 [6-12].

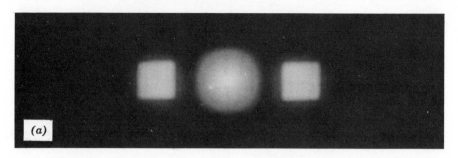

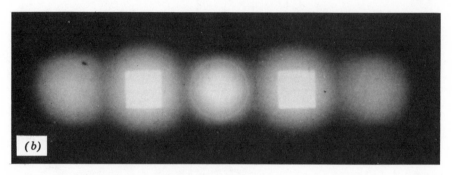

Figure 6-12 Photographs of image plane of a hologram recorded on 649-F with a reference-object wave intensity ratio K of 1.4 to 1. (*a*) Average transmittance of 0.5; (*b*) average transmittance of 0.24 [6-12].

same direction as the primary image. The phase of **V** simply determines the location of the images. Dropping the **V** and the 2 gives

$$\mathbf{U}\mathbf{U}^*\mathbf{U} = [\mathbf{h}(x-a,y;d) + \mathbf{h}(x+a,y;d)]$$
$$\times [\mathbf{h}(x-a,y;-d) + \mathbf{h}(x+a,y;-d)]$$
$$\times [\mathbf{h}(x-a,y;d) + \mathbf{h}(x+a,y;d)]. \qquad (6\text{-}9)$$

The right half of (6-9) becomes

$$3\mathbf{h}(x-a,y;d) + 3\mathbf{h}(x+a,y;d) + \mathbf{h}\left(x-a,y;\frac{d}{2}\right)\mathbf{h}(x+a,y;-d)$$
$$+ \mathbf{h}(x-a,y;-d)\mathbf{h}\left(x+a,y;\frac{d}{2}\right). \qquad (6\text{-}10)$$

The first two terms describe waves that are the same as the primary waves described by the linear terms. Expansion of the third term in exponential form gives

$$\mathbf{h}\left(x-a,y;\frac{d}{2}\right)\mathbf{h}(x+a,y;-d)=\exp\left[\frac{ik}{2d}(x^2-6ax+a^2+y^2)\right] \quad (6\text{-}11)$$

which can be written as

$$\exp\left\{\left[\frac{ik}{2d}(x-3a)^2+y^2\right]\right\}\exp\left[\frac{-ik}{2d}(8a^2)\right] \quad (6\text{-}12)$$

which describes a spherical wave from a point at $x=3a$ and d from the hologram. The exponential needed to complete the square is simply a constant. The fourth term of (6-10) describes a wave from a point at $x=-3a$. Consequently, even though there were only two-point objects, the nonlinearities of the recorder cause four images to appear—the two primary images and two false images. Noise of this type is referred to as intermodulation noise because it arises when one signal (modulation of the carrier) interacts with another to produce noise.

The occurrence of false image points may cause no problems in a display, but if the hologram is used to store digital data, errors can occur. Figure 6-13 shows an array of data points (bright spots represent ones and dark locations zeros, e.g.) recorded on a hologram. The image of Fig. 6-13a was formed with a hologram having a reference-to-object-beam intensity of 29. The image of Fig. 6-13b was formed with a hologram having an intensity ratio of unity. For comparison, the intensity autocorrelation

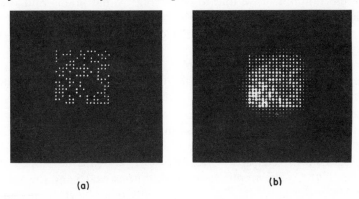

(a)   (b)

Figure 6-13 Holographic image of an array of data points. (a) Reference-object wave intensity ratio K is 29; (b) $K=1$ [6-13].

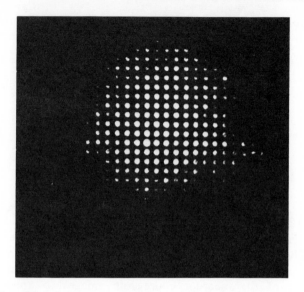

Figure 6-14   Intensity autocorrelation of the array of Fig. 6-12*a* [6-13].

function for the array is shown in Fig. 6-14. This is the distribution added to the image when nonlinearities become important. Kozma et al. found that when the ratio of reference-beam-to-object-beam intensity is greater than 10, the film grain noise becomes more important than intermodulation noise [6-13].

Total signal-to-noise ratio.   The total noise in a reconstructed wave is the sum of the random noise due to scattering by the grains and the intermodulation noise due to nonlinearities. We have seen that one way in which noise can be reduced is to decrease the ratio $|U|^2/|V|^2$. Unfortunately, this also decreases the signal level because the fringes on the hologram have lower visibility, making the hologram less efficient. The parameter of interest, then, is the ratio of the image intensity I_i to the noise intensity I_n—a S/N ratio.

If the only noise is due to scatter, a plot of log (I_i/I_n) versus log K is a straight line [6-14]. This is illustrated in Fig. 6-15. The object was a point 50 mm from the hologram and the reference was a uniform plane wave. The hologram was recorded on Kodak 649-F and was developed in D-19 for 5 min at 20°C. The amplitude transmittance was 0.5 and the area of the hologram was 4.5 mm^2. Because there is only one object, there can be no intermodulation noise and all noise is due to scattering.

When nonlinear effects come into play, the plot of I_i/I_n versus K takes

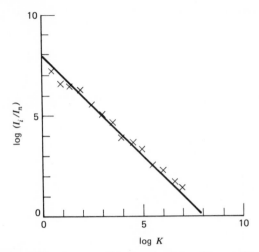

Figure 6-15   Intensity S/N ratio versus K for a point object. The amplitude transmittance on 649-F was $t = 0.5$ [6-14].

the form shown in Fig. 6-16. In this case, the object was a random array of points. The smooth curve gives the calculated values [6-14].

We have seen that scatter as a function of the amplitude transmittance follows the form given by Fig. 6-9. When nonlinear effects can be ignored, the curve of I_i/I_n versus t is similar. However, the noise due to nonlinearities reduces the ratio I_i/I_n for low values of transmittance. Figure 6-17 shows the variation in I_i/I_n as a function of t for a hologram formed with $K = 1$.

Some of the noise associated with holograms recorded on film is undoubtedly caused by phase modulations in the emulsion. As discussed in Chapter 3, phase modulation with a high index of modulation causes higher-order images and distortions of the first-order images. Experimentally, phase effects are difficult to separate from those due to the nonlinear effects just described. If the modulator produces only phase modulation, the amplitude effects do not enter in, and the intermodulation noise is due solely to a large index of phase modulation and energy detection.

6.4   Phase Modulators

In Chapter 3 we saw that spatial phase modulation of a wave can produce a reconstructed wave. Let us now consider various modulators. The phase of the wave can be changed in three basic ways: the thickness of a transparent medium can vary, producing a variation in optical path length; the index of refraction of the transparent medium can be modulated to

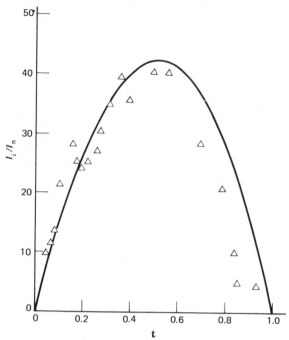

Figure 6-16 Signal-to-average-noise ratio versus amplitude transmittance **t** for a digital array of small squares with $K=20$; Δ indicates measured data [6-14].

produce changes in the phase front of a transmitted wave; the surface of a reflector can be modified to produce different path lengths across the aperture. Often a single medium can produce all three effects, but we shall consider each effect separately.

Surface variations. The thickness variations of a transparency produce spatial variations in the phase of an illuminating wave. Photographic emulsions have variations in thickness because of the removal of the undeveloped silver halides during fixing. The emulsion then collapses in those regions. This type of phase modulation is not particularly effective because the phase shift obtained is a highly nonlinear function of the spatial frequency of the signal [6-15]. There is a strong peak in phase shift when the period of the spatial frequency is approximately the thickness of the emulsion and a sharp drop in phase shift for shorter periods. This is illustrated in Fig. 6-18.

Another means of phase modulation by thickness variation uses thermoplastics. In this process, an electric charge distribution is applied to one side of the thermoplastic as in a xerographic process and the other side is in contact with a conductor. When the plastic is heated, the force due to

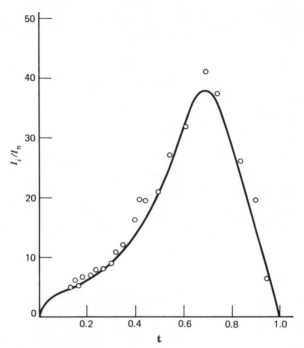

Figure 6-17 Signal-to-average-noise ratio versus **t** for a diffuse object with $K = 1$ [6-14].

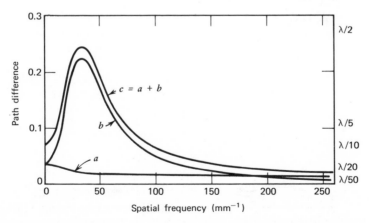

Figure 6-18 Optical path differences in passing through Agfa Agepan FF film developed in Agfa-Gevaert G3p and bleached in Kodak R21. (*a*) Path differences due to refractive index variations; (*b*) path differences due to surface relief; (*c*) total path differences [6-15].

the charges deforms the soft plastic. Upon cooling, the deformations are frozen into the plastic. This process has been demonstrated to be useful in holography [6-16].

Other implementations use a uniform electric charge, which is then modified by the varying conductivity of a photoconductor [6-17]. Refer to Fig. 6-19, which shows the steps in using this type recorder. First, a uniform charge is placed on the thermoplastic. Exposure to light increases the conductivity of the photoconductor so that the charges move as shown in the figure. A second charging increases the electric field in the exposed region. Heating of the thermoplastic to its softening point (60–100°C) allows deformations due to the attracting charges. After cooling, the result is a permanent phase transparency. Erasure is achieved by heating to a higher temperature long enough for the conductivities of the photoconductor and thermoplastic to increase and the charges to neutralize. The

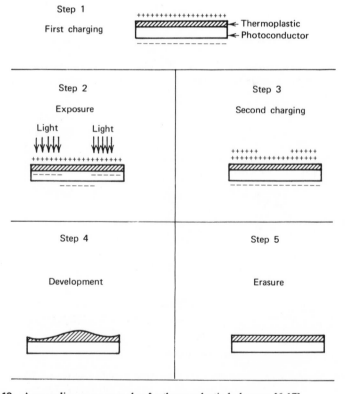

Figure 6-19   A recording erasure cycle of a thermoplastic hologram [6-17].

surface tension of the thermoplastic removes the surface deformations. The resolution obtainable in this process is as high as 1500 cycles/mm and the sensitivity comparable to that of 649-F.

Thermoplastic phase recorders have a spatial frequency band pass similar to that of bleached emulsions in which the emulsion thickness variations are of primary importance. Figure 6-20 shows the diffraction efficiency (which of course is a function of thickness variations in the thermoplastic) as a function of spatial frequency of the recorded grating and the angle between the interfering waves [6-17]. The thickness of the photoconductor is 2.5 μm and the thickness of the thermoplastic is 1 μm. The diffraction efficiency is lower for a diffuse object.

It is possible to use oil films on which a scanning electron beam places a charge on the oil surface. The attraction of the charge to its mirror image on the other side of the film deforms the surface in accordance with the pattern laid down by the electron beam. This scheme has been successful in projecting closed circuit television programs onto large screens [6-18].

Holograms can be embossed into plastic by means of a master copy. This has the advantage that many copies of a hologram can easily be

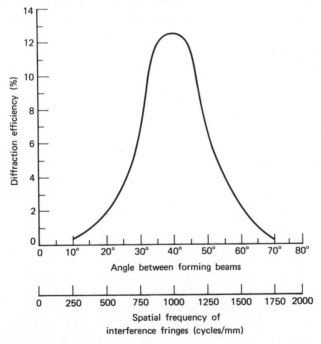

Figure 6-20 Diffraction efficiency of a thermoplastic hologram showing the band-pass effects of the thermoplastic [6-17].

produced and is being used by those interested in commercial uses of holograms [6-19].

All devices employing surface changes to produce phase shifts in transmitted waves can also be used in a reflection mode. The wave reflected from the distorted surface is also phase modulated. The reflectivity of the surface may be low and the diffraction efficiency small in some cases. If a hologram recorded on photographic plate is viewed in reflection, an image can be seen. If the surface of the emulsion is coated with a reflector, the image becomes brighter. Some surface deformation devices such as membrane modulators are intended to be used only in reflection [6-20]. These devices have reflecting membranes which are deformed by electric charges.

Index variations. There are many materials that undergo changes in the index of refraction when exposed to light. We consider photographic emulsions in detail and then briefly review the characteristics of some other recording media.

When the silver grains are formed during photographic exposure, the index of refraction of the medium changes. Because the change in index is proportional to the exposure, the phase transmittance of the transparency is proportional to the exposure in the same manner as the density varies with exposure. This phase-modulating mechanism, as well as the variations in thickness, operates even when amplitude modulation is desired. The effects of thickness variations can be removed by use of a liquid gate, which is a container formed of two flat windows filled with a fluid having an index of refraction equal to that of the emulsion. When the hologram is immersed in such a device, the effect of the thickness variation is eliminated.

If the emulsion is bleached, the attenuation of the transparency can be reduced to the point that phase modulation due to index changes dominates any residual amplitude modulation. The bleaching changes the silver to a transparent silver salt. The description of the mechanism by which phase modulation can reconstruct the desired wave is given in Chapter 3.

The brightness of the image from a bleached hologram is greater than from a nonbleached one, but so is the scatter. Some of the scatter is caused by stress relief in the emulsion, and some by the surface variations of the emulsion [6-21]. Figure 6-21 illustrates the scatter from a bleached hologram and its reduction by the use of a simple liquid gate formed by pressing mineral oil between the emulsion surface and a glass plate.

Another means of reducing the scatter is to use what is called *reversal bleaching* [6-22]. In normal bleaching processes, the emulsion is thicker where the index of refraction is greater. The flare light about an object

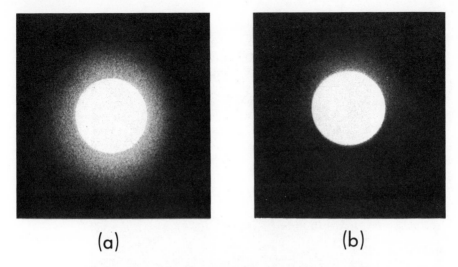

(a)   (b)

Figure 6-21 Photograph of the holographic images of a diffusely illuminated disk. (*a*) Normal bleaching process; (*b*) normal bleaching and index matching liquid [6-22].

recorded in the nonlinear region of an emulsion and with a low value of reference-to-object intensity ratio is brighter when a bleached hologram is used. This is because the flare light is caused by the multiplication of the first order image by $|U|^2$ which is predominately a low-frequency distribution, and the variation in thickness of the emulsion has a peak response in the low-frequency region. The effect of the index variation adds to this efficient modulation of the low frequencies. In reversal bleaching, the index of refraction is greatest when the emulsion is *thinnest*. This causes a partial cancellation of the low spatial frequency modulation and dramatically reduces the flare about the image. This is shown in Fig. 6-22.

Dichromated gelatin can be used to produce phase modulation. The presence of a dichromate such as $(NH_4)_2Cr_2O_7$ in a gelatin causes cross-linking of the gelatin molecules when the gelatin is exposed to light [6-22, 6-23]. This cross-linking, known as hardening, makes the exposed gelatin insoluble in water. Consequently, the unexposed gelatin can be washed away, forming a phase transparency.

Another method is to preharden the entire gelatin before exposure [6-24]. Then no gelatin is dissolved in water, but the gelatin absorbs water in varying amounts, depending on the degree of hardening induced by exposure to light. The harder the gelatin, the less water absorbed. If the gelatin is dried rapidly, it does not return to its original shape, but remains expanded. The result is an efficient hologram with very little scatter.

Figure 6-22 Photograph of the holographic image of a diffusely illuminated disk with reversal bleaching [6-22].

Diffraction efficiency. Phase modulators prove to be more efficient in terms of the portion of incident illumination that is diffracted to form the desired image. Amplitude modulation using absorption can theoretically diffract only 6.25% of the incident energy into an image. Experimentally, the number is about 4%. These numbers are for sinusoidal interference patterns. The patterns appearing on general holograms have lower diffraction efficiencies. A sinusoidal phase hologram, on the other hand, can have diffraction efficiencies as high as 33% (33.9% theoretically). Again, when general interference patterns, rather than sinusoidal ones, are considered, the efficiency is lower. Other limitations are set by intermodulation noise produced when the index of phase modulation is high. This is discussed in Chapter 3.

6.5 Volume Recorders and Modulators

When surface or thin modulators are employed, either amplitude or phase modulation is used to impress both the amplitude and the phase information upon the wave. When a volume recorder is employed, amplitude modulation can be used to impress amplitude variations upon the wave and phase modulation can be used to produce variations in the phasefront. Consequently, it is possible to form only one image. The conjugate wave is not needed to make the wave distribution produced by the hologram purely real, as in the case of only amplitude modulation, or purely imaginary, as in the case of only phase modulation. See Chapter 3 for a further discussion of the types of modulation.

First, we examine photographic emulsions. The effects of nonlinearities are briefly considered, and the effects of emulsion shrinkage outlined. Other volume recorders are mentioned. An examination of the efficiency of volume holograms shows that some types can approach 100% efficiency in reconstructing plane waves.

Photographic emulsions. We have seen that photographic emulsions have variations in both density and index of refraction which depend on the exposure. Consequently, regions of high field strength in the standing wave patterns such as illustrated in Figs. 3-7 and 3-8 produce regions of increased density and refractive index. These can act as modulated diffraction gratings, modulated mirrors, or a combination of both, depending on the tilt of the surfaces with respect to the direction of propagation of the illuminating wave. The description of a hologram in terms of a generalized grating and as a mirror with modulations of reflectivity and shape is given in Chapter 3. The case of surfaces of differing index or density inclined at an arbitrary angle with respect to the surface of the emulsion can be described using the coupled wave theory for thick holograms developed by Kogelnik [6-25]. The results of this theory are considered later in this section.

One difficulty with photographic emulsions is that when developed, fixed, and dried, they tend to shrink. This is no problem in the case of holograms where the diffracting surfaces are normal to the surface of the emulsion, because the shrinkage does not affect the spacing of the recorded fringes. When the surfaces are inclined, rather noticeable effects occur. For example, in the case of a reflecting hologram, shrinkage of the emulsion causes the spacing of the reflecting surfaces to decrease, thereby changing the wavelength at which constructive interference produces maximum reflectivity. Appendix 3 gives some techniques used to prevent shrinkage of an emulsion. Holograms stored in crystals do not have problems with shrinkage, but the recording sensitivity is quite low in comparison with a photographic emulsion.

Blazed holograms. Blazed holograms are an approximation to holograms having a single reflecting surface with variations in the surface shape and reflectivity. They can be formed by recording the hologram in photoresist, a material that is selectively dissolved by its developer, depending on the exposure [6-26]. The surfaces of standing wave maxima, illustrated in Fig. 6-23, reflect the read-out wave and directly phase-modulate the wave. Because of the inclination of the reflectors, most of the energy appears in one order. When overcoated with aluminum, an efficiency greater than 73% can be achieved for plane wave images. The efficiency is lower for general objects. Another limiting factor is that variations in the amplitude of the object wavefront appear as variations in the phase, distorting the reconstructed wavefront [6-27]. The greater the efficiency, the greater the distortion.

Coupled mode theory. The coupled mode theory can be used to describe the operation of a volume wavefront modulator [6-25]. This theory accounts for the effect produced when the waves go through many surfaces of absorption or refractive index change. It also allows for absorptive loss in the hologram and the reduction in illuminating wave amplitude as it passes through the surfaces. The procedures used in applying the theory to volume modulators are outlined here, and more detail is given in reference [6-28], but a thorough treatment is available only in the original [6-25].

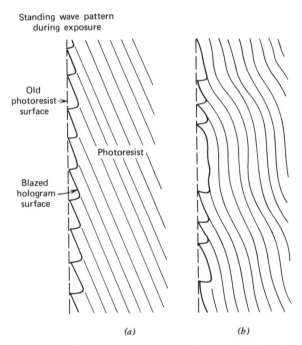

(a)   (b)

Figure 6-23 Formation of a blazed hologram on the surface of a photoresist material for (a) the wavefield of two collimated beams and (b) a more arbitrary wavefield [6-26].

The wave equation in scalar form is

$$\nabla^2 \mathbf{E} + \gamma_g^2 \mathbf{E} = 0 \tag{6-13}$$

where $\mathbf{E}$ is the complex amplitude of the y component of the electric field, and γ_g is the propagation constant in the grating. Both $\mathbf{E}$ and γ_g are functions of x and z. If the medium has loss,

$$\gamma_g^2 = \frac{\omega^2}{c^2}\epsilon_r + i\omega\mu\sigma \tag{6-14}$$

where c is the velocity of light in free space, ϵ_r is the relative dielectric constant, μ is the permeability of the medium, ω is the angular frequency, and σ is the conductivity. The relative dielectric constant and conductivity are assumed to vary sinusoidally as

$$\epsilon_r = \epsilon_{r0} + \epsilon_{r1}\cos(k_g \cdot r) \tag{6-15}$$

and

$$\sigma = \sigma_0 + \sigma_1 \cos(k_g \cdot r) \tag{6-16}$$

where the subscript zero denotes the average value, the subscript 1 denotes the modulation, k_g is a grating vector normal to the fringes, and r is a direction vector. The magnitude of the grating vector is given by

$$|k_g| = \frac{2\pi}{\lambda_g} \tag{6-17}$$

where λ_g is the period of the grating. The equations (6-14), (6-15), and (6-16) can be combined as

$$\gamma_g^2 = \beta_0^2 + 2i\alpha_0\beta_0 + 2\kappa\beta_0(e^{ik_g \cdot r} + e^{-ik_g \cdot r}) \tag{6-18}$$

where the average propagation constant β_0 and the average attenuation α_0 are defined as

$$\beta_0 = k\sqrt{\epsilon_{r0}}, \qquad \alpha_0 = \frac{\mu c\sigma_0}{2}\sqrt{\epsilon_{r0}}. \tag{6-19}$$

The *coupling constant* κ is

$$\kappa = \frac{1}{4}\left[\frac{\epsilon_{r1}}{\epsilon_{r0}}\beta_1 + i2\alpha_1\right] \tag{6-20}$$

where β_1 and α_1 are defined as in (6-19) but with ϵ_{r1} in β_1 and σ_1 in α_1. The coupling constant describes the coupling between the read-out wave **W** and the image wave **S**. If $\kappa = 0$, there is no coupling, and consequently no diffraction to produce an image wave.

The total electric field in the medium consists of the read-out wave and the image wave. Solutions of the coupled wave equations take the form

$$\mathbf{W}(z) = w_1 \exp(\gamma_1 z) + w_2 \exp(\gamma_2 z) \qquad (6\text{-}21)$$

$$\mathbf{S}(z) = s_1 \exp(\gamma_1 z) + s_2 \exp(\gamma_2 z) \qquad (6\text{-}22)$$

where the w_i and s_i are constants that depend on the boundary conditions and the γ_i are complex propagation constants. Figure 6-24 illustrates the manner in which the read-out wave decays and the image wave grows within transmission and reflection holograms.

The solution of the coupled wave equations can be applied to thick transmission or reflection holograms using either dielectric or absorption modulation. Kogelnik's paper considers holograms having arbitrary inclination of the surfaces within the hologram, but the examples listed here have diffracting surfaces either normal (transmission gratings) or parallel (reflection gratings) to the outer surface of the hologram. A further simplification incorporated here is that the holograms are assumed to be either purely absorptive ($\epsilon_{r1} = 0$) or purely dielectric ($\sigma_1 = 0$).

Thick dielectric holograms. In the case of a pure dielectric hologram of a plane wave, the efficiency can theoretically be 100%. The efficiency of a

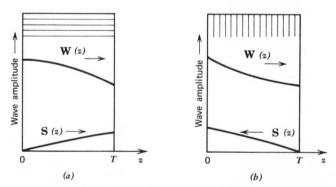

Figure 6-24 Wave propagation in (a) transmission and (b) reflection holograms of thickness d. The read-out wave W decays as it propagates to the right. In (a) the signal wave S grows in amplitude as it moves to the right and in (b) it grows as it moves to the left [6-25]. (Copyright 1969 by American Telephone and Telegraph Co.; reprinted by permission.)

thick, dielectric, transmission grating is given by

$$\eta = \frac{\sin^2(\iota_t^2 + \Upsilon_t^2)^{1/2}}{(1 + \Upsilon_t^2/\iota_t^2)}. \tag{6-23}$$

The variable ι_t is a phase modulation index for dielectric transmission holograms and is defined by

$$\iota_t = \frac{\pi \mathbf{n}_1 T}{\lambda_a \cos \phi_0} \tag{6-24}$$

where $\mathbf{n}_1$ is the amplitude of the sinusoidal variation in index of refraction of the medium. That is,

$$\mathbf{n} = \mathbf{n}_0 + \mathbf{n}_1 \cos \mathbf{k}_g \cdot \mathbf{r}, \tag{6-25}$$

where $\mathbf{n}_0$ is the average index of refraction of the medium. The thickness of the hologram is T, the wavelength of the illumination in air is λ_a, and ϕ_0 is the Bragg angle as measured between the surface of refractive index maxima and the direction of propagation in the medium.* It is assumed that the angle of illumination is close to the Bragg angle. The other variable, Υ_t, is a normalized measure of the deviation of the illumination from optimum:

$$\Upsilon_t = \Delta\left(\frac{2\pi \mathbf{n}_0}{\lambda_a}\right) T \sin \phi_0 \tag{6-26}$$

where the subscript t again denotes a dielectric transmission hologram. The variable Δ is either the deviation of the illumination angle from the Bragg angle or a normalized deviation in wavelength from that used in recording the hologram [6-25, 6-28]. In the latter case,

$$\Delta = \frac{\Delta\lambda_a}{2\mathbf{n}_0\lambda_g \cos \phi_0} = \frac{\Delta\lambda_a}{\lambda_a} \tan \phi_0. \tag{6-27}$$

Notice that effects of variations in the spacing of the surfaces, λ_g, due to emulsion shrinkage can also be taken into account. If $\Upsilon_t = 0$, $\eta = \sin^2 \iota_t$.

Figure 6-25 illustrates the manner in which the efficiency diminishes with changes in either incident angle or wavelength for various effective thicknesses. The value η_0 is the maximum efficiency for the given value of ι. The maximum efficiencies for the three values of ι_t are $\iota_t = \pi/4$, $\eta_0 = 50\%$; $\iota_t = \pi/2$, $\eta_0 = 100\%$; and $\iota_t = 3\pi/4$, $\eta_0 = 50\%$.

*The Bragg angle is the angle at which the waves reflected from each surface of refractive index or absorption maxima add in phase.

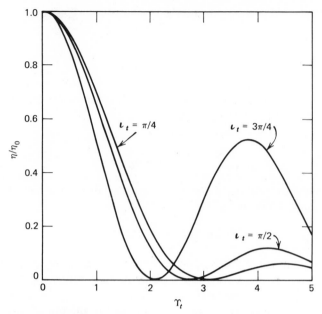

Figure 6-25 Relative efficiency of a thick, dielectric, transmission hologram as a function of thickness, Bragg angle, angle of incidence, and wavelength. See text for definitions of ι_t and Υ_t [6-25]. (Copyright 1969 by American Telephone and Telegraph Co; reprinted by permission.)

$\eta_0 = 50\%$; $\iota_t = \pi/2$, $\eta_0 = 100\%$; and $\iota_t = 3\pi/4$, $\eta_0 = 50\%$.

The dip to zero in the efficiency makes thick holograms suitable for recording multiple images and displaying one at a time. All that is necessary is to record the wavefronts with reference waves at slightly different angles. The Bragg angles for each hologram are adjusted so that when one angle of illumination produces a maximally bright image, the adjacent image is quenched. If, during recording, the object wave is introduced at the same angle each time with the hologram rotated after each exposure, the images sequentially appear in the same location when the hologram is rotated during illumination.

When the surfaces of refractive index maxima are parallel to the surface of the hologram, the efficiency is given by

$$\eta = \left\{ 1 + \frac{1 - \Upsilon_r^2 / \iota_r^2}{\mathrm{sh}^2 (\iota_r^2 - \Upsilon_r^2)^{1/2}} \right\}^{-1} \qquad (6\text{-}28)$$

where sh is the hyperbolic sine and the subscript r denotes reflection hologram. The values of Υ_r and ι_r are given by

$$\iota_r = \frac{\pi \mathbf{n}_1 T}{\lambda_a \cos \psi_0} \tag{6-29}$$

and

$$\Upsilon_r = \Delta \left(\frac{2\pi \mathbf{n}_0}{\lambda_a} \right) T \cos \phi_0$$

$$= \frac{\Delta \lambda}{\lambda_a} \left(\frac{2\pi \mathbf{n}_0}{\lambda_a} \right) T \sin \phi_0, \tag{6-30}$$

where ψ_0 is the angle between the normal to the surface of refractive index maxima and the direction of propagation in the medium. A plot of the relative efficiency as a function of ι_r and Υ_r is given in Fig. 6-26.

Thick absorption holograms. The absorption hologram is one in which the changes in index of refraction are neglected and variations in the absorption of the material are taken as being responsible for the diffraction effects.

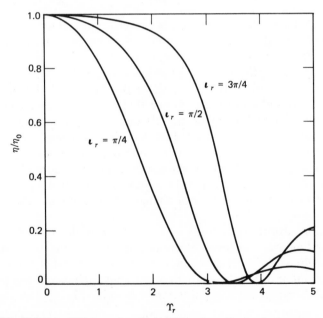

Figure 6-26 Relative efficiency of a thick, dielectric, reflection hologram as a function of thickness, Bragg angle, angle of incidence, and wavelength. See text for definitions of ι_r and Υ_r [6-25]. (Copyright 1969 by American Telephone and Telegraph Co.; reprinted by permission.)

The efficiency in the case of a transmission grating (diffracting surfaces normal to the hologram surface) and Bragg angle incidence of the read-out wave is given by

$$\sqrt{\eta} = \text{sh}\left(\frac{\mathfrak{D}_1}{2}\right)\exp(-\mathfrak{D}_0) \qquad (6\text{-}31)$$

where $\mathfrak{D}_0$ is a measure of the average density of the grating,

$$\mathfrak{D}_0 = \frac{\alpha_0 T}{\cos\phi_0}, \qquad (6\text{-}32)$$

and $\mathfrak{D}_1$ is the amplitude of the density modulation,

$$\mathfrak{D}_1 = \frac{\alpha_1 T}{\cos\phi_0}. \qquad (6\text{-}33)$$

The attenuation α_0 is the average attenuation of the medium and α_1 is the amplitude of the sinusoidal variations in attenuation that produce the grating effect. The maximum possible diffraction efficiency of 3.7% occurs when α_1 has its maximum value of α_0 and

$$\mathfrak{D}_0 = \ln 3. \qquad (6\text{-}34)$$

Figure 6-27 shows the square root of the efficiency as a function of the amplitude of the absorption modulation α_1 and for various values of $\alpha_0/\alpha_1 = \mathfrak{D}_0/\mathfrak{D}_1$. The square root of efficiency is used to show the regions where the output amplitude of the grating varies linearly with modulation. The dashed curves of constant $\mathfrak{D}_0$ show the action of the grating for constant background absorption. Notice that a linear response and maximum efficiency for each value of modulation is obtained when $\mathfrak{D}_0 \approx 1$.

If the Bragg condition is not satisfied, the efficiency decreases as shown in Fig. 6-28. It is assumed that the maximum depth of modulation is used. That is, $\mathfrak{D}_0 = \mathfrak{D}_1$. The efficiency normalized to the maximum efficiency for each value of $\mathfrak{D}_0 = \mathfrak{D}_1$ is plotted as a function of changes in angle of incidence of the read-out wave or of changes in wavelength. Compare with Fig. 6-25.

A reflection absorption grating is formed when the two plane waves during recording are introduced from opposite sides of the hologram. If we assume that the surfaces of absorption maxima run parallel to the surface of the hologram and that the changes in refractive index are negligible, the efficiency with Bragg angle illumination is described by

$$\sqrt{\eta} = -\frac{\mathfrak{D}_1}{2}\left[\mathfrak{D}_0 + \left(\mathfrak{D}_0^2 - \frac{\mathfrak{D}_1^2}{4}\right)^{1/2}\coth\left(\mathfrak{D}_0^2 - \frac{\mathfrak{D}_1^2}{4}\right)^{1/2}\right]. \qquad (6\text{-}35)$$

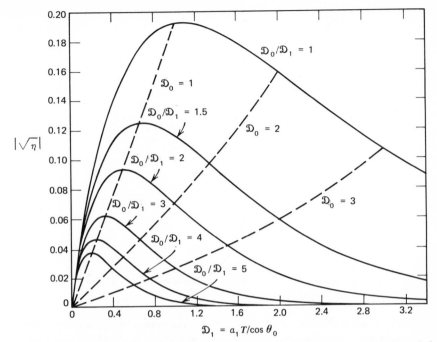

Figure 6-27 Efficiency of a thick, absorptive transmission hologram as a function of the modulation $\mathfrak{D}_1 = \alpha_1 T/\cos\theta_0$ for various modulation depths $\mathfrak{D}_1/\mathfrak{D}_0$ and various bias levels $\mathfrak{D}_0 = \alpha_0 T/\cos\theta_0$ [6-25]. (Copyright 1969 by American Telephone and Telegraph Co.; reprinted by permission.)

The efficiency is a maximum when the modulation of the absorption constant is equal to the bias absorption and the values of $\mathfrak{D}_0$ and $\mathfrak{D}_1$ approach infinity. The value of η in this case is 7.2%. This indicates that for high efficiency, absorption *reflection* holograms should have a high photographic density. Figure 6-29 shows the variation in the square root of the efficiency (a measure of the *amplitude* of the image wave) as a function of the modulation $\mathfrak{D}_1$ for various values of the average loss constant $\mathfrak{D}_0$ and for various modulation depths. Notice the linear variation of $\sqrt{\eta}$ with $\mathfrak{D}_1$ for constant $\mathfrak{D}_0$. This linear variation has been experimentally verified by Friesem and Walker [6-29].

The sensitivity of the reflection grating to deviations of the illumination angle from the Bragg angle and to changes in λ is shown in Fig. 6-30. The maximum depth of modulation, $\mathfrak{D}_1 = \mathfrak{D}_0$ is assumed. The efficiency is normalized by the maximum efficiency for each value of $\mathfrak{D}_1 = \mathfrak{D}_0$. Note the broadening of the curve as $\mathfrak{D}_0$ increases and compare with Fig. 6-28.

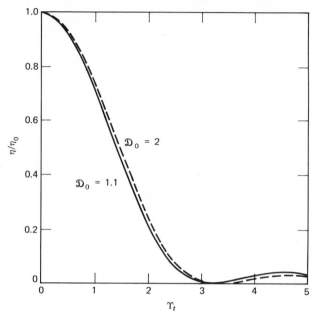

Figure 6-28 Relative efficiency of a thick, absorptive transmission hologram as a function of thickness, Bragg angle, angle of incidence, and wavelength [6-25]. (Copyright 1969 by American Telephone and Telegraph Co.; reprinted by permission.)

Effect of emulsion thickness in amplitude modulation holograms. It has been shown that if the reference-to-object-wave intensity ratio K is large, the diffraction efficiency of an absorption hologram is not a function of the emulsion thickness [6-30]. This is due to the way in which the absorption coefficient and density are related to emulsion thickness. If the density is to be fixed, the absorption coefficient must decrease as the thickness increases. Therefore, the modulation can be expressed in terms of density only. Figure 6-31 shows the manner in which diffraction efficiency varies with thickness of an experimental, fine-grain emulsion. The ratio K was 68 and the spatial frequency of the recorded grating 1000 cycles/mm. For the thickness dependence to be small, the hologram must be illuminated at the Bragg angle. The absorption, hence diffraction efficiency, increases with thickness initially because a given exposure will produce a greater density if the emulsion is thicker.

Experimental measures of angular sensitivity. High angular sensitivity in a hologram is a disadvantage if a wide viewing angle is desired in a display. If, however, the object is to store as many holograms as possible on a

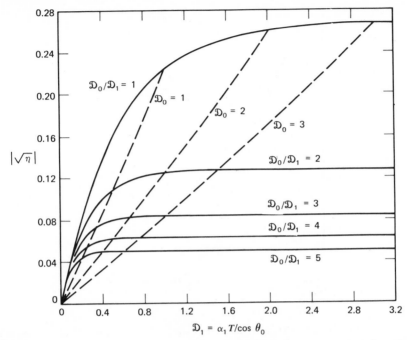

Figure 6-29 Efficiency of a thick, absorptive reflection hologram as a function of the modulation $\mathfrak{D}_1 = \alpha_1 T/\cos\theta_0$ for various modulation depths $\mathfrak{D}_1/\mathfrak{D}_0$ and various bias levels $\mathfrak{D}_0 = \alpha_0 T/\cos\theta_0$ [6-25]. (Copyright 1969 by American Telephone and Telegraph Co.; reprinted by permission.)

noninterfering basis, as in data storage holograms, a high angular sensitivity is desired.

We have seen how the diffraction efficiency varies as a function of the amount of deviation from the Bragg angle for different values of degree of modulation, hologram thickness, and Brewster's angle. Figure 6-32 shows a plot of measured decreases in efficiency as a function of the deviation from the Bragg angle. The hologram was recorded using Kodak 649-F spectroscopic plates which have emulsions 15 μm thick. Similar measurements made using photochromic glass 1587.5 μm thick produced the curve of Fig. 6-33. In both cases, the reference and signal plane waves were incident at 30° on opposite sides of the normal to the hologram surface. The angle between the two waves in air thus was 60°. Note the change in the units of the abscissa from degrees to minutes of arc.

Figure 6-32 indicates that if multiple images are to be stored in 649-F, there would have to be at least a 10° shift in angle between recordings.

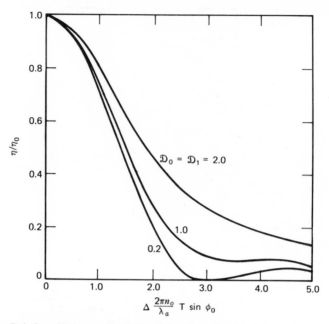

Figure 6-30   Relative efficiency of a thick, absorptive reflection hologram as a function of thickness, Bragg angle, angle of incidence, and wavelength [6-25]. (Copyright 1969 by American Telephone and Telegraph Co.; reprinted by permission.)

Consequently, only a few images can be recorded without interference. As can be seen from Fig. 6-33, however, a thicker medium can record two orders of magnitude more images—in theory, that is. In practice, the possibility of saturation of the medium limits either the total number of images that can be recorded or the exposure of each. The latter causes a decrease in the diffraction efficiency for each image. Another factor to be

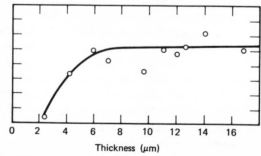

Figure 6-31   Relative diffraction efficiency of an amplitude modulation hologram as a function of emulsion thickness. The circles denote experimental points [6-30].

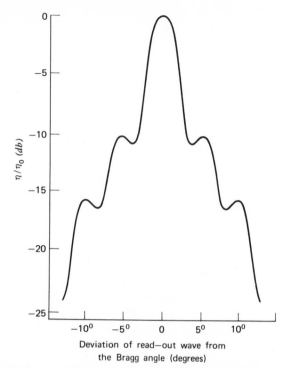

Figure 6-32 Relative efficiency as a function of deviation from the Bragg angle of an absorptive transmission hologram recorded on Kodak 649-F plates with emulsion thickness of 15 μm [6-29].

considered is the increased exposure for most thicker materials. For example, the exposure for the photochromic glass of Fig. 6-33 was 10^4 times that for the photographic plate of Fig. 6-32.

Other treatments of recordings in thick media are given in references [6-31] and [6-32]. Others will be referred to in later chapters.

Efficiency. As can be seen, the efficiency of a hologram depends on the type of modulation (absorption or dielectric), the mode of use (transmission or reflection), and the angle of incidence of the waves. Let us assume that the angle of incidence of the read-out wave is the Bragg angle. We can then summarize the results as in Table 6-2. Remember that these values are for plane wave object and reference waves. That is, the holograms are gratings. More general object waves result in lower efficiencies.

Cross sections of thick emulsions. Cross sections of amplitude and phase modulation holograms have been recorded using a transmission electron

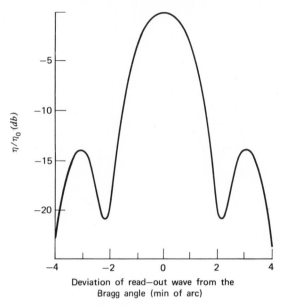

Figure 6-33 Relative efficiency as a function of deviation from the Bragg angle for a hologram recorded in photochromic glass 1587.5-μm thick [6-29].

microscope [6-33]. The holograms were formed when two plane waves of equal intensity interfered on the recording material. The wavelength used was 632.8 nm. Figure 6-34 shows a cross section of an amplitude modulation hologram made with Kodak 649-F sheet film. When the emulsion was bleached, the grains were enlarged as shown in Fig. 6-35. For comparison, a cross section of a bleached hologram made with Agfa 14C75 appears in Fig. 6-36. This shows the effect of reduced resolution and reduced emul-

TABLE 6-2   THEORETICAL EFFICIENCIES FOR VARIOUS HOLOGRAMS

Thickness	Modulation	Mode of Use	Maximum Efficiency %
Thin	Absorption	Transmission	6.25
Thin	Refractive index	Transmission	33.9
Thick	Absorption	Transmission	3.7
Thick	Refractive index	Transmission	100.0
Thick	Absorption	Reflection	7.2
Thick	Refractive index	Reflection	100.0

Figure 6-34 Transmission electron microscope view of a cross section of an amplitude modulation hologram recorded on Kodak 649-F film [6-33].

sion thickness. It is not so effective when used as a volume recorder, but is more sensitive. If Fig. 6-36 is viewed at a small angle from the page, the fringe pattern can be seen. The magnification of all three figures is X 4600.

6.6 Characterization of Transparencies

There are many ways of describing the effect of a transparency on an illuminating wave. One of the older methods is to analyze the effect of the slope of the H & D curve. A more useful technique in optical information processing is the use of a plot of amplitude transmittance as a function of

Figure 6-35   Transmission electron microscope view of a cross section of a phase modulation hologram recorded on Kodak 649-F film [6-33].

exposure. An amplitude transfer function proves to be more useful in bleached or reflection holograms, and two methods of presenting this function are discussed.

Photographic gamma.   The slope of the straight-line or linear region of the H & D curve is referred to as gamma. Unfortunately, gamma is also almost universally used to represent the mutual coherence function. In this text, γ represents the slope of the H & D curve, and γ represents the normalized mutual coherence function. Study of Fig. 6-4 shows that a film with a large γ rapidly goes from low to high density with a relatively small change in exposure. Such a film is a *high-contrast* film. Film with a low value of γ is *low-contrast* film. Some typical curves are shown in Fig. 6-37; all emulsions are sensitive to the same wavelengths. The resolution is lowest with 103-F and increases from I-F to V-F and is highest with 649-F. The same developer, D-19, was used for all emulsions.

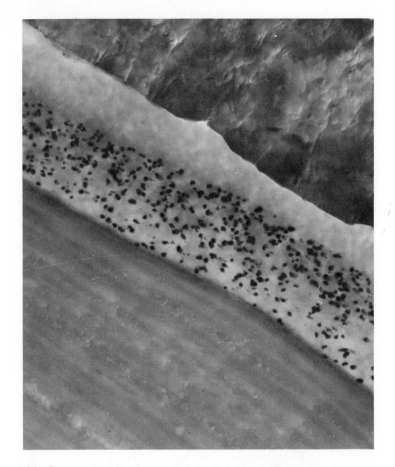

Figure 6-36 Cross section of a phase modulation hologram on Agfa 14C75 film [6-33].

The developing time influences the γ of the transparency and the developing chemical can shift its value drastically. Figure 6-38 illustrates the effect of time of development and type of developer upon γ. The two developers D-19 and D-8 are high-contrast developers. Low-contrast developers such as D-165 can reduce γ so that a large range of intensity can be recorded.

The linear region of the H & D curve (Fig. 6-4) can be described by

$$\mathfrak{D} = \gamma(\log \mathscr{E} - \log \mathscr{E}_0) \qquad (6\text{-}36)$$

where $\log \mathscr{E}_0$ is where the straight line, if extended, would intersect the log

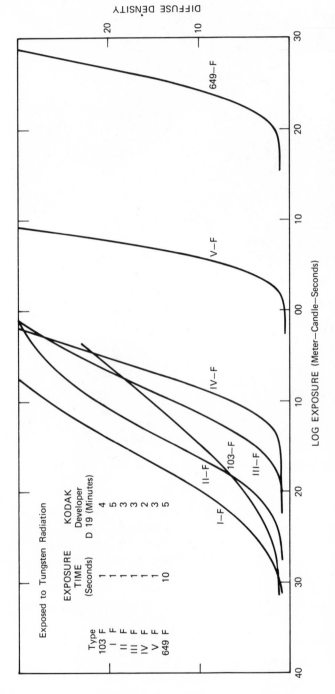

Figure 6-37 H&D curves for seven Kodak emulsions. [From *Kodak Plates and Films for Science and Industry* (Eastman Kodak Co., 1967, p. 15d)]. (Reproduced with permission from a copyrighted Eastman Kodak publication.)

166

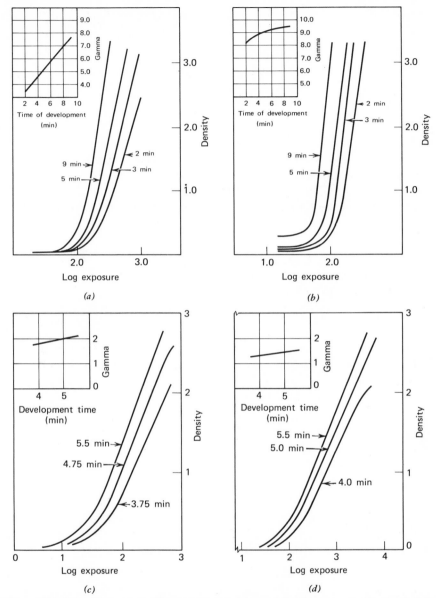

Figure 6-38 The effects of development time and type of developer on γ and the H & D curve of Kodak 649-F materials. (*a*) D-19, (*b*) 1:2 solution of D-8, (*c*) 1:3 solution of D-165, and (*d*) 1:4 solution of D-165. [After *Kodak Plates and Films for Science and Industry* (Eastman Kodak Co., 1967), reproduced with permission from a copyrighted Eastman Kodak publication, and D. P. Jablonowski, R. A. Heinz, and J. O. Artman, *Appl. Opt.* **10**, 1988 (1971)].

167

$\mathcal{E}$ axis. Use of (6-1) and rewriting the difference of the logarithms of (6-36) yield

$$\log\left(\frac{1}{\tau}\right) = \log\left(\frac{\mathcal{E}}{\mathcal{E}_0}\right)^{\gamma}, \qquad (6\text{-}37)$$

so that

$$\tau = \mathcal{E}_0^{\gamma}\mathcal{E}^{-\gamma} = C\mathcal{E}^{-\gamma} \qquad (6\text{-}38)$$

where C is a constant. Notice that the intensity transmittance is not linearly proportional to the incident exposure or intensity but that the recording is highly nonlinear. Let the exposure $\mathcal{E}$ be written in terms of the intensity I and the time Δt. Equation (6-38) can then be written as

$$\tau = KI^{-\gamma} \qquad (6\text{-}39)$$

where the constant K incorporates C and $(\Delta t)^{-\gamma}$.

The results above apply to a negative transparency. To indicate this, the subscript n can be added to (6-39), giving

$$\tau_n = K_n I^{-\gamma}. \qquad (6\text{-}40)$$

A positive transparency can be obtained by illuminating the negative with uniform intensity. The intensity distribution illuminating the second emulsion is

$$I = I_0 \tau_n \qquad (6\text{-}41)$$

where I_0 is the uniform illumination. If the second recording is made in the linear region of the H & D curve,

$$\tau_p = C_p (I_0 \tau_n \Delta t)^{-\gamma_p} \qquad (6\text{-}42)$$

where (6-41) was placed into (6-39), the exposure time is Δt, and C_p is the constant for a positive process. If I_0, Δt, C_p, and K_n are all incorporated into another constant K_p, (6-42) can be written as

$$\tau_p = K_p I^{\gamma_n \gamma_p}. \qquad (6\text{-}43)$$

We see that a linear mapping of intensity onto a transparency is possible only if the product of the gammas of the two operations is unity.

In optical processing, we are normally concerned with the amplitude transmittance of a transparency rather than the intensity transmittance. Because the index of refraction and the thickness of the emulsion vary across the transparency after exposure and development, the amplitude transmittance **t** can be complex. These variations in the phase of **t** are related to the signal. Other phase variations are caused by variations in the thickness of the acetate or glass base on which the emulsion is coated. Consequently, the amplitude transmittance cannot be taken as simply the square root of the intensity transmittance. However let us assume that the phase effects can be eliminated or ignored and define **t** as the positive square root of **t**. We then find that equations (6-39) and (6-43) yield

$$t_n = K_n^{1/2} I^{-\gamma_n/2} = K_n^{1/2} |U|^{-\gamma_n} \qquad (6\text{-}44)$$

and

$$t_p = K_p^{1/2} I^{\gamma_n \gamma_p/2} = K_p^{1/2} |U|^{\gamma_n \gamma_p}. \qquad (6\text{-}45)$$

Equation (6-45) indicates that to obtain an amplitude transmittance that is linearly proportional to the incident intensity on the original emulsion, the product $\gamma_n \gamma_p$ must be 2. This is one way in which the assumption that film is a square law detector can be justified. It is not necessary to cause $\gamma_p \gamma_n$ to be 2 over the linear regions of the H & D curves, however. Regions of the curves can be chosen where the slope is 2. If off-axis holograms are used, it is not necessary to obtain a positive. A negative γ of 2 simply causes the phase of the off-axis waves of the hologram to be shifted by 180°.

The amplitude transmittance versus exposure curve. A more useful curve is a plot of amplitude transmittance as a function of exposure. Kozma pointed out that the amplitude transmittance can vary linearly with exposure even if the overall γ of the developing process is not 2 [6-34]. Figure 6-39 shows some typical curves for 649-F. Obviously, a region can be found where the amplitude transmittance varies linearly with exposure. The regions of linearity on these curves correspond to regions near the toe of the H & D curves. Consequently, holograms exposed to a linear region on the $t - \mathcal{E}$ curve appear to be underexposed. The emulsion can be preexposed to ensure operation in the linear region of the $t - \mathcal{E}$ curve if the amplitude of the reference wave is not sufficient to determine the average exposure. If the reference intensity is, for example, ten times the intensity of the object or information wave, the intensity fluctuations of the object wave will have little effect on the total intensity. The average exposure can then be made to fall in the center of the linear region by selection of the exposure time.

In using the $t - \mathcal{E}$ curve, transfer characteristics can be drawn in the same way that transfer characteristics are obtained for nonlinear electrical components. Figure 6-40 shows a typical $t - \mathcal{E}$ curve and a *sinusoidal* exposure. The variations in exposure cause corresponding variations in **t**,

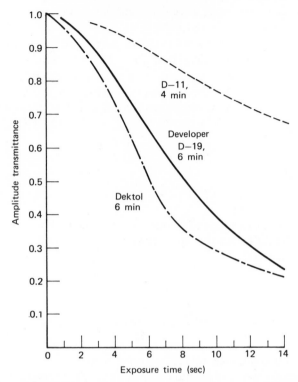

Figure 6-39 Amplitude transmission versus exposure curves for Kodak 649-F film. The intensity was constant at 0.44 mW/cm². [From A. A. Friesem and J. S. Zelenka, *Appl. Opt.* **6**, 1755–1759 (1967)].

but because of the nonlinearity of the transfer function, $\mathbf{t}$ is not sinusoidal. The exact form of $\mathbf{t}(x)$ can be found by expanding $\mathbf{t}(\mathcal{E})$ in a Taylor series about the average exposure $\mathcal{E}_0$. We then have

$$\mathbf{t}(\mathcal{E}) = \mathbf{t}_0 + \frac{d\mathbf{t}}{d\mathcal{E}}\bigg|_{\varepsilon_0}(\mathcal{E}-\mathcal{E}_0) + \frac{1}{2}\frac{d^2\mathbf{t}}{d\mathcal{E}^2}\bigg|_{\varepsilon_0}(\mathcal{E}-\mathcal{E}_0)^2 + \dots. \qquad (6\text{-}46)$$

Because both U and V are constants,

$$\mathcal{E}-\mathcal{E}_0 = 2UV\Delta t\cos(2\zeta x), \qquad (6\text{-}47)$$

$$\mathbf{t}(x) = \mathbf{t}_0 + 2\frac{d\mathbf{t}}{d\mathcal{E}}\bigg|_{\varepsilon_0}\Delta t\, UV\,\cos(2\zeta x)$$

$$+ 2\frac{d^2\mathbf{t}}{d\mathcal{E}^2}\bigg|_{\varepsilon_0}(\Delta t)^2 U^2 V^2\cos^2(2\zeta x) + \dots. \qquad (6\text{-}48)$$

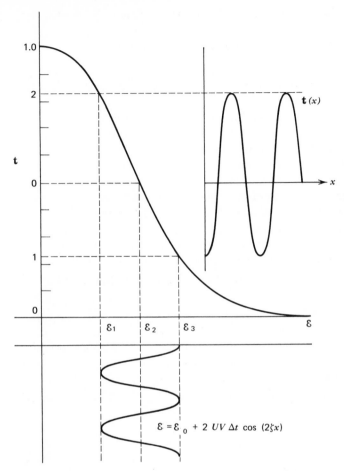

Figure 6-40   Transfer characteristics of a transparency near the linear region of the $t-\varepsilon$ curve.

We can see that, in addition to the $UV \cos(2\zeta x)$ term that can produce the desired reconstructed wave and its conjugate, there are other waves produced by the $\cos^2(2\zeta x)$ term. These waves will propagate along the same axis as the illuminating wave and at an angle equal to twice the angle at which the desired wave propagates. If a third-order term exists in (6-48), it will produce waves at three times the angle of the desired wave and at the same angle as the desired wave. This will cause distortion of the image. No image distortion is caused by the second-order term because **U** and **V** are constants. Consequently, their effect is to simply change ε_0. If **U** and **V** do not represent constants, the second-order term also causes image distortion as discussed in the section "Distortions Due to Nonlinearities." The results discussed here apply *only* for a sinusoidal variation in exposure, which can be caused by the interference of two plane waves.

The amplitude transfer function.   The $t-\mathscr{E}$ curve is more appropriate than the H & D curves for holography, but its primary use is with holograms in which the thickness of the emulsion is neglected and the illuminating wave is only amplitude modulated. Another characterization of the modulating material (it need not be photographic film) is the amplitude of the reconstructed wave as a function of the amplitude of the incident wave and the exposure. The function describing the amplitude of the reconstructed wave as a function of the exposure and the ratio of signal to reference wave amplitude has been called the *amplitude transfer function* (ATF) by Upatnieks and Leonard [6-35]. In the plots of ATF, the square root of the hologram diffraction efficiency is used rather than the amplitude of the reconstructed wave. Because the efficiency, η, is defined as the ratio of the energy in the reconstructed wave to that in the illuminating wave, $\sqrt{\eta}$ gives a *normalized* measure of the amplitude of the reconstructed wave.

Figure 6-41 illustrates the variation of $\sqrt{\eta}$ with the ratio of object-to-reference-wave amplitudes $|U/V|$ with the exposure of the reference wave $\mathscr{E}_r$ as a parameter. A spatial frequency of 11.5 cycles/mm was recorded on Agfa 10E75 with an 18-sec exposure and a 12-min development in Methinol-U. The diffraction grating thus recorded has a diffraction efficiency η which depends on the ratio of the amplitudes of the reference

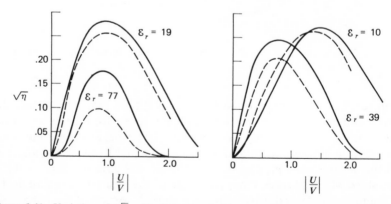

Figure 6-41   Variation of $\sqrt{\eta}$ with the ratio of object to reference wave amplitude and the exposure of the reference wave, $\mathscr{E}_r$, in ergs/cm². The grating spatial frequency is 11.5 cycles/mm. The solid lines are calculated curves and the dashed lines show the measured values [6-35].

and object waves and on the exposure time. Notice the shift in optimum object-to-reference amplitude ratio as the exposure increases. This is not easy to interpret because the parameter is the exposure of the *reference* wave, not the total exposure. Neither can a direct comparison with Fig. 6-27 be made because of the different parameters. However, one can determine which conditions give both high efficiency and a linear variation in amplitude. The ATF curves do not show the effect of the resolution of the recorder on the efficiency, but curves can be plotted for different spatial frequencies. Figure 6-42 shows ATF curves for a spatial frequency of 60 cycles/mm recorded on Agfa 10E75 with an 18-sec exposure and development in Methinol-U for 12 min. Notice that the efficiency is diminished for the larger angle of diffraction associated with the higher spatial frequency. This is due to the reduced MTF of the film.

Another way of presenting the curves of diffraction efficiency was described by Lin [6-36]. The visibility of the fringes, rather than the ratio of wave amplitudes was chosen as a parameter. Another difference is that Lin plotted $\sqrt{\eta}$ against υ, the visibility of the fringes rather than the ratio of the interfering waves. The same type of information can be obtained. For example, the range of υ, hence the ratio $|\mathbf{V}/\mathbf{U}|$, over which the holographic recording of $\mathbf{U}$ is linear can be found and the maximum efficiency selected as a function of exposure. Figures 6-43 through 6-45 show the $\sqrt{\eta} - \upsilon$ curves for 649-F, bleached 649-F, and dichromated gelatin. The normal to the plate bisected the 45° angle between the two plane waves. The 649-F was exposed to 632.8-nm wavelength light and the dichromated gelatin to 488-nm wavelength light. Development of the

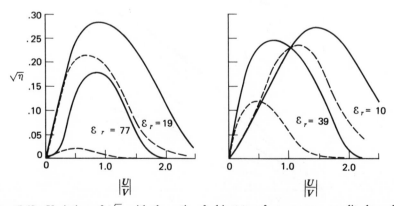

Figure 6-42 Variation of $\sqrt{\eta}$ with the ratio of object-to-reference wave amplitude and the exposure of the reference wave, $\mathcal{E}_r$, in ergs/cm². The grating spatial frequency is 60 cycles/mm. The solid lines are calculated curves and the dashed lines show the measured values [6-35].

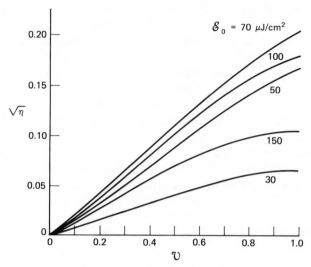

Figure 6-43 $\sqrt{\eta}$ –υ characteristics of Kodak 649-F emulsion developed in D-19 [6-36].

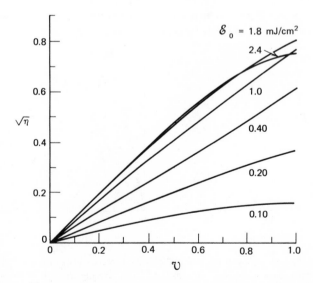

Figure 6-44 $\sqrt{\eta}$ –υ characteristics of bleached Kodak 649-F emulsion developed in D-76 and bleached with potassium ferricyanide bleach bath [6-36].

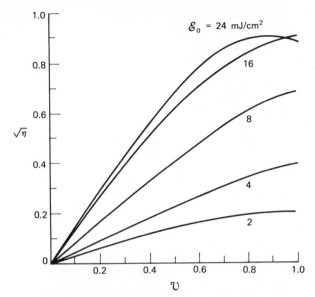

Figure 6-45 $\sqrt{\eta} - \upsilon$ characteristics of dichromated gelatin film [6-36].

nonbleached 649-F was with D-19. The bleached 649-F was developed with D-76 and bleached using a potassium ferricyanide bleach [6-36, 6-37]. The dichromated gelatin was prepared in accordance with the procedure given in [6-24]. Note the change in scale of the exposure from microjoules to millijoules.

In sum, the ATF or $\sqrt{\eta} - \upsilon$ curves present more information than the $t - \mathcal{E}$ curves, but they require more data to be plotted. None of the curves can accurately predict the efficiency of a hologram made with a diffusely reflecting object rather than with a plane wave object, but an estimate can be made. The $t - \mathcal{E}$ curve is sufficient for amplitude modulation holograms when the thickness of the emulsion is not a factor. Either the ATF or the $\sqrt{\eta} - \upsilon$ curve is preferred for bleached or reflection holograms.

PROBLEMS

6-1.   Consider the $t - \mathcal{E}$ curve of Fig. 6-10. Write an analytical expression for t as a function of $\mathcal{E}$ for the linear region. Assume that a transparency recorded in the linear region is then used in a contact print to obtain a positive. Obtain an analytical expression for t_p as a function of $\mathcal{E}$.

6-2.   Assume that a thin absorption hologram is recorded by interference of two uniform plane waves of equal amplitude. Further assume that the amplitude transmittance is linearly proportional to the exposure. If a uniform plane wave illuminates the hologram, what percentage of the incident energy is found in the first-order image? Give your answer in terms of the variables involved.

6-3.   Assume that a thick dielectric transmission hologram diffracts 100% of the incident energy into a plane wave image ($\iota = \pi/2$). The Bragg angle in air is 30°, $n_0 = 1.5$, $\lambda_a = 500$ nm, and $T = 15\,\mu$m. Find the incident angle closest to the Bragg angle that will produce zero energy diffracted to the image.

6-4.   Obtain (6-10) from (6-9).

REFERENCES

6-1.   J. F. Hamilton, "The Photographic Grain," *Appl. Opt.* **11**, 13–21 (1972).

6-2.   C. E. K. Mees and T. H. James, *The Theory of the Photographic Process* (Macmillan, New York, 1966).

6-3.   A. Vander Lugt and R. H. Mitchel, "Technique for Measuring Modulation Transfer Functions of Recording Media," *J. Opt. Soc. Am.* **57**, 372–379 (1967).

6-4.   J. W. Goodman, "Film-Grain Noise in Wavefront-Reconstruction Imaging," *J. Opt. Soc. Am.* **57**, 493–502 (1967).

6-5.   E. L. O'Neill, *Introduction to Statistical Optics* (Addison-Wesley Publishing Co., Reading, Mass., 1963), Ch. 7.

6-6.   S. A. Benton and R. E. Kronauer, "Properties of Granularity Weiner Spectra," *J. Opt. Soc. Am.* **61**, 524–529 (1971).

6-7.   P. E. Castro, J. H. B. Kemperman, and E. A. Trabka, "Alternating Renewal Model of Photographic Granularity," *J. Opt. Soc. Am.* **63**, 820–825 (1973).

6-8.   Klaus Biedermann, "The Scattered Flux Spectrum of Photographic Materials for Holography," *Optik* **31**, 367–389 (1970).

6-9.   Howard M. Smith, "Light Scattering in Photographic Materials for Holography," *Appl. Opt.* **11**, 26–32 (1972).

6-10.   D. H. R. Vilkomerson, "Measurements of the Noise Spectral Power Density of Photosensitive Materials at High Spatial Frequencies," *Appl. Opt.* **9**, 2080–2087 (1970).

6-11.   O. Bryngdahl and A. Lohmann, "Nonlinear Effects in Holography," *J. Opt. Soc. Am.* **58**, 1325–1334 (1968).

6-12.   J. W. Goodman and G. R. Knight, "Effects of Film Nonlinearities on Wavefront-Reconstruction Images," *J. Opt. Soc. Am.* **58**, 1276–1283 (1968).

6-13.   A. Kozma, G. W. Jull, and K. O. Hill, "An Analytical and Experimental Study of Nonlinearities in Hologram Recording," *Appl. Opt.* **9**, 721–731 (1970). Corrections: **9**, 1947 (1970).

6-14.   Wai-Hon Lee and Milton O. Greer, "Noise Characteristics of Photographic Emulsions Used for Holography," *J. Opt. Soc. Am.* **61**, 402–409 (1971).

6-15. H. Hannes, "Interferometric Measurements of Phase Structures in Photographs," *J. Opt. Soc. Am.* **58**, 140–141 (1968); "Interferometrische Messungen an Phasenstrukturen für die Holographie," *Optik* **26**, 363–380 (1967).

6-16. John C. Urbach and Reinhard W. Meier, "Thermoplastic Xerographic Holography," *Appl. Opt.* **5**, 666–667 (1966).

6-17. L. H. Lin and H. L. Beauchamp, "Write-Read-Erase in Situ Optical Memory Using Thermoplastic Holograms," *Appl. Opt.* **9**, 2088–2092 (1970).

6-18. E. Baumann, "The Fischer Large-Screen Projection System (Eidophor)," *J. SMPTE* **60**, 344–356 (1953).

6-19. E. Bartolini, W. Hannan, D. Karlsons, and M. Lurie, "Embossed Hologram Motion Pictures for Television Playback," *Appl. Opt.* **9**, 2283–2290 (1970).

6-20. Kendall Preston, Jr., "The Membrane Light Modulator and its Application in Optical Computers," *Opt. Acta* **16**, 579–585 (1969).

6-21. K. S. Pennington and J. S. Harper, "Techniques for Producing Low-Noise, Improved Efficiency Holograms," *Appl. Opt.* **9**, 1643–1650 (1970).

6-22. R. L. Lamberts and C. N. Kurtz, "Reversal Bleaching for Low Flare Light in Holograms," *Appl. Opt.* **10**, 1342–1347 (1971).

6-23. T. A. Shankoff, "Phase Holograms in Dichromated Gelatin," *Appl. Opt.* **7**, 2101–2105 (1968).

6-24. L. H. Lin, "Hologram Formation in Hardened Dichromated Gelatin Films," *Appl. Opt.* **8**, 963–966 (1969).

6-25. H. Kogelnik, "Coupled Wave Theory for Thick Hologram Gratings," *Bell Syst. Tech. J.* **48**, 2909–2947 (1969).

6-26. N. K. Sheridon, "Production of Blazed Holograms," *Appl. Phys. Lett.* **12**, 316–318 (1968).

6-27. D. Kermish, "Wavefront-Reconstruction Mechanism in Blazed Holograms," *J. Opt. Soc. Am.* **60**, 782–786 (1970).

6-28. R. J. Collier, C. B. Burckhardt, and L. H. Lin, *Optical Holography* (Academic Press, New York, 1971), Ch. 9.

6-29. A. A. Friesem and J. L. Walker, "Thick Absorption Recording Media in Holography," *Appl. Opt.* **9**, 201–214 (1970).

6-30. H. M. Smith, "Effect of Emulsion Thickness on the Diffraction Efficiency of Amplitude Holograms," *J. Opt. Soc. Am.* **62**, 802–806 (1972).

6-31. P. J. van Heerden, "Theory of Optical Information Storage in Solids," *Appl. Opt.* **2**, 393–400 (1963).

6-32. E. N. Leith, A. Kozma, J. Upatnieks, J. Marks, and N. Massey, "Holographic Data Storage in Three-Dimensional Media," *Appl. Opt.* **5**, 1303–1311 (1966).

6-33. M. Akagi, T. Kaneko, and T. Ishiba, "Electron Micrographs of Hologram Cross Sections," *Appl. Phys. Lett.* **21**, 93–95 (1972).

6-34. A. Kozma, "Photographic Recording of Spatially Modulated Coherent Light," *J. Opt. Soc. Am.* **56**, 428–432 (1966).

6-35. J. Upatnieks and C. D. Leonard, "Linear Wavefront Reconstruction from Nonlinearly Recorded Holograms," *Appl. Opt.* **10**, 2365–2367 (1971).

6-36. L. H. Lin, "Method of Characterizing Hologram Recording Materials," *J. Opt. Soc. Am.* **61**, 203–208 (1971).

6-37. C. B. Burckhardt and E. T. Doherty, "A Bleach for High-Efficiency Low-Noise Holograms," *Appl. Opt.* **8**, 2479–2482 (1969).

7

Spatial Filtering

We found in Chapter 5 that the spatial frequency spectrum of a wave amplitude distribution in the front focal plane of a lens appears in the back focal plane. We also saw that, with the use of a second lens, the spectrum can be transformed to reproduce the original distribution—an image. The spatial spectrum is readily available for modification, allowing the spectrum to be altered and the altered spectrum to be retransformed to yield an altered image. The important aspect is that the spatial frequency content of an image is easily changed. In addition to the two-lens system shown in Fig. 7-1, the single-lens system of Fig. 7-2 can be used to allow the alteration of the spatial frequencies of an image. The $1/i\lambda f$ terms found in Chapter 5 are dropped in both figures. In all figures, unless otherwise indicated, a point source and a collimator should be assumed so that the illumination of the transparency is with a uniform plane wave.

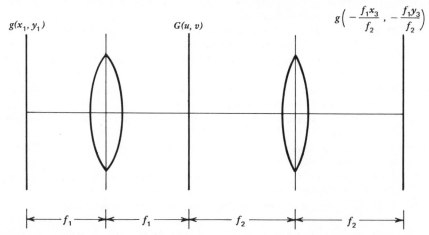

Figure 7-1 Two-lens imaging system allowing operation on the spatial frequency spectrum.

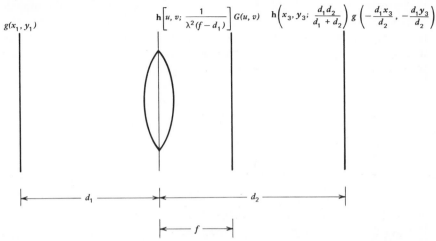

Figure 7-2 Single-lens imaging system allowing operation on the spatial frequency spectrum.

The primary topics of this chapter are the techniques of altering the spatial spectrum of a spatial signal, an analysis of the effects, and a discussion of some of the problems involved. The subject of spatial filtering is introduced by first considering only amplitude modification of the spectrum. Because the phase is not changed, this class of spatial filtering is often called *intensity spatial filtering*, even if the filters are described in terms of their effect on the amplitude of the spectrum. If the phase as well as the amplitude are altered (other than insertions of π radian shift), the operation is referred to as *complex spatial filtering*. After the basic theory is treated, some of the practical difficulties are discussed, and some approaches being taken to overcome the difficulties are outlined.

7.1 Intensity Filtering

The spatial frequency spectrum of the amplitude of a wave distribution is obtained by using an optical system such as shown in Fig. 7-1. The spectrum of the object and image is present in the back focal plane of the first lens. Consequently, the spectrum is readily available for operations in the frequency domain. By placing blocks or aperture stops in the transform plane, certain of the spatial frequency components of the image can be eliminated or reduced in amplitude.

For example, consider the case where a transparency of amplitude transmittance

$$t(x,y) = 1 + \tfrac{1}{2}\cos\frac{2\pi x}{\sigma} + \tfrac{1}{2}\cos\frac{2\pi x}{2\sigma} \qquad (7\text{-}1)$$

is placed in the front focal plane of lens L_1 in Fig. 7-3 and is illuminated with a uniform, normally incident plane wave. The factor of $\frac{1}{2}$ is necessary to assure that $t(x,y)$ is never negative. The cosine terms in the distribution $t(x,y)$ can be written in terms of exponentials before the Fourier transform is performed, allowing the use of the shift theorem, to obtain

$$\mathbf{T}(u,v) = \left[\delta(u) + \tfrac{1}{4}\delta\left(u - \frac{1}{\sigma}\right) + \tfrac{1}{4}\delta\left(u + \frac{1}{\sigma}\right) + \tfrac{1}{4}\delta\left(u - \frac{1}{2\sigma}\right)\right.$$

$$\left. + \tfrac{1}{4}\delta\left(u + \frac{1}{2\sigma}\right)\right]\delta(v) \qquad\qquad (7\text{-}2)$$

where $\mathbf{T}(u,v)$ is the Fourier transform of $t(x,y)$. The distribution in the transform plane is $\mathbf{T}(u,v)$ convolved with the Fourier transform of the pupil function, but for the present purposes we neglect the effect of the lens size. The 1 and the exponential components of the cosines of (7-1) represent plane waves as shown in Fig. 7-3. The plane waves are focused to the points indicated by the arrows in the transform plane. These now serve as point sources producing waves that are collimated by the lens L_2. Because of the offset in the position of the point sources, the resulting plane waves are tilted in the same manner as the original waves representing $t(x,y)$. If the focal lengths f_1 and f_2 are equivalent, the tilt of the waves is exactly the same as in the object plane, and they interfere to produce the

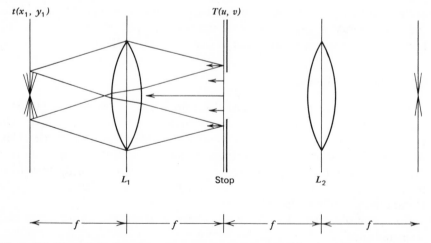

Figure 7-3 Two of the five plane wave components generated by the transparency $t(x_1,y_1)$ are blocked by the low-pass stop in the spatial spectral plane.

cosine distributions of (7-1). If, however, a stop or iris is placed in the transform plane as shown in Fig. 7-3, the points associated with the higher-frequency cosine are eliminated, thereby removing the $\cos 2\pi x/\sigma$ distribution from the image. Such a stop is a *low-pass spatial filter*. A block to eliminate the zero-frequency* distribution is a *high-pass spatial filter*. Two such filters are shown in Fig. 7-4. The cutoff of the filter is, of course, determined by its size. The filters of Fig. 7-4 have *no* effect on the phase of the spectrum. A band-stop filter could be used to remove the $\cos 2\pi x/2\sigma$ component, resulting in an image having twice the frequency variations of the one produced when a low-pass filter is used.

A square wave can be thought of as being composed of a number of sinusoids. Figure 7-5 shows the spatial spectral analysis of a grating that can be described by

$$\mathbf{t}(x,y) = \left[\Pi\left(\frac{x}{a}\right) \otimes \mathrm{III}\left(\frac{x}{2a}\right)\right]\Pi\left(\frac{x}{b}\right)\Pi\left(\frac{y}{d}\right), \qquad (7\text{-}3)$$

where $b \gg a, d \gg a$. Figure 7-5*b* is a photograph of the distribution in the transform plane.

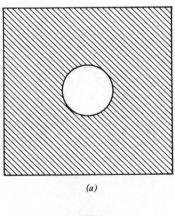

(a)

(b)

Figure 7-4 (*a*) A low-pass spatial filter. (*b*) A high-pass spatial filter with the same cutoff frequency.

*Not dc, please; dc stands for direct current. We have no current, but we do have a distribution having a spatial frequency of zero cycles per unit length.

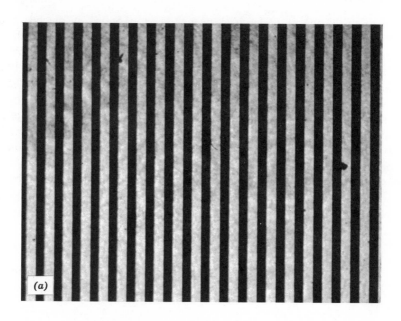

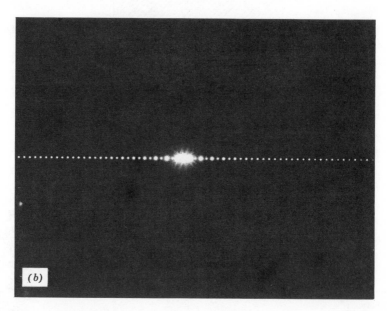

Figure 7-5 (a) Photograph of a transparency described by $t(x,y) = [\text{II}(x/a) \otimes \text{III}(x/2a)]$ $\text{II}(x/b)\text{II}(y/d)$, $b \gg a, d \gg a$. (b) The transform of $t(x,y)$ showing the spatial frequency components of the square wave.

182

An image of a grid can be described in terms of two square waves—one running in the x direction and the other in the y direction. The spectrum of such an image is shown in Fig. 7-6b. All spatial variations in the x direction can be removed by use of a spatial filter, in the spectrum plane, which is transparent only near $u = 0$. Such a filter is shown superimposed on the spectral distribution in Fig. 7-6c. Figure 7-6d shows the result of this type of filtering. The variations in the y direction could have been removed in a similar manner. From these results we can guess that if the filter is transparent only near $u = 0$ *and* $v = 0$, only the zero-frequency portion of the image remains. Figure 7-7 illustrates how a pinhole can be used to remove the diffraction rings due to dust in the optical system and provide a uniform illuminating or reference wave.

Comparison of temporal and spatial filters. When an electrical filter is constructed for a temporal signal, the amplitude and phase of the transfer function are not independent. If the circuit is a minimum-phase circuit, the amplitude and phase functions are related by the Hilbert transform. The amplitude and phase transmission of a spatial filter, on the other hand, can be specified and controlled independently. It is primarily this difference in temporal and spatial filters that makes an exact analogy between the two impossible. One example of the care necessary in making analogies is that of a low-pass filter. The low-pass spatial filter of Fig. 7-4a has uniform amplitude transmission and no phase variation across the aperture. Consequently, if the input to a system such as is shown in Fig. 7-3 is a square spatial pulse, the output will be a *symmetrical* pulse with rounded edges such as shown in Fig. 7-8b. However, the output of a temporal minimum-phase low-pass filter would be the *asymmetrical* output as shown in Fig. 7-8d. The reason is that the phase multiplier of the output spectrum causes the inverse Fourier transformation to yield a completely different function.

The asymmetry of the output of the electrical circuit can also be related to the directionality of time. In the spatial case, the function can extend positively or negatively in space. In the temporal case, causality must be obeyed.

Sampled images. As an example of a sampled image let us consider the transform of a halftone photograph such as appears in a magazine or a newspaper. The photograph is made up of a number of dots as illustrated in Fig. 7-9. If the size of the dots is assumed to be very small, the image can be described by

$$g(x,y)\mathrm{III}\left(\frac{x}{\sigma}\right)\mathrm{III}\left(\frac{y}{\sigma}\right). \tag{7-4}$$

If the finite size of the halftone dots is taken into account, the shah

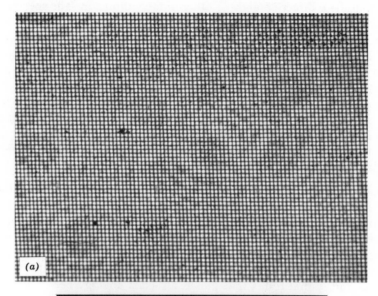

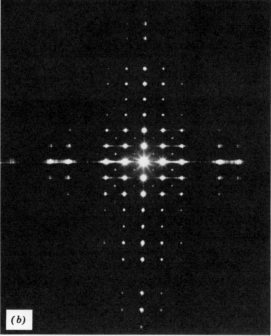

Figure 7-6 (*a*) Image of a wire grid. (*b*) Spatial spectrum of a wire grid.

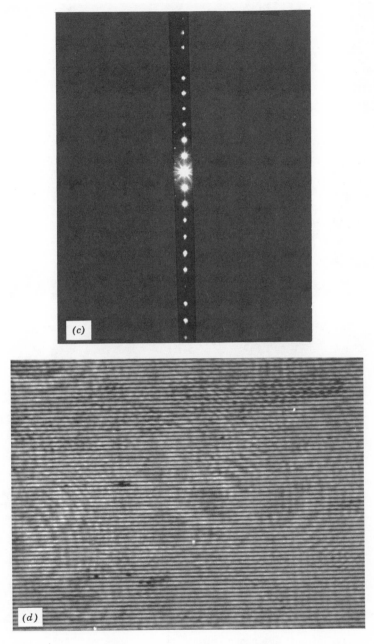

Figure 7-6 (c) Slit filter passing only $u = 0$ region of the spectrum. (d) The effect of the slit filter on the image distribution.

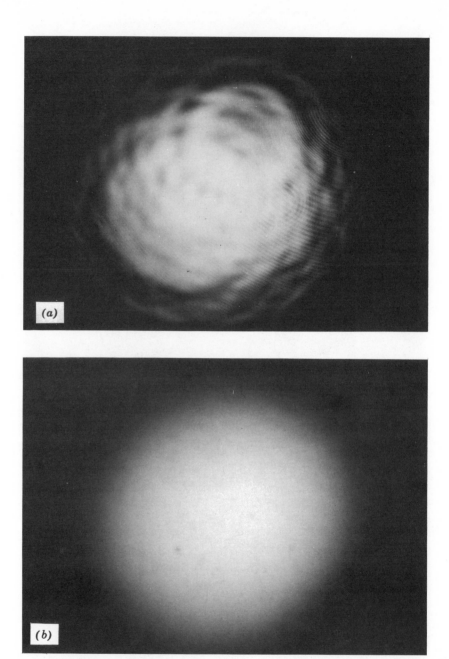

Figure 7-7 (*a*) Intensity distribution obtained by spreading a wave from a laser with a microscope objective. (*b*) Intensity distribution of the same wave after a pinhole is placed in the focus of the microscope objective.

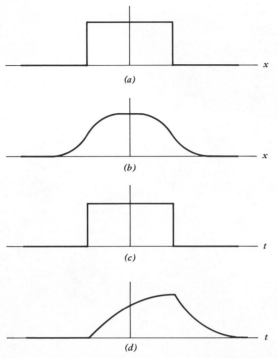

Figure 7-8 Comparison of effects of optical spatial and electrical temporal low-pass filters. (*a*) Square pulse input to imaging system. (*b*) Symmetrical distribution after low-pass filtering, using a uniform phase filter. (*c*) Square pulse temporal input to electrical system. (*d*) Asymmetrical output of a minimum phase low-pass filter.

functions of (7-4) must be convolved with a circle function. If a transparency having an amplitude transmittance $\mathbf{t}(x,y)$ equal to (7-4) is illuminated by a normally incident plane wave, the spectrum of the resulting distribution is

$$[\mathbf{t}(x,y)] = \sigma^2 \mathbf{G}(u,v) \otimes [\text{III}(\sigma u)\text{III}(\sigma v)]. \qquad (7\text{-}5)$$

The use of a lens system as described above provides the transform in the back focal plane of the lens. Notice that (7-5) is of the form of a replicated function. Consequently, we conclude that the effect of the halftone sampling is to replicate the spectrum of the nonsampled image. If the sampling rate is sufficiently high, a system with a stop such as shown in Fig. 7-3 can be used to block all but one of the replicated spectra. This spectrum can than be retransformed by the second lens of Fig. 7-3 to produce an image free of the halftone dots (see problem 2-12).

Figure 7-9   A section of a halftone photograph.

If the sampling rate is not at least double the highest frequency component of the sampled distribution, the replicated spectra will overlap and separation cannot be achieved. Consequently, the original distribution cannot be retrieved. If the spectra overlap and the stop of Fig. 7-3 is used, the adjacent spectra are not rejected completely if all of the central spectrum is to be passed. Figure 7-10 illustrates the total spectrum obtained when the sampling rate is insufficient. Consequently, one must choose between losing some higher-frequency components of the image or keeping them along with the reversed tails of the adjacent spectra. When energy from the high-frequency regions appears as lower frequencies in this manner, *aliasing* is said to occur.

If a slit rather than a circular hole is used as a stop, the image is produced with lines running normal to the slit. This is due to the interference of the waves from the individual spectra. The phenomenon is similar to that involved with the removal of a portion of the image of the wire grid.

Differentiation and integration. We have seen that, by use of the appropriate imaging system, the spatial frequency spectrum is available and is easily modified. Let us now consider the effect of placing, in the transform plane, a transparency having a transmittance

$$\mathbf{T}(u,v) = i2\pi u. \tag{7-6}$$

The result is that the transform of the image is multiplied by $i2\pi u$. It can easily be shown by writing the derivative in terms of its limits [7-1] that

$$\mathcal{F}\left[\frac{\partial}{\partial x} \mathbf{g}(x,y) \right] = i2\pi u \mathbf{G}(u,v). \tag{7-7}$$

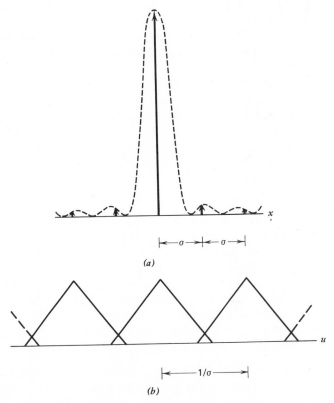

(a)

(b)

Figure 7-10 (a) A function sampled at intervals greater than one-half the period of the highest frequency component. (b) The resulting overlap of the spectra.

Consequently, rather than an image appearing in the x_3, y_3 plane of Fig. 7-1 or 7-2, the derivative of the image distribution with respect to x appears. The reader may not be accustomed to seeing the $i2\pi$ in (7-7), but it arises because of the symmetrical definition of the transform and its inverse.

We can just as easily obtain the derivative in the y direction, second and higher-order derivatives, or integrals. The integral of the image distribution is obtained by using a transparency with a transmittance proportional to $1/u$, $1/v$, or both. This makes the coherent imaging system a very versatile analog computer. It has the same advantages and disadvantages of most analog systems. It has a large resolution, that is, many data points on $\mathbf{g}(x,y)$ can be used, but it is difficult to achieve high accuracy. A little thought will show how any nonuniformity of the illuminating wave or nonlinearity in the film recording of the transparency will produce an error in the wave representing $\mathbf{t}(x,y)$. Further problems are caused by lens aberrations and the output device—film if a photograph of the result is desired, or an array of detectors if an electrical signal is needed. Another advantage of optical processing is that the optical system has the capability of handling either two-dimensional information or multiple channels of one-dimensional data [7-2].

The spatial filters described above are not simply intensity filters. The phase must be shifted by 180° at the origin to give the negative values of u; for example, when the transparency $\mathbf{t}(u,v) = i2\pi u$ is used. This can be done by providing a transparent object with a π radian shift in optical thickness at the origin. Quite often, it is not the derivative that is sought but simply an enhancement of the high-frequency detail of an image. To obtain this, the phase shift is not needed.

Edge enhancement. The edges and other regions of rapid change in an image are enhanced by differentiation, but differentiation is not necessary. A simple blocking of the low-frequency components is helpful. A phenomenological description of the process is that large areas having relatively constant transmittances produce transmitted waves that focus on the block. On the other hand, the edges and other regions of rapid change in transmittance diffract energy around the high-pass spatial filter (block). The energy diffracted around the block appears on the image plane, allowing the regions with high frequency variations to be imaged. Figure 7-11 shows the effect of such a high-pass filter on an image.

Signals in noise. Assume that a signal $1 + \cos(2\pi x/\sigma)$ and spatial noise $\mathbf{n}(x,y)$ are combined on a single transparency such that

$$\mathbf{t}(x,y) = 1 + \cos\left(\frac{2\pi x}{\sigma}\right) + \mathbf{n}(x,y). \qquad (7\text{-}8)$$

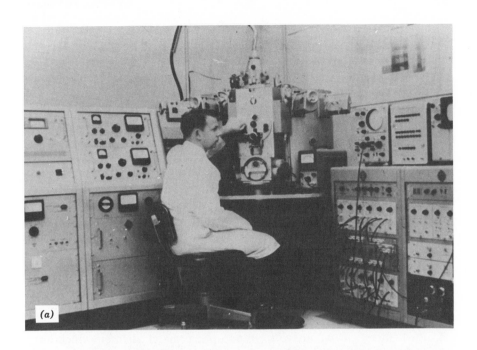

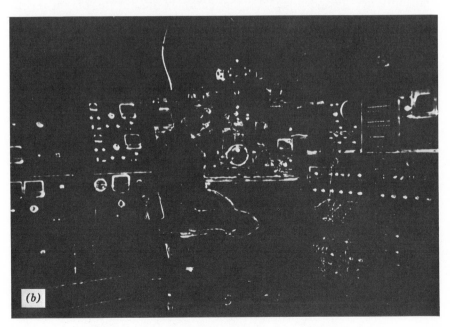

Figure 7-11 (*a*) Image before filtering. (*b*) Image after use of high-pass spatial filter.

Quite often the noise is not additive but multiplicative. To illustrate the optical processing, however, we shall assume additive noise. We have seen that the transform of the signal is three $[J_1(\rho)/\rho]$ type distributions located at $u = 0, 1/\sigma$, and $-1/\sigma$, where the Bessel function distribution is the transform of the lens or transparency aperture, whichever is smaller. The spectrum of the noise is spread over the entire spectral plane. Consequently, if we know the spectrum of the signal, we can place three holes in an opaque screen in the spectral plane to pass the spectrum of the signal. A portion of the noise spectrum is passed, but most of the noise energy is blocked. The result is an image with a greatly increased signal-to-noise ratio [7-3]. One way of describing the filter is to say that the filter is *matched* to the spectrum of the known signal. It can be shown that a matched filter is the optimum linear filter for detecting the presence of a signal in additive, stationary noise [7-4]. The matched filter should have a transmittance proportional to the *complex conjugate* of the spectrum of the signal and inversely proportional to the spectral density of the noise.

7.2 Complex Filters

Matched filters.   If we are to form an optical matched filter, which has a transmittance proportional to the complex conjugate of the signal spectrum, we must preserve the phase of the spectrum. In general, the spectrum of a spatial distribution will be complex because the Fourier transform of an asymmetrical function is complex [7-5]. Vander Lugt has shown that the phase can be preserved by the addition of a reference wave [7-6]. In essence, a hologram is made of the spatial spectrum as shown in Fig. 7-12.

The reference wave could have an arbitrary amplitude and phase distribution. We shall consider the consequences of this later, but we assume, for the present discussion, that the reference wave is a uniform plane wave propagating at an angle θ with respect to the optical axis of the lens. The total distribution in the transform plane is

$$\mathbf{A}\left(\frac{x_2}{\lambda f}, \frac{y_2}{\lambda f}\right) + A_0 e^{i\alpha x_2} \tag{7-9}$$

where $\mathbf{A}(x_2/\lambda f, y_2/\lambda f)$ is the Fourier transform of $\mathbf{U}(x_1, y_1)$, A_0 is a real constant, and $\alpha = 2\pi \sin\theta/\lambda$. The recording of the distribution of (7-9) with an energy detector results in

$$|\mathbf{A}|^2 + A_0^2 + \mathbf{A}A_0 e^{-i\alpha x_2} + \mathbf{A}^* A_0 e^{i\alpha x_2}. \tag{7-10}$$

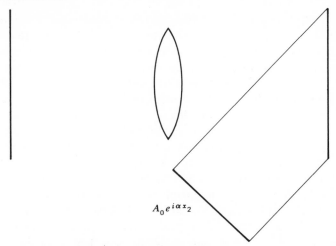

$A_0 e^{i\alpha x_2}$

Figure 7-12 Recording the complex spectrum.

We can see from (7-10) that not only is the complex spectrum recorded but also the conjugate. If the recording is made into a transparency having a transmittance proportional to the function in (7-10), we obtain waves along the axis and at the angles $\pm \theta$ from the optical axis.

The illumination of the recorded spatial filter can be achieved as shown in Fig. 7-13. If a point source is placed in the front focal plane of the lens, a plane wave illuminates the filter, giving rise to, among others, a wave represented by $\mathbf{A}(x_2/\lambda f, y_2/\lambda f)$. This wave is transformed by the second lens to provide an image of the original distribution. Because of the point

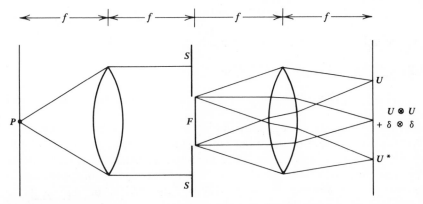

Figure 7-13 Illumination of a filter to determine the impulse response. P—point source, S—aperture stop, F—filter, U—image, U^*—conjugate image, $U \otimes U$ and $\delta \otimes \delta$—on-axis correlation distributions.

source in the input plane, the resulting image is the *impulse response* of the filter. Note that the spectrum of the impulse is flat as indicated by the plane wave in the spatial spectral domain. In addition to the primary image, an inverted conjugate image is obtained. (Recall that $\mathcal{F}[A^*]$ $=U^*(-).$) On axis, we find optical distributions proportional to $A_0^2\delta(x,y)$ and $U(x,y) \star U(x,y)$. Figure 7-14 shows a filter and the impulse response for the character T. The edge enhancement is due to the overexposure of the center portion of the filter, attenuating the low spatial frequencies.

Overexposure of the central region and extension of the dynamic range can be accomplished for certain filters by using a very weak reference beam and bleaching the filter [7-7]. This approach results in a filter with diffraction efficiency nearly independent of exposure over a range of $200:1$. Other approaches use a modulating grating in the input plane to cause many replicated spectra to appear in the transform plane [7-8, 7-9]. These spectra have differing energy levels and hence cause different exposures of the array of filters. The same grating is placed in the input plane when the filters are used so that energy is diffracted to all filters. The output is the superimposed filtered inputs. The grid structure of the output can be removed by placing another grating in the output plane and reimaging through a low pass spatial filter [7-9].

Convolution and correlation. Let us assume that a filter was recorded using $H(u,v)$, the spectrum of $h(x,y)$, and that the filter is placed in the transform plane of Fig. 7-13. Rather than a point source, a transparency having a transmittance proportional to $g(x,y)$ is placed in the input plane and is illuminated with a uniform plane wave. The distribution just to the right of the filter is described by

$$G(u,v)\left[A_0^2 + |H|^2 + A_0 H e^{-i\alpha\lambda fu} + A_0 H^* e^{i\alpha\lambda fu}\right]. \qquad (7\text{-}11)$$

The second lens Fourier-transforms this distribution to give

$$A_0^2 g(x,y) + g(x,y) \otimes [h(x,y) \star h^*(x,y)]$$

$$+ A_0 g(x,y) \otimes h(x+f\sin\theta, y) + A_0 g(x,y) \star h^*(x-f\sin\theta, y). \qquad (7\text{-}12)$$

The first term gives an image of the input $g(x,y)$ on-axis, and the second adds the convolution of the input with the autocorrelation of the function for which the filter was formed. The next two terms are of greater interest. The third term of (7-12) shows that, centered at $x = -f\sin\theta$, the *convolution* of $g(x,y)$ with $h(x,y)$ is produced. The last term represents the *correlation* of $g(x,y)$ with $h(x,y)$ centered at $x = f\sin\theta$. Figure 7-15 shows the correlation and on-axis distributions in the output or correlation plane. Figure 7-15 also shows one of the more interesting applications of matched

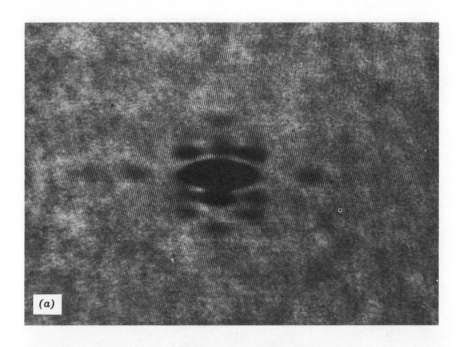

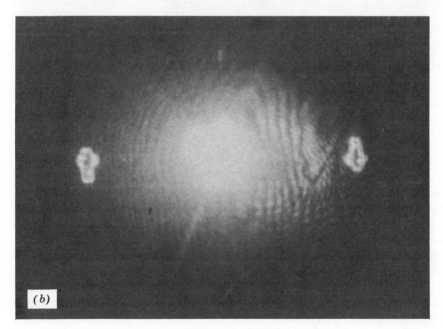

Figure 7-14 (*a*) Filter for the character *T* and (*b*) its impulse response.

Figure 7-15 Output of an optical correlating system: *A*—on-axis distributions; *B*—correlation of *T* with the word *spatial filters*.

spatial filtering—character recognition. More of the aspects of character recognition are discussed in Section 7.3.

The mathematical description of correlation and convolution was given in Chapter 2. It is helpful, however, to have a physical description of the optical processes involved. Figure 7-16 shows the detail of a spatial filter for the character *O*. Notice the fringes produced by the interference between the wave describing the transform of *O* and the reference wave. Now consider the effect of placing a plane wave illuminated transparency of an *O* in the input plane of a system as shown in Fig. 7-13. No matter where the *O* is in the plane, the transform is centered on the axis in the filter plane. Because the transform causes energy to fall exactly where the recorded fringes lie (the fringes appear where the transform contained energy during the recording process), energy is diffracted off-axis to be

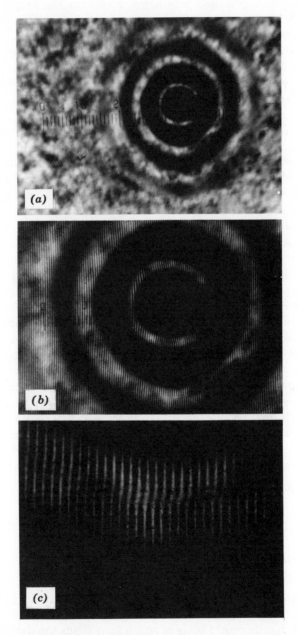

Figure 7-16 Details of the complex filter for the character O. (a) X 50 magnification, 20 μm/div, (b) X 100 magnification, 10 μm/div, (c) X 400 magnification, 2.5 μm/div. Note the shift in fringe position at the phase reversals.

focused by the second lens producing a spot in the output plane. The description above ignores the film saturation obvious in Fig. 7-16; we shall consider such effects in the next section.

Obviously, energy is diffracted into the region displaying convolution as well as into the region showing correlation. The difference is that in one case, the wave is multiplied by its conjugate, giving a distribution with a uniform phasefront. That is, a plane wave appears behind the filter directed toward the correlation region. If the wave were a *uniform* plane wave, it would be focused to a spot by the lens. Because it has an amplitude distribution described by $|\mathbf{H}|^2$, it is transformed to the auto-correlation of $\mathbf{h}$. On the other hand, the wave giving rise to the convolution is not multiplied by its conjugate and is not converted into a plane wave. Consequently, it cannot be focused into anything resembling a point by the second lens; the convolution of a function with itself yields a smeared or spread-out distribution.

We have seen that illumination of the spatial filter with a duplicate of the reference wave yields the original wave (impulse response). Similarly, if the filter is illuminated by the conjugate of one of the stored distributions, we retrieve a plane wave. The plane wave is not a uniform plane wave as was the reference, but it could be made to so if the filter were of the form $\mathbf{H}^*/|\mathbf{H}|^2$. The reconstruction of a plane wave produces a focused point in the output plane. Notice that the plane wave reference could have been provided by placing a point source alongside the input in the filter recording process. The offset in the input plane would have given the necessary tilted plane wave in the transform plane. If a filter output other than a correlation or a point is desired, *that* output could be placed in the input plane during the filter recording process so that its transform served as the reference for the recording of the transform of the function to which the filter is to respond. To provide an undistorted output, the denominator of the filter representation should be $|\mathbf{H}|^2$, yielding a filter transfer function of

$$\frac{\mathbf{H}^*\mathbf{Q}}{|\mathbf{H}|^2} \qquad (7\text{-}13)$$

where $\mathbf{Q}$ is the transform of the desired output $\mathbf{q}$. For example, a filter could be made where an input β caused B to appear in the output plane.

Computer-generated spatial filters. Greater flexibility in specifying the filter response can sometimes be achieved by calculating and plotting the desired filter distribution. Photoreduction then yields the desired filter. The computer-generated filters need not look the same as the filters obtained photographically. In fact, they most often have no shades of gray in the transparency but have a binary transmittance. The mode of achieving amplitude and phase control is also often different. Computer-generated spatial filters are not treated further in this chapter, but are discussed as special forms of computer-generated holograms in Chapter 9.

7.3 Pattern Recognition

The possibility of optically producing a correlation of one two-dimensional function with another is the basis of a coherent optical pattern recognition technique. This section treats aspects of a complex spatial filtering system that arise especially in pattern recognition. Of particular interest are techniques for increasing the degree of discrimination between similar patterns and means of providing real-time operation. Incoherent optical correlation systems can be used, but their outputs have a low S/N ratio.

Recognition system. One system that could be conceived involves an input presenting a transparency containing information to be read—a page of print, for example; an arrangement for placing different filters in the filter plane; and an array of detectors or a scanning detector to detect the presence of a correlation indication. For example, the filter for the letter a would produce correlation indications everywhere the a appeared in the input text. The detector system would sense the presence of the correlation indications and send the locations of all a's to a memory for storage. The filter for b would then come into place and the locations of all b's detected. The process would continue until all characters and symbols were processed.

The processing could be done, at least in part, in parallel rather than in series. One filter could be made to detect both a's and b's, for example. In this case, the correlation indications for the a's and the correlation indications for the b's would appear in different regions of the output. This is easily done by introducing the reference for the a and the b at different angles. (See problem 7-6.) The number of sequential operations could be halved if all filters operated on two symbols, but the number of detectors (or detector scan time) would be doubled.

Energy normalization. One of the problems arising in the character recognition system is best described in terms of the energy that a filter causes to go into a correlation spot. Again, thinking of the filter as a modulated grating, we recall that an O, for example, causes energy to fall on the portions of the filter for an O over which fringes were recorded. A portion of this energy is deflected to form a correlation spot. If parallel operations are being performed, where, for example, the filter for a period is also illuminated, the cross-correlation spot resulting from the O being correlated with a period may be as bright as the one for a period with itself. Consequently, some normalization scheme must be used, or a logic system must be used to process the output data.

The use of the noise spectral density in the denominator of the representation of the filter performs a degree of normalization. Strictly speaking, it represents only random noise. However, characters other than the one being sought can be thought of as noise. It probably reduces confusion if such signals are referred to as clutter, in analogy to the radar problem where one person's signal is another person's noise. For example, for an airborne radar intended to detect other aircraft the return from the ground is clutter or noise from which any signal must be extracted. On the other hand, the ground return is the signal for a ground-mapping radar. Similarly, an *a* is the signal when we are looking for it and is clutter when we are looking for a *b*.

By dividing the transfer function of a filter for one character by the spectral density of all other possible characters, we can increase the S/N ratio in the output plane. In addition to the use of such a filter, the discrimination of spatial filters can be varied by using modified matched filters.

Discrimination enhancement.   One method of increasing the ratio between the intensities of autocorrelation and cross-correlation indications is to recognize that all characters will have a zero-frequency component and to make filters that ignore the contribution of the zero-frequency energy. Compare, for example, the transform pattern for the *T*, Fig. 7-14, and for the *O*, Fig. 7-16. If the low-frequency regions of both are removed, they have very little in common. If a filter for an *O* is illuminated by the spectrum of a *T*, almost no energy will be deflected to the correlation plane if the low-frequency region of the filter is blocked. Figure 7-17 illustrates the result, where part *a* shows the outputs of filters for *T* and *O* when a *T* is the input. Part *b* shows the same outputs when a low-frequency block is used. The *difference* between the output of the two is unchanged, but the *total* energy in both correlations is reduced and the *ratio* of the output of the *T* filter to the output of the *O* filter is increased.

If we analyze separately the effect of the matched filter and the high-pass filter, we can conclude that the effect of the high-pass filter is to cause what is equivalent to the correlation of one edge-enhanced character with another. Consider the effect of edge-enhancement upon the problem of the period and the *O*, mentioned previously.

Normal film causes an effect similar to that of a block because of the saturation of the film near the center of the focus (the low spatial frequency region of the transform). Film has a limited dynamic range over which it can record light distributions and the high frequencies of a

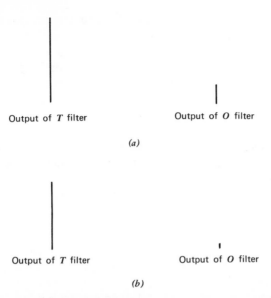

Output of T filter   Output of O filter

(a)

Output of T filter   Output of O filter

(b)

Figure 7-17 The output levels of filters for T and O when a T is the input. (a) All frequency regions preserved. (b) A high-pass filter is superimposed on both matched filters.

distribution normally have less energy than the low frequencies. Consequently, if the high frequencies are to be recorded, the low-frequency region of the transform is overexposed. The film can act as a variable center frequency bandpass filter if the frequencies of interest are optimally exposed. Then the lower frequencies will *normally* be overexposed and the higher frequencies *normally* underexposed. Obviously, if the signal being used to make the filter has an especially strong high-frequency component, the general rule is not obeyed. Of course, if a sharp cutoff bandpass filter or a filter of special characteristics is needed, a separate filter can be clamped to the matched filter.

Another variable in the production of matched spatial filters is the ratio of the reference wave amplitude to the amplitude of the distribution in the spectral plane. The visibility of the recorded fringes, hence their efficiency as gratings in deflecting energy to the correlation region, is a maximum when the two amplitudes are equal. Therefore, in recording a spatial filter, attention must be given to the ratio of reference to transform wave amplitude. If a region of the spectrum can be picked as being more important than another, the reference level can be adjusted to be the same as the amplitude of that region of the spectrum. The exposure of the film is then adjusted accordingly. One way of easily doing this is to view the transform with a microscope. The reference level can then be adjusted while alternately blocking the reference and the transform. This allows easy visual comparisons of the levels in the region of interest.

The discrimination between two similar signals such as an *O* and a *Q* can be increased by, after making the filter for one, double exposing the filter with the transform of the other *without the use of a reference*. This will darken the film in the regions common to both spectra. The effect is similar to that of removing the common low frequencies of all signals — the difference between the auto and cross-correlations is the same, but the ratio is drastically changed. In essence, the filter now looks only for *differences* between the transforms of the *O* and the *Q*. This argument can be extended to take into account all similar signals. The result is similar to the division of the filter transfer function by the spectral density of the clutter.

Classification filtering. Rather than modification of a filter to reject similar but different signals, a filter can be designed to respond to the common regions of the spectra of similar signals so that any one of a class or group of signals will respond. For example, automobiles in a photograph could be detected, rather than automobiles of a certain model, by building a filter to respond to rectangles of a certain size and shape. The probability of error would likely increase.

7.4 Practical Difficulties

We have considered some of the concepts of spatial filtering and the techniques used in pattern recognition. When these are applied to a practical system, additional problems arise. The difficulties are primarily involved with the changing and positioning of the filter, and the input and output processes. Some other problems such as changes in the scale or orientation of the pattern are common to any pattern recognition system.

Filter positioning. If a system is planned such that the filters must be changed, the positioning of the filter becomes more important. The position of the filter is not more critical, but the fact that it must be positioned frequently can cause mechanical difficulties. An intuitive feel for the effect can be obtained by visualizing the filter being illuminated by the spectrum from which it was made. If the filter is now moved transverse to the optical axis, the bright regions no longer optimally illuminate the grating-like structure of the filter. The level of the correlation indications decreases. On the other hand, the levels of correlation indications for the incorrect patterns may increase. If the mispositioning is longitudinal along the optical axis, the effect is not quite so severe. One effect is a changing of the scale of the transform that illuminates the filter. Another effect is that, because the filter is no longer in the transform plane, shifts in position of an input no longer produce only shifts in phase in the filter plane. Consequently, a change in longitudinal position of the filter causes a displacement of the transforms of input signals not on the optical axis. The amount of displacement is proportional to the distance that the signal is off-axis.

Vander Lugt [7-10] has studied the effects of small displacements of spatial filters on the system performance. He has shown that the performance of the system, defined as a normalized S/N ratio, varies as

$$\text{sinc}^2 \frac{\Delta u L}{4\pi} \qquad (7\text{-}14)$$

where L is the linear dimension of the signal and Δu is the displacement in the u,v plane. To obtain (7-14), homogeneous additive noise and a signal of the form $\text{II}(x/L)$ was assumed. Figure 7-18 shows the performance of a system for various values of the effective $F^{\#}$ of the system. The effective $F^{\#}$ is determined by the size of the signal and the focal length of the lens. The signal size enters the calculations because, when collimated illumination is used, the portion of the lens used is determined by the signal size.

One of the more interesting results is that when the noise spectral density is nonuniform the tolerable filter displacement is considerably

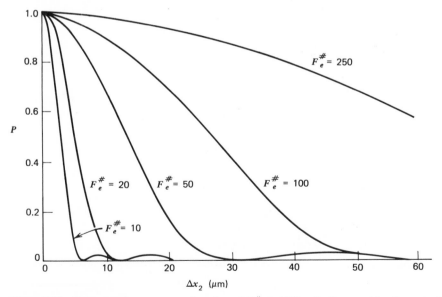

Figure 7-18 System performance as a function of $F^{\#}$ and filter displacement for the case of uniform noise [7-10].

reduced. Vander Lugt's analysis uses the background spectral density characteristic of aerial photographs as a case of a nonuniform spectral density. Figure 7-19 shows the results of that analysis. Notice that the difference between the tolerances for the two different types of noise increases as the $F^{\#}$ increases.

The tolerance on longitudinal filter displacement is an order of magnitude looser than on transverse displacement [7-10].

Scale and orientation. There are cases where it is desirable to make a filter for a pattern of one size and then use it to recognize or detect similar patterns of another size. If the pattern differs in size by only a small amount, there is a small reduction in the system performance. An experimental rule of thumb is that size variations of up to 10% are tolerable. The exact tolerance depends on the degree of discrimination desired.

A technique for obtaining a variable-scale Fourier transform is described in Chapter 5. The use of this technique in making a filter is illustrated in Fig. 7-20 where lens L_1 is assumed to be illuminated by a uniform plane wave. This filter is used in the variable-scale system shown in Fig. 7-21. A change in the location of the input results in a scale change in the transform plane. A 2:1 range of scales can easily be handled by such a system.

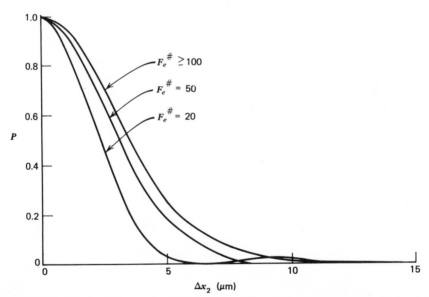

Figure 7-19 System performance as a function of $F^{\#}$ and filter displacement for the case of nonuniform noise [7-10].

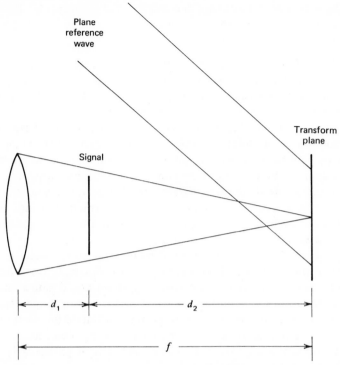

Figure 7-20   The recording of a spatial filter for use in a variable-scale system.

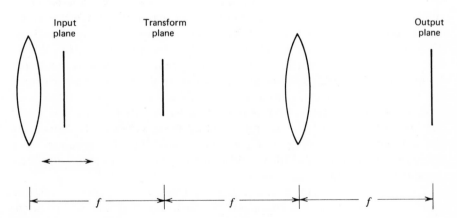

Figure 7-21   A variable-scale spatial filtering system.

205

The effect of orientation depends on the pattern. An O, for example, presents no problem, while a Y does. A rule of thumb is that a 10° rotation can usually be tolerated.

Read-in. It must be remembered that the discussions thus far have either explicitly stated or implicitly assumed that the input to the spatial filtering system is a transparency. If we attempt to use light reflected from a piece of paper on which the pattern is printed, we obtain not the transform of the pattern, but the transform of the pattern modified by an essentially random phase modulation caused by the surface variations of the paper. A pattern painted on a mirror would be acceptable but the paper causes the transform of the pattern to be convolved with the transform of the phase distribution due to the paper's surface. The result is that the transform of the pattern is completely changed. The information is still present, however, and techniques have been developed for using such incoherently illuminated inputs. The S/N ratio is low.

For high S/N, a transparency having no superfluous random phase modulation is needed. This means that, for optimum operation, all inputs should be transferred to a transparency and a liquid gate should be used if the transparency is on a base that has variations in thickness.

The input need not spatially amplitude-modulate the wave; spatial phase modulation can be used. The possibility of using phase modulation extends the number of materials that can be used. This is particularly important in achieving real-time operation.

Read-out. The read-out of the correlation distribution is another operation with potential for reducing the S/N ratio. The size of the detector should be matched to the correlation spot in the output plane. If the detector is too large, more than the peak of the correlation signal is detected, and the S/N decreases. On the other hand, if the detector is too small, the entire peak will not be detected, also reducing the S/N. Unfortunately, even with a well-matched detector, problems arise. We do not know where the correlation spot is to appear. It is easy to locate the regions of high correlation visually, but imagine the process as performed by an array of detectors or a scanning detector such as a vidicon. A portion of the correlation spot may be seen by one detector, or scan, and another portion seen by a second detector, or scan. We could reduce the detector size and spacing or have more scans, but this requires more detectors or more bandwidth in addition to some logic circuitry to determine if one spot has already been counted.

The general problem has not been solved. In the special case of character recognition, the characters lie on a line that can be specified.

Even then, for such things as computer printout, the characters do not always lie on the line. Some of these problems can be ameliorated by using specially weighted filters to take into account the probability distribution describing the location of the character about the line [7-11]. There is still work to be done before a spatial filtering system can read standard fonts at an error rate acceptable for computer use. The possibilities of parallel processing and high data rates keeps the interest high, however.

7.5 Real-Time Techniques

The primary problem in achieving real-time operation is the lack of a good real-time transparency. If such can be found, the filtering techniques previously discussed could be used, but there are other approaches that may become more feasible. The subject of transparencies is discussed briefly here, but because most developmental work has been done in conjunction with holography, transparencies are treated in more detail in Chapter 6 and in the chapters on holography.

Transparencies. The greatest block to realizing a real-time system is the input. Not enough data are available in transparency form. Even data in that form have a film base which, without a liquid gate, introduces random phase modulation of the input wave. The ideal input mechanism is a transparency having a transmittance that can be modulated by either an electrical signal or another optical distribution. Many electrooptic materials change transmittance when a voltage is applied. The problem is that it is difficult to obtain high resolution and high sensitivity simultaneously. The sensitivity problem is even more acute in materials that change transmittance at one wavelength when illuminated with another wavelength. Some materials have easily modulated transmittances, but they are diffusing transparencies—the phase is randomly modulated.

Phase effects in themselves are not harmful; it is the random phase modulation that degrades system performance. We have seen that phase modulation, rather than amplitude modulation, can be used to impress the information on the wave. Phase modulation can be rather effective in spatial filtering, and practical real-time phase modulators may be easier to achieve than real-time amplitude modulators.

Spatial heterodyning. If a real-time transparency is available, a process that could be called *spatial heterodyning* is worth considering. Spatial heterodyning is a process of multiplying the spectra of two signals directly and then transforming the product rather than recording a filter using one signal and then illuminating the filter with the spectrum of the other signal [7-12]. Figure 7-22 shows a possible mechanism using a transparency

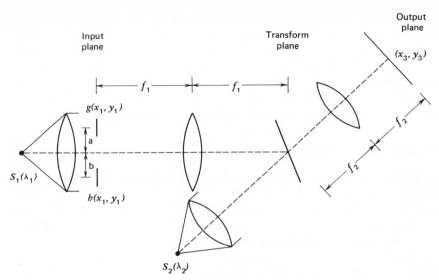

Figure 7-22 The spectra can be multiplied directly and the product transformed to obtain the correlation.

having a transmittance at a wavelength λ_1 which can be modulated by illumination with energy at a wavelength λ_2. The source S_1 emitting λ_1 illuminates transparencies of the functions $\mathbf{g}(x,y)$ and $\mathbf{h}(x,y)$. The transform distributions illuminate the transform plane transparency with energy at wavelength λ_1 causing its transmittance at λ_2 to vary. The resulting wave with wavelength λ_2 is Fourier-transformed by the lens L_2, and, as we shall see, a correlation of $\mathbf{g}(x,y)$ with $\mathbf{h}(x,y)$ is obtained in the output plane.

The distribution in the transform plane of Fig. 7-22 is described by

$$\mathcal{F}\left[\mathbf{g}(x-a,y)+\mathbf{h}(x+b,y)\right]=e^{-i2\pi au}\mathbf{G}(u,v)+e^{i2\pi bu}\mathbf{H}(u,v). \quad (7\text{-}15)$$

Presumably, any transparency would be energy sensitive, meaning that the transmittance would vary proportionally with the square of (7-15). The transmittance of the transparency can be written as

$$\mathbf{t}(u,v)=|\mathbf{G}|^2+|\mathbf{H}|^2+\mathbf{G}\mathbf{H}^*\exp\left[-i2\pi(a+b)u\right]+\mathbf{G}^*\mathbf{H}\exp\left[i2\pi(a+b)u\right]$$

$$(7\text{-}16)$$

where, to be more accurate, there should be a constant of proportionality in (7-16). The illumination of the transparency $\mathbf{t}(u,v)$ with a plane wave of wavelength λ_2 and the transformation with the lens of focal length f_2 yields

$$\mathbf{g}(x',y') \star \mathbf{g}^*(x',y') + \mathbf{h}(x',y') \star \mathbf{h}^*(x',y') + \mathbf{g}(x'+a+b,y')$$

$$\star \mathbf{h}^*(x',y') + \mathbf{g}^*(x',y') \star \mathbf{h}(x'-a-b,y') \qquad (7\text{-}17)$$

where

$$x' = \frac{\lambda_1 f_1}{\lambda_2 f_2} x_3, \qquad y' = \frac{\lambda_1 f_1}{\lambda_2 f_2} y_3. \qquad (7\text{-}18)$$

The result is that the correlation of $\mathbf{g}$ with $\mathbf{h}$ appears off-axis by an amount equal to $(\pm \lambda_2 f_2 / \lambda_1 f_1)(a+b)$ and the autocorrelations appear on-axis. It is interesting that, with this technique, no convolutions appear.

Another, more important, difference between the spatial filtering and spatial heterodyning operations is that there is no critical positioning problem in the spatial heterodyning system. Once the system is aligned optically, new inputs can be placed in the system with no fear of misalignment in the region where the spectra are multiplied. The location of the correlation indications is dependent on the location of the input, just as in the case of a spatial filtering system.

The spatial heterodyning technique works with film, but this requires the development of film for each correlating operation.

Hybrid optical-electronic systems. There are a number of hybrid systems that attempt to use the better characteristics of both optical and electronic systems. Most of these employ a scanning optical sensor to feed data into a computer which performs the correlation. These systems, while effective, are slow and require a considerable amount of computer capability. Improvements are being made, however, and systems of this type are becoming more practical. The discussion of such systems properly belongs in a treatise on computer-based pattern recognition, and are not treated here. The one hybrid system discussed is based on spatial heterodyning.

Figure 7-23 shows a hybrid system using the spatial heterodyning approach [7-13]. The inputs are illuminated by a uniform plane wave. The lens L_1 produces the transform in the plane of the vidicon. The vidicon, being an energy detector, gives an output similar to (7-16). The primary difference is that u is replaced by st where s is the scan velocity and v is replaced by lt where l is the line rate of the vidicon. Consequently, the spatial signal becomes a temporal signal. Before going into the mathematics or the physical description of the system, consider the black box description. The vidicon produces a temporal signal, and the associated electronics removes the sync and blanking pulses and integrates over one vidicon frame time. The spectrum analyzer now transforms the signal from

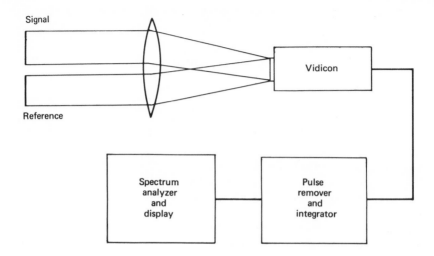

Figure 7-23 A hybrid optical-electronic spatial heterodyning system.

the time domain to the frequency domain, an operation analogous to the operation of the last lens of Fig. 7-22. The result is a correlation indication on the display of the spectrum analyzer. Figure 7-24 shows the result of using one *a* as an input and the word *spatial* as the other. All characters lie on the same line. The display shown in the upper portion of the figure indicates the correlation of the two inputs. The word *spatial* is shown in the lower portion of the figure to indicate the shapes of the characters. The horizontal axis on the spectrum analyzer is, of course, frequency.

The correlation indication shows only one coordinate of the output signal. If the other coordinate location is desired, it could be obtained by use of a different display system. How this can be done is easier to see after the operation is analyzed further.

One method of describing the operation of the system is to consider the fringes formed by interference between the waves described by the transforms of the two functions. The photograph of Fig. 7-16 illustrates the type of fringes formed even though those of Fig. 7-16 were formed by interference between a plane wave and the transform of an *o*. The vidicon scan normal to the fringes produces a signal frequency dependent upon the scan rate and the fringe spacing. The spectrum analyzer isolates this signal. Because the fringe spacing is dependent on the separation of the reference signal and the input signal, the frequency displayed on the analyzer is indicative of the horizontal distance between the input signal and the

Figure 7-24 Detection of the *a*'s in the word *spatial* by the use of a hybrid optical-electronic system [7-10].

reference signal. The strength of the signal is dependent on the visibility of the fringes and the area over which the fringes appear. Consequently, interference patterns between like spectra produce a stronger signal than interference patterns between unlike spectra.

The spectrum analyzer is centered on the frequency formed by interference between the reference and the center character. If a character is off the line of characters parallel to the vidicon scan, the interference fringes will be skewed with respect to the scan. In addition to the frequency component due to the scan, there will be a frequency component due to the line rate. This lower frequency can be detected by a spectrum analyzer centered at a frequency related to the line rate rather than the scan velocity. This results in two coordinate locations of the signal. With multiple signals, ambiguities become a problem.

Another hybrid optical-electronic system obtains the first Fourier transform optically using a spatial heterodyning technique and the second Fourier transform electronically with a digital computer [7-14]. The vidicon is still employed, but an analog-to-digital (A/D) converter is used to provide the appropriate input to the computer.

7.6 Incoherent Systems

An incoherent system is one that does not record or use the phase of the signals. Another definition is that incoherent illumination is used. There are numerous incoherent systems, many following the correlation description mechanically. That is, some systems involve the moving of an image over a transparency to give a correlation of the image with the distribution on the transparency. The shifting in

$$\mathbf{g}(x,y) \otimes \mathbf{h}(x,y) = \int \mathbf{g}(w,s)\mathbf{h}(x-w, y-s)\, dw\, ds \qquad (7\text{-}19)$$

is done by physical motion where the shifting was not necessary in techniques involving the multiplication of transforms. Such physical motion correlation schemes are adequate for some purposes and are in use. They are slow and quite often have a low S/N.

It is possible to have an incoherent system that does not require motion. One such system is shown in Fig. 7-25 [7-15]. The extended source, S, illuminates the transparency having intensity transmittance τ_1 such that the distribution caused by τ_1 is projected onto the transparency τ_2. Consider, first, a single point on the illuminating source, with coordinates x_s, y_s. The spherical wave from this point is collimated and illuminates the transparency $\tau_1(x, y)$. The plane wave produced by this point projects the distribution τ_1 onto τ_2 but with a displacement because of the off-axis position of the illuminating point. Other illuminating points give projections with different displacements, and the extended detector, D, integrates all the energy to yield a correlation. We can obtain a mathematical representation geometrically. Assuming that the point source is as shown in Fig. 7-25, we see that τ_2 is illuminated with the distribution propor-

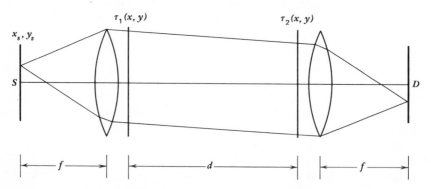

Figure 7-25 A system for providing correlation without motion. S–illumination source, D–extended detector [7-15].

tional to

$$\tau_1\left[x+\left(\frac{d}{f}\right)x_s, y+\left(\frac{d}{f}\right)y_s\right]. \qquad (7\text{-}20)$$

This yields an intensity distribution proportional to the product of τ_1 and τ_2. The lens collects the light over all x and y resulting in an intensity distribution in the focus of the lens which is described by

$$I=\iint \tau_1\left[x+\left(\frac{d}{f}\right)x_s, y+\left(\frac{d}{f}\right)y_s\right]\tau_2(x,y)\,dx\,dy. \qquad (7\text{-}21)$$

If, for some value of x_s and y_s, τ_1 is the same as τ_2, the light from τ_1 is passed optimally by τ_2 and a bright spot appears in the location (x_s,y_s) in the detector plane. A reduced similarity in the two distributions gives lower transmission and a dimmer spot, indicating a lower degree of correlation.

The resolution of a system such as described above is limited by the diffraction occurring between τ_1 and τ_2. If the pattern on the transparency τ_1 has fine detail, the resulting diffraction will prevent a facsimile of τ_1 from being projected onto τ_2. For the system to work well, the approximation of a geometrical projection from one distribution to another must hold. Another difficulty is that complex functions are difficult to represent with a nonnegative distribution such as the intensity transmittance.

A similar approach is used to obtain incoherent Fourier transforms [7-16]. The difference is that the object distribution is projected onto a mosaic of grids with different orientations and spacing. The result is that the energy passing through the grid is proportional to the amplitude of the object's spatial frequency component having the same period. The phase of the spatial frequency component can be found by obtaining the sine and cosine transform. Consequently, a distribution is obtained that has information concerning the spatial frequency distribution of the object.

Another approach to pattern recognition using incoherent illumination uses holography to make a lens with a special impulse response [7-14]. The impulse response of an imaging system is the image it forms of a point object. In calculating the image distribution, the impulse response is convolved with the object distribution. Consequently, if the impulse response can be made to be the reversed conjugate of the pattern to be recognized, the distribution in the image plane would be the autocorrelation of the input pattern. This approach has been used, in conjunction with feature extraction, to build a real-time system with a reasonably high S/N in the output [7-17]. The system is restricted to one input character at a time, but in many cases, a single input would be preferred with a system using matched spatial filters.

PROBLEMS

7-1. A television image consisting of modulated lines can be represented by

$$\left\{ \left[g(x,y) \otimes \text{II}\left(\frac{x}{a}\right) \right] \text{III}\left(\frac{x}{\sigma}\right) \right\} \otimes \text{II}\left(\frac{x}{a}\right)$$

where $g(x,y)$ describes the intensity distribution of the image, σ is the line spacing, and a is the linewidth. Recall that

$$g(x) \otimes h(x)|_{x=0} = \int g(u) h(u)\, du,$$

indicating that the line distribution is dependent on the integral of the signal over the linewidth. Consequently $\text{III}(x/\sigma)$ picks out the

$$\int_{m\sigma - a/2}^{m\sigma + a/2} g(x,y)\, dx$$

terms. What is the spatial frequency spectrum of the image? If the image is recorded on film so that

$$t(x,y) = \left\{ \left[g(x,y) \otimes \text{II}\left(\frac{x}{a}\right) \right] \text{III}\left(\frac{x}{\sigma}\right) \right\} \otimes \text{II}\left(\frac{x}{a}\right),$$

how could the scan lines be removed?

7-2. How could cosine modulation of an image distribution be used to store and retrieve more than one image on a single piece of photographic film? Each image extends over the entire film. How could the images be retrieved?

7-3. Determine the effect of removing the zero-frequency component from the image of a transparency having a transmittance of

$$t(x,y) = \frac{1}{\sigma} \text{II}\left(\frac{x}{a}\right) \otimes \text{III}\left(\frac{x}{\sigma}\right).$$

Sketch the distribution $t(x,y)$ for $\sigma/2 < a < \sigma$. Assume that a spatial filtering and imaging system such as shown in Fig. 7-1 or 7-2 is used to remove the zero-frequency component of the spectrum (the average amplitude of the image). Sketch the resulting intensity distribution in the image plane. What happens when $\sigma = 2a$?

7-4. What is the result of removing all components of the spectrum of the distribution

$$\left(\frac{1}{2a}\right)\mathrm{II}\left(\frac{x}{a}\right)\otimes\mathrm{III}\left(\frac{x}{2a}\right)$$

except the zero-frequency component and the $3/2a$ cycle/unit length component? Sketch the resulting image and compare with

$$\mathrm{II}\left(\frac{x}{a}\right)\otimes\mathrm{III}\left(\frac{x}{2a}\right).$$

7-5. A transparency of a page of text $a \times a$ in size and illuminated by a uniform plane wave is used as an input to a spatial filtering system. The focal length of all lenses is f. At what angle must the reference wave be introduced during the recording of a filter for a character so that when the filter is used, the off-axis correlation indications of the character and the distribution centered on the optical axis do not overlap? (Figure P7-5.)

7-6. Assume that a transparency of a page of text $a \times a$ in size is used as an input to a spatial filtering system. It is desired that the filter be designed to detect two characters at a time. If the focal length of all lenses is f, at what angles must the reference wave be introduced during the two exposures to avoid overlap of the correlation indications of the two characters or overlap of either with the on-axis distribution? Assume that both of the correlation regions are to be centered along the x-axis.

7-7. If a two lens imaging system is used where the input, one focal length before a lens, is normally illuminated with a uniform plane wave and the second lens is two focal lengths from the first, the image appears one focal length behind the second lens. Now, if the location of the input is changed, the image location changes. From

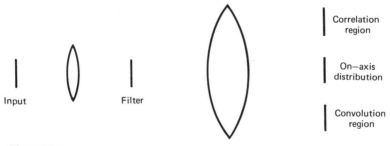

Figure P7-5

this we might conclude that the correlation plane of the system shown in Fig. 7-21 would not be one focal length behind lens L_2. Why would this conclusion be incorrect? Note that the filter is constructed with the input as shown in Fig. 7-20.

7-8. Where is the *convolution* plane of a system such as shown in Fig. 7-21?

7-9. A complex spatial filter is to be formed by placing a transparency against the transforming lens. Where should a point *reference* (real or virtual) be placed to cancel the quadratic phase error in the filter?

7-10. A spatial filtering experiment is being performed using the system shown in Fig. P7-10. Because the transform of $t(x,y)$ can be seen in plane P_2 for any value of d_1, the transparency $t(x,y)$ is placed close to lens L_1 to conserve space. A spatial filter for the function $g(x,y)$ is available from another experiment in which $g(x,y)$ was in the front focal plane of a lens and is placed in the focal plane P_2. Lens L_2 is to Fourier-transform the product distribution in plane P_2 and produce $t(x,y) \otimes g(x,y)$ in plane P_3. The correlation, however, does not appear in plane P_3. Why not?

7-11. A complex spatial filter is recorded for the signal $g(x,y) = \delta(x + b,y) - \delta(x,y)$. Show that when the filter is used in conjunction with an input transparency having a transmittance $h(x,y)$, the amplitude distribution of the image in the convolution plane is proportional to $(\partial/\partial x)\, h\,(x,y)$, the derivative of the input [7-18].

7-12. A complex spatial filter is formed using the unit step function as an input. Show that this filter will produce an image in the convolution plane proportional to

$$\int_{-\infty}^{x} \int_{-\infty}^{x} h(w,s)\, dw\, ds$$

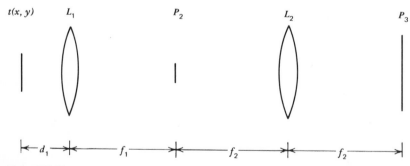

Figure P7-10

where $\mathbf{h}(w,s)$ is the amplitude distribution in the input or object plane [7-18].

7-13. Program on a computer the calculation of $|\text{sinc}(x/b)\otimes\text{II}(x/a)|^2$ and plot for $b/a = 1, 0.5, 0.1, 0.05, 0.01$. This will give the intensity distribution of the image of a slit of width a when a low-pass spatial filter having a cutoff at $u = \pm 1/2b$ is placed in the transform plane.

7-14. Describe the differences between the distributions in the correlation plane of a matched spatial filtering system when a negative and when a positive recording of the matched filter is used.

7-15. What is the *intensity* distribution in the spectral plane of Fig. 7-22 if the transparencies g and h are d_1, not f_1, in front of the transforming lens? Compare the results to those obtained when $d_1 = f_1$.

7-16. A spatial filter is recorded in plane P_2 of the system shown for problem 7-10 except that a reference wave is added. In making the filter, $d_1 = 0$. The entire system is then used where an input transparency $\mathbf{t}(x,y)$ appears in the input plane and the filter is placed in plane P_2. What must be the value of d_1 to cause (*a*) the correlation and (*b*) the convolution plane to appear in plane P_3?

REFERENCES

7-1. R. N. Bracewell, *The Fourier Transform and its Applications* (McGraw-Hill Book Co., New York, 1965), p. 117.

7-2. L. J. Cutrona, E. N. Leith, C. J. Palermo, and L. J. Porcello, "Optical Data Processing and Filtering Systems," *IRE Trans. Inf. Theory* **6**, 386–400 (1960).

7-3. E. L. O'Neill, "Optical Filtering in Optics," *IRE Trans. Inf. Theory* **2**, 56–65 (1956).

7-4. See, for example, G. L. Turin, "An Introduction to Matched Filters," *IRE Trans. Inf. Theory* **6**, 311–329 (1960).

7-5. See [7-1], p. 14.

7-6. A. Vander Lugt, "Signal Detection by Complex Spatial Filtering," *IEEE Trans. Inf. Theory* **10**, 139–145 (1964).

7-7. S. I. Ragnarsson, "A New Holographic Method of Generating a High Efficiency, Extended Range Spatial Filter with Application to Restoration of Defocussed Images," *Physica Scripta* **2**, 145–153 (1970).

7-8. G. B. Brandt, "Spatial Frequency Diversity in Coherent Optical Processing," *Appl. Opt.* **12**, 368–372 (1973).

7-9. J. W. Goodman and H. B. Strübin, "Increasing the Dynamic Range of Coherent Optical Filters by Means of Modulating Gratings," *J. Opt. Soc. Am.* **63**, 50–58 (1973).

7-10. A. Vander Lugt, "The Effects of Small Displacements of Spatial Filters," *Appl. Opt.* **6**, 1221–1225 (1967).

7-11. W. T. Cathey, "Probability Weighting of Spatial Filters," *J. Opt. Soc. Am.* **61**, 478–482 (1971).

7-12. J. E. Rau, "Comparison of Coherently Illuminated Images," *J. Opt. Soc. Am.* **56**, 541A (1966), and "Detection of Differences in Real Distributions," *J. Opt. Soc. Am.*

56, 1490–1494 (1966); C. S. Weaver and J. W. Goodman, "A Technique for Optically Convolving Two Functions," *Appl. Opt.* **5**, 1248 (1966).

7-13.   J. E. Rau, "Real-Time Complex Spatial Modulation," *J. Opt. Soc. Am.* **57**, 798–802 (1967).

7-14.   C. S. Weaver, S. D. Ramsey, J. W. Goodman, and A. M. Rosie, "The Optical Convolution of Time Functions, " *Appl. Opt.* **9**, 1672–1682 (1970).

7-15.   L. S. G. Kovásznay and A. Arman, "Optical Autocorrelation Measurements of Two-Dimensional Random Patterns," *The Rev. Sci. Inst.* **28**, 793–797 (1957).

7-16.   L. Leifer, G. L. Rogers, and N. W. F. Stephens, "Incoherent Fourier Transformation: A New Approach to Character Recognition," *Opt. Acta* **16**, 535–553 (1969).

7-17.   W. T. Maloney, "Lensless Holographic Recognition of Spatially Inchoherent Patterns in Real Time," *Appl. Opt.* **10**, 2127–2131 (1971).

7-18.   S. K. Yao and S. H. Lee, "Spatial Differentiation and Integration by Coherent Optical-Correlation Method," *J. Opt. Soc. Am.* **61**, 474 (1971).

8

Analysis of
Imaging Systems

The analysis in Chapter 5 determines the conditions under which an image is obtained with coherent illumination. This chapter treats the imaging system as a linear system for which a spatial frequency transfer function can be obtained. We shall see that the transfer function of the imaging system is dependent on the type of illumination, and a transfer function will be found for coherent, incoherent, and partially coherent illumination. The angular spectrum of plane waves is used to determine the transfer function for coherent, plane wave illumination, and the same angular spectrum analysis is then extended to diffuse and incoherent illumination.

8.1 Coherent Plane Wave Illumination

Spatial coherence. In the discussion of imaging systems, it is the spatial coherence that is important, not the temporal coherence. It makes no difference if the phase of the illuminating wave changes with time as long as the same phase appears across the plane of the object. Consequently, we shall not specify the temporal coherence, but only the spatial coherence. Coherent illumination in this context means spatially coherent illumination and incoherent illumination refers to spatially incoherent illumination. Quite often the two types are distinguished by reference to *point source* or *extended source* illumination, where the term extended source refers to a collection of independent point sources. That is, an extended plane wave would not be considered as an extended source because it can be focused to a point.

Angular spectrum. One of the ways of visualizing the effect of the lens aperture on an image is to write the object distribution as an angular spectrum of plane waves and follow each plane wave through the system. This is very similar to the technique used in Chapter 1 to derive the diffraction formula and is based on the work of Abbe, Rayleigh, and Porter [8-1, 8-2, 8-3]. Basically, the Fourier transform pairs

$$\mathbf{A}_1(u,v) = \int \int \mathbf{U}_1(x,y)\exp[-i2\pi(xu+yv)]\,dx\,dy \qquad (8\text{-}1)$$

and

$$\mathbf{U}_1(x,y) = \int \int \mathbf{A}_1(u,v)\exp[i2\pi(xu+yv)]\,du\,dv, \qquad (8\text{-}2)$$

where

$$u = \frac{\cos\theta}{\lambda}, \qquad v = \frac{\cos\xi}{\lambda}, \qquad (8\text{-}3)$$

relate the amplitude distribution to the angular spectrum. The angles θ and ξ are the angles between the *normal* to a plane wave and the x-axis and the y-axis, respectively. Equation (8-2) indicates that a sum of plane waves of amplitude and phase as specified by $\mathbf{A}_1$ and tilt as indicated by the exponential term can give the distribution $\mathbf{U}_1$. In determining the transfer function of a lens, we can simply follow the angular spectral components through the system.

Spatial frequency cut off of a lens.   Any distribution can be made up of an angular spectrum of *infinite* plane waves. This is a valid mathematical representation and also gives meaningful results when the waves are followed through an imaging system. Some objects, however, are better suited to slightly different approaches. After considering the general approach, we shall treat two special cases.

One satisfactory approach is to follow the plane wave components in pairs, as shown in Fig. 8-1. The spatial frequency cutoff of the lens can be described as occurring when the plane wave pair representing a certain spatial frequency no longer overlap to form fringes in the image plane. The plane waves become spherical waves in the image plane of a single-lens imaging system, but the result is similar. We can see from Fig. 8-1 that the smaller the lens, the smaller the region of overlap because the lens truncates the plane wave. For the moment, neglect any diffraction of the plane wave components. It is also obvious that as the angle between the plane wave components increases, the region of overlap diminishes. At some point, the fringes cease to exist. This is an abrupt transition, leading to the sharp spatial frequency cutoff of a lens. We shall see that the extent of the region of overlap, l, is

$$l = \frac{d_i}{d_o}(D - 2d_o\cot\theta) \qquad (8\text{-}4)$$

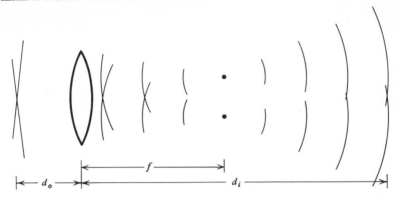

Figure 8.1 Two plane wave components passing through an imaging system.

where d_i and d_o are the image and object distances shown in Fig. 8-1 and D is the lens diameter. The region of wave overlap vanishes, removing the associated spatial frequency component from the spatial spectrum when

$$\cot\theta = \frac{D}{2d_o}. \tag{8-5}$$

For values of θ near $\pi/2$, $\cot\theta \approx \cos\theta$ and division of (8-5) by λ gives the cutoff frequency

$$u_c = \frac{D}{2\lambda d_o}. \tag{8-6}$$

Consequently, on the optical axis, a lens of diameter D has a unity spatial frequency transfer function up to the spatial frequency $D/2\lambda d_o$. At that point, the transfer function falls to zero. A similar equation is valid for v_c. Figure 8-2 shows the transfer function of a perfect circular lens of diameter D.

The cutoff given by (8-6) is valid for the spatial frequencies of the object. If we are concerned with the spatial frequency distribution of the image, the cutoff frequency is given by

$$u_{ci} = \frac{D}{2\lambda d_i} \tag{8-7}$$

The relation between u_{ci} and u_{co}, the cutoff frequency in terms of the object spatial frequency, is simply

$$\frac{u_{co}}{u_{ci}} = \frac{d_i}{d_o} = M \tag{8-8}$$

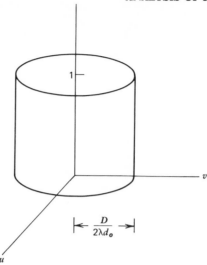

Figure 8-2   The transfer function of a perfect circular lens of diameter D.

where M is the magnification of the image. Equation (8-8) indicates that if, for example, the image is smaller than the object, a sinusoidal feature of the object will show up as a higher-frequency spatial frequency in the image. Consequently, if the feature is to be removed, the spatial frequency cutoff of the image must be higher than the cutoff of the object frequencies. This is accomplished in (8-7) by the division by d_i rather than d_o.

Geometrical analysis using plane waves.   It is interesting that the plane wave angular spectral components can be followed through the imaging system *geometrically* and yield the same results as a diffraction formula analysis using the original distribution [8-4–8-6]. Figure 8-3 shows the figure used to derive (8-4). The infinite plane waves are truncated by the lens, are focused in the focal plane, and propagate to the image plane where they overlap. From the figure, we find that

$$\frac{w}{d_i - f} = \frac{D}{f} \tag{8-9}$$

$$\frac{w - l}{d_i} \approx \tan 2\chi, \tag{8-10}$$

and, for small χ,

$$\tan 2\chi \approx 2\cot\theta. \tag{8-11}$$

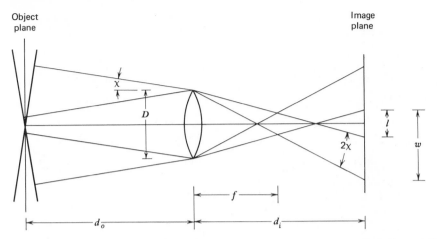

Figure 8-3 Geometry used in tracing the plane wave components to find the cutoff frequency.

Solving for l and using the relation

$$\frac{d_i - f}{f} = \frac{d_i}{d_o} \qquad (8\text{-}12)$$

yields (8-4).

The question naturally arises as to the effect of diffraction of the plane wave by the lens aperture. The effect is a second-order one, blurring the sharp cutoff. Figure 8-4 shows the effect of a decreasing lens diameter on the image of a square wave ruling. Notice how the area of overlap decreases as the lens diameter decreases. Beyond the spatial frequency cutoff for the fundamental spatial frequency, the waves no longer overlap on axis and no semblance of an image remains. The effect of diffraction upon the plane wave components is clearly shown.

Strictly speaking, the waves truncated by the lens should be replaced by a new set of infinite plane waves that propagate to the image plane. The result is simpler, however, if we realize that the effect of the finite aperture is to replace each plane wave by a narrow bundle of plane waves traveling, on the average, in the same direction as the original plane wave. In the plane of the aperture, they add to form a truncated plane wave; in the focal plane, they are focused to form the Airy pattern rather than a spot. However, by neglecting the bundle and assuming that the truncated waves propagate as infinite waves, a first-order result is obtained that gives the cutoff frequency.

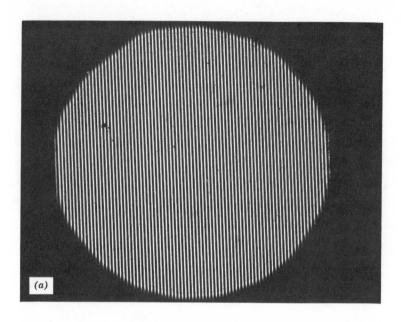

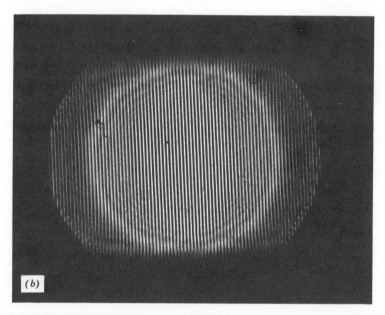

Figure 8-4 Effect of decreasing lens diameter on the plane wave components of the image of a square wave. (a) to (d): Decreasing lens diameter.

224

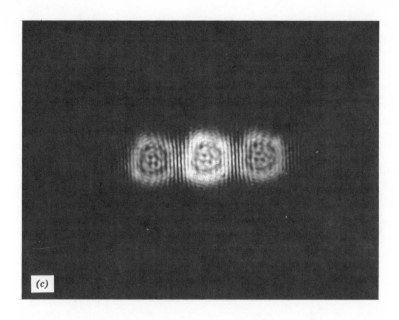

(c)

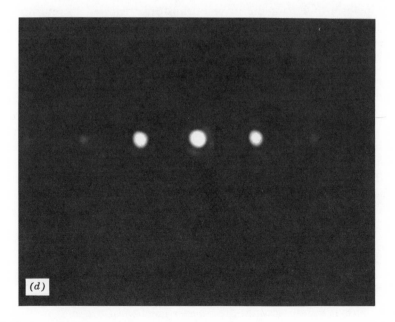

(d)

Figure 8-4   (*Continued*)

The neglect of the diffraction effect upon the truncated plane waves can be related to the dropping of the phase terms of equation (5-76) restated below:

$$\mathbf{U}_i(x_i,y_i) = \frac{1}{M}\mathbf{h}(x_i,y_i;d_i)\left\{ \tilde{\mathbf{P}}\left(\frac{x_i}{M},\frac{y_i}{M}\right) \otimes \left[\mathbf{U}_o\left(-\frac{x_i}{M},-\frac{y_i}{M}\right)\mathbf{h}\left(\frac{x_i}{M},\frac{y_i}{M};d_o\right)\right]\right\}.$$

$$(8\text{-}13)$$

The pupil function of the imaging system is P, its transform is $\tilde{P}$, and the magnification is denoted by M. The Fourier transform of (8-13) yields the relation between the object and image angular spectra,

$$\mathbf{A}_i(u,v) = -M^3(\lambda^2 d_i d_o)\mathbf{h}\left(u,v;-\frac{1}{\lambda^2 d_i}\right) \otimes \left\{ \mathbf{P}(-Mu,-Mv)\right.$$

$$\left. \times \left[\mathbf{A}_o(-Mu,-Mv)\otimes\mathbf{h}\left(u,v;-\frac{1}{\lambda^2 d_o}\right)\right]\right\}. \qquad (8\text{-}14)$$

The negative sign in the argument of $\mathbf{P}$ appears because we have used the direct transform of $\tilde{\mathbf{P}}$, not the inverse transform required to return to $\mathbf{P}$. That is, $\tilde{\mathbf{P}} = \mathbf{P}(-)$. Recall that a convolution with the impulse response expresses the diffraction formula:

$$\mathbf{U}_2 = \mathbf{U}_1 \otimes \mathbf{h}. \qquad (8\text{-}15)$$

Consequently, (8-14) says that it is *not* the angular spectrum of $\mathbf{U}_o$ that is multiplied by $\mathbf{P}$, but that it is a diffraction-smeared form of the angular spectrum that multiplies $\mathbf{P}$. Similarly, the convolution of the entire term in the braces with $\mathbf{h}(u,v;-1/\lambda^2 d_i)$ indicates that the spectrum passing through the pupil $\mathbf{P}$ undergoes further diffraction to yield $\mathbf{A}_3$, the spectrum of the image. The neglect of the diffraction of the plane wave components is indicated by the dismissal of the $\mathbf{h}$'s in (8-13), resulting in

$$\mathbf{U}_i(x_i,y_i) = \frac{1}{M}\tilde{\mathbf{P}}\left(\frac{x_i}{M},\frac{y_i}{M}\right)\otimes \mathbf{U}_o\left(-\frac{x_i}{M},-\frac{y_i}{M}\right) \qquad (8\text{-}16)$$

and

$$\mathbf{A}_i(u,v) = M^3\mathbf{P}(-Mu,-Mv)\mathbf{A}_o(-Mu,-Mv). \qquad (8\text{-}17)$$

Equation (8-17) indicates that the *transfer function* of the imaging system is simply the pupil function modified by the magnification. The u's and v's of (8-17) refer to the spatial frequencies of the image. The spatial frequencies of the image can be specified as

$$u_i = \frac{x_2}{\lambda d_i},$$

(8-18)

and the spatial frequencies of the object as

$$u_o = M u_i = \frac{x_2}{\lambda d_o},$$

(8-19)

where x_2 is the coordinate in the lens plane. The cutoff is found by replacing x_2 by its maximum value, $D/2$.

Coherent imaging of transparent objects smaller than the lens. If the object on a transparency is smaller than the lens, the truncated plane argument used above is applicable.* The difference is that the lower frequency plane waves are truncated at the object rather than at the lens plane. As a result, the region of overlap is a constant until the frequency is reached at which the lens begins to truncate the plane waves. Then (8-4) holds. Consequently, the frequency at which cutoff on the optical axis occurs is found in the same manner as when the object is represented by infinite plane waves.

Comparison of effects of lens and focal plane stops. We found in Chapter 5 that spatial frequency components of an object wave could be removed by operations in the focal plane. Figure 8-1 shows how each frequency component is focused in the transform plane. An aperture in this plane can serve as a low-pass spatial filter removing all spatial frequency components above the cutoff frequency. We have just seen that a finite lens has a similar effect. The effects, however, are not identical. The transform or frequency plane filter allows the passed frequency components to appear *over the entire image*, and the blocked frequencies appear nowhere in the image. On the other hand, when the finite lens diameter starts affecting the image, the spatial frequency components are removed from the outer portion of the image first.

Oblique illumination. It is interesting that fringes can be seen off the optical axis of Fig. 8-4c, whereas they cannot be seen on-axis. This is illustrated in Fig. 8-5 which shows the regions of overlap of the three components of an image of a cosinusoidal grating. The lens is sufficiently small to cut off the frequency of the image on the optical axis. Off-axis,

---

*It is the *object* size, not the size of the transparency, that is important. Note that the small blemishes on the grating of Fig. 8-4 are imaged with higher quality than the grating.

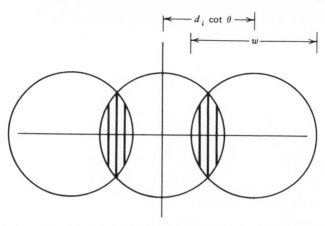

Figure 8-5 Fringes can sometimes be seen off the optical axis even when they are not visible on axis.

however, fringes can still appear. We shall find that they are not of the same form as the fringes that form on-axis at lower spatial frequencies, but the period is the same. The size of the three image components of Fig. 8-5 is determined by the size of the lens or the transparent object, whichever is smaller. The center-to-center separation of the components is determined by the difference in spatial frequencies represented. From the geometry of Fig. 8-3 we find that the center of a component is at $d_i \cot \theta$ or approximately $\lambda d_i u$. The extent over which a frequency component appears in the image is, if limited by the lens,

$$w = \frac{d_i - f}{f} D = \frac{d_i}{d_o} D. \tag{8-20}$$

Consequently, the fringes vanish on-axis when

$$\lambda d_i u > \frac{w}{2} = \frac{d_i}{2 d_o} D$$

or

$$u > \frac{D}{2 \lambda d_o}. \tag{8-21}$$

We can see from Fig. 8-5 that for the fringes to vanish everywhere in the image plane,

$$\lambda d_i u > w \tag{8-22}$$

or

$$u > \frac{D}{\lambda d_o}. \tag{8-23}$$

Remember that the resolution or spatial frequency resolving power of an imaging system is specified *on the optical axis*. Consequently, if oblique illumination is used to shift the fringes onto the axis, by definition the resolution of the system has been increased. The expression in (8-23) indicates that the cutoff frequency for oblique illumination can be up to twice the cutoff frequency for normally incident plane wave illumination. To shift the off-axis overlapping components to be on-axis, the normal of the illumination wave should be at an angle θ from the optical axis. The result is a shifting of the transfer function of Fig. 8-2 one-half diameter to one side. There is, in addition, a reduction in the modulus of the transfer function (except at zero spatial frequency) which is due to not having as many waves producing a given frequency component.

Notice that the same argument can be applied to the case where the spatial frequency transfer is restricted by a stop in the transform plane. Illumination with an oblique plane wave moves the transform to one side so that the higher frequency components to one side of the axis will be passed. The effect is the same as moving the spectral plane stop off the optical axis. In either case—frequency restriction by the lens or a focal plane stop—the result of oblique illumination is similar to single side band (SSB) modulation. The frequency components one side of the zero (or carrier) frequency are passed while those on the other side are eliminated.

The result of the single side band effect is that even though fringes having the proper spatial period can be formed beyond the normal cutoff frequency of the imaging system, they do not have the same shape as those formed below the cutoff frequency. As an example, consider a transparency having an intensity transmittance of $(\frac{1}{2} + \frac{1}{2}\cos 2\pi x/\sigma)^2$. The amplitude transmittance is

$$\mathbf{t}(x,y) = \tfrac{1}{2} + \tfrac{1}{2}\cos 2\pi \frac{x}{\sigma}. \qquad (8\text{-}24)$$

The distribution of (8-24) can be represented by one plane wave having amplitude $\frac{1}{2}$ and zero tilt; one plane wave with amplitude $\frac{1}{4}$ and phase $2\pi x/\sigma$; and one plane wave with amplitude $\frac{1}{4}$ and phase $-2\pi x/\sigma$. If the imaging system passes all three waves, they combine in the image plane to form $\frac{1}{2} + \frac{1}{2}\cos 2\pi x/\sigma$. (In addition, there is usually the quadratic phase term if a single-lens system is used. In the present discussion, the quadratic phase is seen to arise because spherical waves, not plane waves, are combined in the image plane.) The image intensity is

$$I = \left[\tfrac{1}{2} + \tfrac{1}{2}\cos \frac{2\pi x}{\sigma} \right]^2 = \tfrac{3}{8} + \tfrac{1}{2}\cos \frac{2\pi x}{\sigma} + \tfrac{1}{8}\cos \frac{4\pi x}{\sigma}. \qquad (8\text{-}25)$$

If the imaging system passes only the zero-frequency component and one off-axis wave or side band, the image intensity is given by

$$I_s = \left| \tfrac{1}{2} + \tfrac{1}{4} \exp\left(\frac{i2\pi x}{\sigma} \right) \right|^2 = \tfrac{5}{16} + \tfrac{1}{4} \cos \frac{2\pi x}{\sigma}. \tag{8-26}$$

The two images have fringes with the same periods, as can be seen in Fig. 8-6. The structure of the fringes is shown to differ, however. The effect of the change in the fringe form upon the images formed is an interesting study, but is not needed for our purposes.

Diffusely reflecting object. Thus far we have considered the imaging of a transparency illuminated by a uniform plane wave. If the object is a diffusely reflecting object, exactly the same analysis is applicable *if we are interested in the surface detail.* That is, if the surface that causes the diffuse reflection is considered a part of the image, then the wave from the object is reduced to an angular spectrum of plane waves just as was the wave from the transparency. If the surface is not under study and is of no interest (a sheet of paper on which an image is printed, e.g.), the effect of the surface is best treated as part of the illumination. In the case of a distribution on paper, we could analyze the imaging process as if a transparent medium were illuminated with a wave having a phase distribution identical to that caused by the paper. This case is treated as a variation of the plane wave illumination imaging and is discussed in the next section.

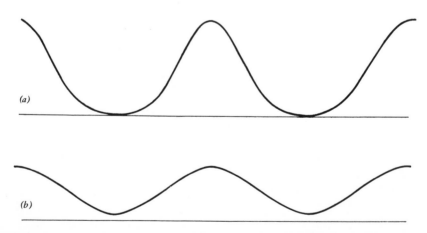

(a)

(b)

Figure 8-6 Comparison of the intensity distributions of the images of $\tfrac{1}{2} + \tfrac{1}{2}\cos 2\pi x/\sigma$ using (a) normally incident and (b) oblique plane wave illumination such that one side order is lost.

8.2 Coherent Complex Wave Illumination

A wave need not be a plane wave to be spatially coherent. A spherical wave or even a diffuse wave with random phase shifts in a plane is spatially coherent as long as the phase is temporally invariant. If the relative phase between two points remains constant, a double pinhole interferometer can be used to produce fringes indicating spatial coherence. The only effect of a phase difference between the two holes is to shift the position of the fringes.

Angular spectrum of the illumination. The illuminating distribution can be assumed to be comprised of an angular spectrum of plane waves which, when added, yield the illuminating distribution. Each component of the angular spectrum illuminates the transparent object and forms an image as described in the preceding discussions. The images formed by all illuminating plane waves are superimposed to give the image formed with a complex illuminating wave. A transfer coefficient of the imaging system is found to be dependent on the angular spectrum of the illumination.

Let us first consider the effect of having just two illuminating plane waves, one propagating along the optical axis and the other tilted slightly. We have seen that with normal plane wave illumination, the area over which a given spatial frequency component of the image appears is a lens-shaped region of the form shown in Fig. 8-7. If the illuminating plane wave propagates at an angle to the optical axis, the distribution shown in Fig. 8-7 is shifted in the image plane accordingly. This has the effect of increasing the area over which the spatial frequency component appears. If

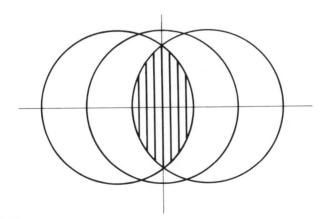

Figure 8-7 Area over which a given spatial frequency component appears in the image when normally incident, plane wave illumination is used.

the illuminating wave has a wide angular spectrum of plane waves, the spatial frequency component can appear over the entire image plane. More importantly, the areas can overlap, producing a brighter fringe pattern than would be produced by a single illuminating wave. The fringes forming the spatial frequency component of the image always have the same origin no matter which illuminating wave is used. That is, the spatial phase of the fringe pattern is the same. The fringes do not shift position because the illumination wave can affect the total phase of the waves forming the fringes, but not the relative phase between them. This can be seen by examining the expressing (8-27) which describes the interference of two plane waves of the spectrum. The tilted illuminating wave produces the same phase effect $e^{i\alpha x}$ on both waves, which then cancels:

$$|e^{i2\pi x/\sigma}e^{i\alpha x} + e^{-i2\pi x/\sigma}e^{i\alpha x}|^2 = 2\cos\left(\frac{2\pi x}{\sigma}\right). \tag{8-27}$$

The fringes formed in the image plane can be thought of as being present over the entire image but visible only through windows such as in Fig. 8-7. The locations of the windows depend on the angles of the individual plane waves comprising the illuminating distribution. If the illuminating plane waves have incidence angles resulting in the overlap of the apertures, the resulting image is brighter at that point. Assuming a fixed object and lens size, the size of the lens-shaped region of Fig. 8-7 is dependent upon the spatial frequency being considered. As the spatial frequency increases, the size of the region decreases until the plane wave components no longer overlap and the region vanishes.

Figure 8-8 shows the areas over which the zero and both side frequencies add to form cosinusoidal fringes. Part a shows the overlapping regions assuming five illuminating plane waves and a spatial frequency being imaged of low enough frequency that the five regions overlap on the optical axis. As the spatial frequency under consideration increases, the area over which the fringes are formed decreases. The center of the fringe areas is determined by the angles of the illuminating plane waves, however. Consequently, as the spatial frequency increases, the number of images of the spatial frequency that contribute to the total image is reduced. Finally, the images no longer overlap and only one illuminating plane wave contributes to the image. The result is that the amplitude of the image spatial frequency components on the optical axis is high for low frequencies and falls off as the spatial frequency of the image increases. The transfer coefficient is no longer flat out to the cutoff frequency. As the number of illuminating plane waves increases, as with diffuse illumination, the steps in the transfer coefficient become smaller and the falloff smoother.

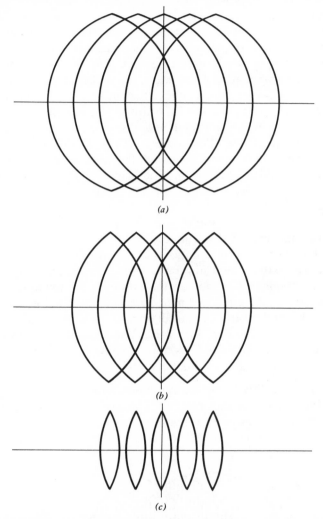

(a)

(b)

(c)

Figure 8-8 The areas of overlapping images due to five illuminating plane waves. (a) Spatial frequency low enough to allow overlap of all five images on axis. (b) Higher spatial frequency causing smaller regions of fringe formation allowing only three images to overlap. (c) Still higher spatial frequency allowing only one illuminating plane wave to cause an image on axis.

233

Another effect occurs with diffuse illumination. Because of the oblique components of the angular spectrum of the illumination, the cutoff frequency of the system is double that of the system illuminated with a single, normally incident plane wave. The preceding discussion applies for the modified fringes which appear except that the overlapping regions caused by each illuminating plane wave are of the form shown in Fig. 8-5 in the case of spatial frequencies above the cutoff u_c for normal plane wave illumination. For spatial frequencies below u_c, the region shown in Fig. 8-9 is used. This region is replicated and displaced as determined by the angular spectrum of the illumination. When all of the contributions, because of the various components of the angular spectrum of the illumination, are added, the transfer coefficient is tapered and the spatial frequency cutoff is twice the cutoff for illumination with a normally incident uniform plane wave.

It is apparent that the transfer coefficient of an imaging system is dependent on the angular spectrum of the illumination. An alternative way of thinking of the process is to notice that the effect of off-axis illumination is to shift the spatial frequency band pass. The strength of the spatial frequency components depends on the power in the illuminating off-axis wave. The sum of the effects of many off-axis illuminating waves can be represented by adding the displaced transfer functions with appropriate levels. This is illustrated in Fig. 8-10. The dashed line represents the ensemble transfer coefficient. This approach could be considered further but is not within the scope of this book [8-7, 8-8].

Speckle. When a diffusely reflecting object is illuminated with a coherent wave, the image appears with speckles. The easiest visualization is in terms of constructive and destructive interference of portions of the wave. The size of the speckle is also related to the resolution of the imaging system.

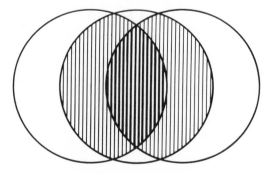

Figure 8-9 The region over which fringes of either shape appear in the image of a spatial frequency.

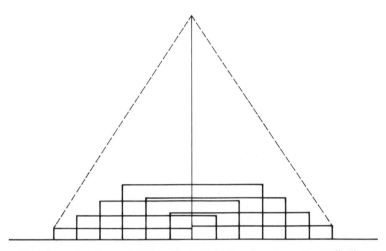

Figure 8-10 The rectangle representing the transfer function for unit amplitude normally incident plane wave illumination is displaced and modified in height according to the tilt and amplitude of the oblique illumination. The dashed line represents the transfer coefficient for diffuse illumination.

Figure 8-11 shows a portion of the surface of an object. If the resolution is sufficiently high, the individual detail of the surface is resolved, and the speckles do not appear. If, however, the resolution is lower, the waves from a number of points are averaged and may cancel. For example, the system giving a resolution spot as shown in Fig. 8-12 for each point on the object will, in general, produce speckle. The contributions from each object point have different phases and will reinforce or cancel each other in the image. It is easy to see that the smaller the resolution spot, the fewer the number that will overlap at a given image point, and the smaller will be any speckle. Figure 8-13 shows the increase in speckle in an image as the resolution decreases. The photographs were made by stopping down the lens of the camera to decrease the resolution in the image. The $F^{\#}$ of the human eye ranges from approximately 10 to 32, depending on the illumination.

8.3 Incoherent Illumination

Again, the property of primary interest is the *spatial* coherence. Incoherent illumination refers to the use of an illumination where there is *no* correlation between any two points on a plane normal to the direction of propagation. Actually, this is an impossible condition to achieve, but it lends itself well to certain forms of analysis. Any form of illumination that approximates this condition is often referred to as incoherent illumination.

Figure 8-11   Surface of constant phase of the object wave.

Incoherent illumination can be obtained with a thermal source that has a large number of radiating atoms having independent phase fluctuations, or the condition of incoherent illumination can be approximated by moving a diffuser in a coherent illuminating beam. We shall start with the moving diffuser technique.

Time-varying diffuse illumination.   If the random diffuser used to provide diffuse illumination in Section 8.2 is moved during the response or integration time of the image detector, we have spatially incoherent illumination.

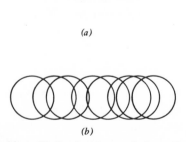

(a)

(b)

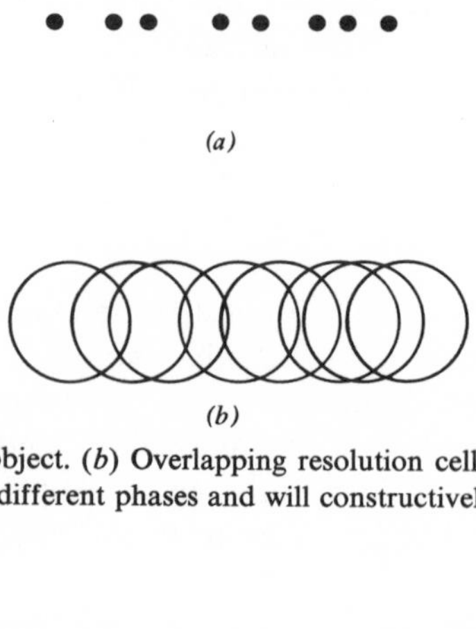

Figure 8-12   (a) Points on object. (b) Overlapping resolution cells of the image. The waves from each object point have different phases and will constructively or destructively interfere in the image plane.

Figure 8-13 Photographs of a coherently illuminated, diffusely reflecting object using different F-numbers (f/D). (a) $F^{\#}=8$; (b) $F^{\#}=16$; $(c)F^{\#}=32$; $(d)F^{\#}=64$.

A double-slit interferometer could form no fringes—they would be washed out by the fringe motion during the exposure. The fringes would move because the relative phase between the two slits is no longer constant but is varying with time.

The transfer coefficient of a diffusely illuminated system changes with the position of the diffuser because the angular spectrum of the illumination changes. If the diffuser is moving, the transfer coefficient is acquired by averaging the time-varying transfer coefficient over the detector integration time. Notice, however, that fringes in the *image* do not move. Alternatively, the transfer function could be obtained from the time average of the angular spectrum of the illumination. In either case, we obtain a transfer function that falls off with spatial frequency and has a cutoff frequency twice that of normally incident plane wave illumination.

Intensity impulse response and transfer function. The image is detected with an energy detector. Consequently, the effect of the system on the spatial frequency components of the intensity distribution is of interest. The intensity in the image plane is described as the time average of the magnitude of the amplitude squared,

$$I_i = \langle U_i U_i^* \rangle. \tag{8-28}$$

When referring to an image, the quantity defined by (8-28) is often called the *irradiance* of the image. We have seen that the image amplitude can be related to the amplitude of the object wave by (8-16). If we write the impulse response of the lens as

$$\mathbf{h}(x_i',y_i') = \tilde{\mathbf{P}}\left[\frac{x_i}{M}, \frac{y_i}{M}\right], \tag{8-29}$$

$$\mathbf{U}_i(x_i',y_i') = \mathbf{h}(x_i',y_i') \otimes \mathbf{U}_o(x_i',y_i') \tag{8-30}$$

where the primed coordinates takes into account the magnification and a reversal of coordinates. The image irradiance can then be written as

$$I_i(x',y') = \Bigg\langle \int\int \mathbf{U}_o(w,s,t)\mathbf{h}(x'-w,y'-s)\,dw\,ds$$
$$\times \int\int \mathbf{U}_o^*(w',s',t)\mathbf{h}^*(x'-w',y'-s')\,dw'\,ds' \Bigg\rangle$$
$$= \int\int\int\int \mathbf{h}(x'-w,y'-s)\mathbf{h}^*(x'-w',y'-s')$$
$$\times \langle \mathbf{U}_o(w,s,t)\mathbf{U}_o^*(w',s',t)\rangle\,dw\,ds\,dw'\,ds' \tag{8-31}$$

where w, s, w', and s' are dummy variables.

In the case of coherent illumination, the distribution across the object has no relative variations with time (spatially coherent) but the absolute phase of the entire image can vary with time. For coherent illumination the variables of $U(x',y',t)$ can be separated as

$$U_o(x',y',t) = U_o(x',y') \frac{U_o(0,0,t)}{\langle |U_o(0,0,t)|^2 \rangle^{1/2}} \qquad (8\text{-}32)$$

where the first term on the right side of the equality is a time-invariant function describing the object wave at one time and the fraction gives the temporal variation on the optical axis normalized by the rms value of the temporal variation. The use of (8-32) in (8-31) yields

$$I_i = \int \int \int \int \mathbf{h}(x'-w, y'-s)\mathbf{h}^*(x'-w', y'-s')U_o(w,s)U_o^*(w',s')$$

$$\left\langle \frac{U_o(0,0,t)}{\langle |U_o(0,0,t)|^2 \rangle^{1/2}} \frac{U_o^*(0,0,t)}{\langle |U_o(0,0,t)|^2 \rangle^{1/2}} \right\rangle dw\,ds\,dw'\,ds' \qquad (8\text{-}33)$$

where the outer time average can only operate on the numerator. The result is that the two integrals can be separated and

$$I_i(x',y') = \left| \int \int \mathbf{h}(x'-w, y'-s)U_o(w,s)\,dw\,ds \right|^2, \qquad (8\text{-}34)$$

which shows that the imaging system is linear in complex amplitude.

The other extreme of coherence is completely incoherent illumination. In this case, not only does the distribution in the object plane have relative variations with time, but it is assumed that there is no correlation between the variations at one point with those at any other point. That is,

$$\langle U_o(w,s,t)U_o^*(w',s',t) \rangle = I_o(w,s)\delta(w-w', s-s'). \qquad (8\text{-}35)$$

There is no correlation except with the distribution at a point and itself. This condition is physically impossible to achieve [8-7]. However, many practical systems can approximate this condition and the mathematical form is easy to use. The use of (8-35) in (8-31) results in

$$I_i(x',y') = \int \int |\mathbf{h}(x'-w, y'-s)|^2 I_o(w,s)\,dw\,ds. \qquad (8\text{-}36)$$

This expression indicates that the incoherently illuminated imaging system is linear in *intensity*, not amplitude, and that the intensity impulse response is $|\mathbf{h}|^2$.

The transfer function can be found by Fourier-transforming the convolution relation (8-36). The spatial frequency spectrum of an intensity distribution is usually normalized by the zero-frequency value of the spectrum. If the spectrum of the intensity is represented by $\mathcal{C}$ and the transfer function by $\mathcal{K}$, we can write the image and object spectra as

$$\mathcal{C}_i(u,v) = \frac{\mathcal{F}[I_i(x',y')]}{\mathcal{F}[I_i(x',y')]_{u=0,v=0}} \tag{8-37}$$

$$\mathcal{C}_o(u,v) = \frac{\mathcal{F}[I_0(x',y')]}{\mathcal{F}[I_o(x',y')]_{u=0,v=0}}, \tag{8-38}$$

and

$$\mathcal{K}(u,v) = \frac{\mathcal{F}[|h(x',y')|^2]}{\mathcal{F}[|h(x',y')|^2]_{u=0,v=0}}. \tag{8-39}$$

Notice that

$$\mathcal{F}[g(x',y')]_{u=0,v=0} = \int\int g(x',y')\,dx'\,dy'. \tag{8-40}$$

Using the functions above, we can write

$$\mathcal{C}_i(u,v) = \mathcal{C}_o(u,v)\,\mathcal{K}(u,v). \tag{8-41}$$

The expression (8-41) indicates that the spatial frequency spectrum of the image intensity is related to the spatial frequency spectrum of the object intensity by the intensity transfer function $\mathcal{K}$. For the incoherent case, the intensity transfer function is related to the amplitude transfer function through (8-39). Use of the convolution theorem gives

$$\mathcal{K}(u,v) = \frac{\mathbf{H}(u,v)\star\mathbf{H}^*(u,v)}{\mathbf{H}(u,v)\star\mathbf{H}^*(u,v)|_{u=0,v=0}} \tag{8-42}$$

where $\mathbf{H}(u,v)$ is the Fourier transform of $\mathbf{h}(x',y')$. Consequently, the intensity transfer function (known as the *optical transfer function*) is the normalized autocorrelation of the coherent amplitude transfer function. The modulus of the optical transfer function $|\mathcal{K}|$ is the *modulation transfer function*. In measurements of the performance of optical systems, it is the modulation transfer function that is usually measured. This avoids the problem of determining the phase of the transfer function for each frequency. (We shall see that if the lens has aberrations, the transfer function can have a negative sign in some regions.)

Equation (8-42) can be written in integral form:

$$\mathcal{K}(u,v) = \frac{\int\int \mathbf{H}(w',s')\mathbf{H}^*(w'-u,s'-v)\,dw'\,ds'}{\int\int |\mathbf{H}(w',s')|^2 dw'\,ds'}$$

(8-43)

Quite often, the change of variables

$$w' = w + \frac{u}{2}, \qquad s' = s + \frac{v}{2}$$

(8-44)

is used, yielding the symmetrical form of the correlation

$$\mathcal{K}(u,v) = \frac{\int\int (w+u/2,s+v/2)\mathbf{H}^*(w-u/2,s-v/2)\,dw\,ds}{\int\int |\mathbf{H}(w,s)|^2 dw\,ds}.$$

(8-45)

In either case, we can find the incoherent transfer function from the coherent transfer function.

Calculation of optical transfer functions. Calculation of the optical transfer function (OTF) for a diffraction-limited lens involves a relatively straightforward correlation as indicated by (8-45). The transfer function for the coherent process is written as

$$\mathbf{H}(u,v) = M^2 \mathbf{P}(-Mu, -Mv)$$

(8-46)

from either (8-17) or the Fourier transform of (8-29). Consequently, we see that the correlation of the pupil function with itself yields the OTF of an incoherently illuminated system. The correlation is easy to obtain for a one-dimensional system or a lens with a square aperture: the $\mathrm{II}(u/u_c)$ $\mathrm{II}(v/v_c)$ function is simply replaced by $\Lambda(u/2u_c)\Lambda(v/2v_c)$ where u_c and v_c are the coherent cutoff frequencies. The resulting incoherent transfer function is shown in Fig. 8-14. The cross section in the $\mathcal{K}-u$ plane is $\Lambda(u/2u_c)$ and the cross section in the u-v plane is a rectangle $4u_c$ by $4v_c$. The u-v cross section indicates the band-pass region of the spatial spectrum.

The case for a circular aperture is easy to visualize but requires slightly more involved calculations to obtain the form of the transfer function. Figure 8-15 shows the correlation to be carried out to obtain the OTF for a circular pupil. It is obvious that $\mathcal{K}$ will fall off monotonically with u and v but to find in what manner, (8-45) must be evaluated. The integral determines the common area of the overlapping circles. This area is easily found geometrically from Fig. 8-16. Only the variation with u need be found because the result is circularly symmetrical. From Fig. 8-16 we can

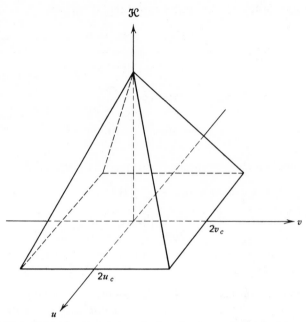

Figure 8-14 Incoherent transfer function for a rectangular lens.

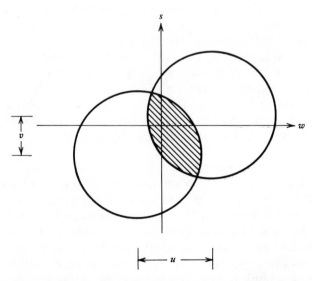

Figure 8-15 Autocorrelation of the pupil function to yield the OTF.

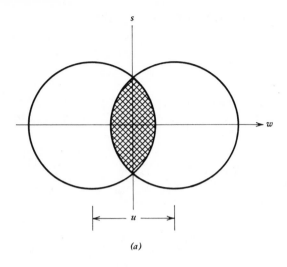

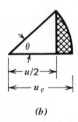

Figure 8-16 Calculation of the OTF of a circular lens. (*a*) Overlapping coherent transfer functions. (*b*) Determination of overlapping area.

also see that the overlapping area can be found for one-quarter of the area and the result quadrupled. The area of the circular sector of Fig. 8-16*b* is

$$\left(\frac{\theta}{2\pi}\right)(\pi u_c^2) = \frac{u_c^2}{2}\cos^{-1}\left(\frac{u}{2u_c}\right) \qquad (8\text{-}47)$$

and the area of the triangle is

$$\frac{1}{2}\left(\frac{u}{2}\right)\sqrt{u_c^2 - \frac{u^2}{4}} \ . \qquad (8\text{-}48)$$

Consequently, the shaded area of Fig. 8-16*b* is

$$\frac{u_c^2}{2} \cos^{-1}\left(\frac{u}{2u_c}\right) - \frac{uu_c}{4}\sqrt{1-\left(\frac{u}{2u_c}\right)^2}.\tag{8-49}$$

The incoherent transfer function is given by four times the result of (8-49) normalized by πu_c^2, the total area of the uniform transfer function of the coherently illuminated lens. Consequently,

$$\mathcal{H}(u,0) = \begin{cases} \dfrac{2}{\pi}\cos^{-1}\left(\dfrac{u}{2u_c}\right) - \dfrac{u}{\pi u_c}\sqrt{1-\left(\dfrac{u}{2u}\right)^2}, & u \leqslant 2u_c \\ \\ 0, & u > 2u_c \end{cases}.$$

The symmetrical transfer function is shown in Fig. 8-17.

8.4 Comparison of Coherent and Incoherent Illumination

There are numerous ways to compare imaging systems—spatial frequency cutoff, two-point resolution, and the Sparrow criterion [8-7] among others. The selection of the best imaging system is usually a function of the criterion used and the object being imaged. In addition to a determination of the resolution of the system, more subjective factors, such as the reaction of a human viewer, may be required. We shall not consider subjective criteria, and the only objective criteria discussed are the sine wave imaging, or spatial frequency cutoff, and two-point resolution criteria.

Spatial frequency cutoff. One common example that shows dramatic differences between images formed with coherent and incoherent illumination is that of a transparent object having an amplitude transmittance of

$$\mathbf{t}(x,y) = \left|\cos\frac{2\pi x}{\sigma}\right|\tag{8-50}$$

and an intensity transmittance of

$$\tau(x,y) = \cos^2\left(\frac{2\pi x}{\sigma}\right).\tag{8-51}$$

We can see, from the previous work, that the spectrum of the image irradiance for a coherent system is given by

$$\mathcal{F}[I_{ic}] = \mathcal{F}[(\mathbf{h}\otimes\mathbf{U}_o)(\mathbf{h}\otimes\mathbf{U}_o)^*] = \mathbf{H}\mathbf{A}_o \star \mathbf{H}^*\mathbf{A}_o^*\tag{8-52}$$

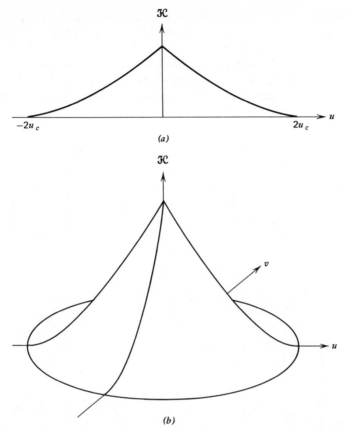

Figure 8-17 Incoherent transfer function of a circular lens. (*a*) Cross section in the $\mathcal{H}$-u plane. (*b*) Three-dimensional sketch of $\mathcal{H}$ as a function of u and v.

and the spectrum of the image irradiance with incoherent illumination is

$$\mathcal{F}[I_{ii}] = \mathcal{F}[\mathbf{hh}^* \otimes \mathbf{U}_o \mathbf{U}_o^*] = (\mathbf{H} \star \mathbf{H}^*)(\mathbf{A}_o \star \mathbf{A}_o^*) \qquad (8\text{-}53)$$

where

$$\mathbf{h}(x,y) = \tilde{\mathbf{P}}\left(\frac{x}{M}, \frac{y}{M}\right) \qquad (8\text{-}54)$$

and

$$\mathbf{H}(u,v) = M^2 \mathbf{P}(-Mu, -Mv). \qquad (8\text{-}55)$$

The use of (8-52) and (8-53) in analyzing the imaging of the object described by (8-50) and (8-51) with coherent and incoherent illumination, respectively, is illustrated in Fig. 8-18. The fundamental frequency of the spectrum of the rectified cosine wave is $2/\sigma$. The higher-order frequencies are not shown in Fig. 8-18. The entire spectrum of the $\cos^2(2\pi x/\sigma)$ signal is shown. Let the lens diameter and image distance be selected so that the coherent spatial frequency cutoff lies between $1/\sigma$ and $2/\sigma$. Assume unity magnification so that $M = 1$. The transfer functions shown are multiplied

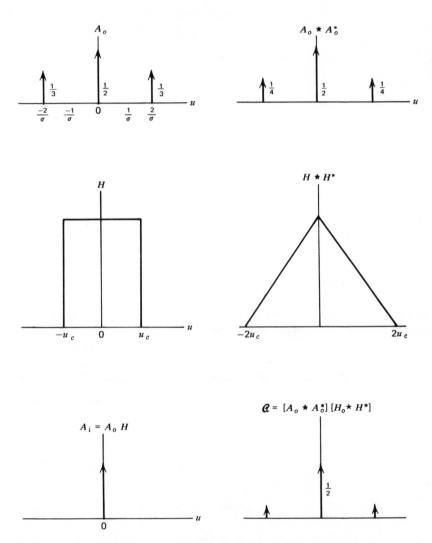

Figure 8-18 Comparison of coherent and incoherent imaging of $|\cos(2\pi x/\sigma)|$.

by the spectra, and the image spectra are found from (8-52) and (8-53). We find that, for the coherent case, no image is obtained. For the incoherent case, an image is obtained, but at reduced contrast because the levels of the higher frequency terms are reduced. This can be explained in terms of the angular spectrum of the incoherent illumination [8-8].

If the amplitude transmittance had been $\cos(2\pi x/\sigma)$ rather than the absolute value of that function, the intensity distribution of the coherently formed image would have had all frequencies without attenuation. The proof of this is left as an exercise for the reader. Such a transparency is not easy to obtain, a phase grating being required to provide the 180° shift every half cycle. The point is, no blanket statement can be made comparing coherent and incoherent imaging; the results depend on the object. In the examples just considered, coherent imaging gave no image in one case and a perfect image in another. Incoherent imaging gives identical images in both cases.

Two-point criterion. A two-point object also has dramatic differences between its coherent and incoherent images. In measuring resolution using two points, the minimum separation at which the image can be seen to be formed of two objects must be found. In the case of incoherent imaging, the illumination of the two points is assumed to be mutually incoherent, no matter how closely the objects are spaced. However, the illumination is coherent over an individual point. Consequently, the image is an incoherent superposition of coherent point images. That is, the irradiances of the point images add, but each point image is an Airy pattern caused by the finite lens aperture. The two points are said to be resolved when the center of the Airy disk caused by one object falls in the null of the Airy pattern caused by the second object. This is the *Rayleigh criterion*. The point separation at which this occurs is

$$s_R = 1.22 \frac{\lambda d_i}{D} \tag{8-56}$$

where s_R is the Rayleigh separation, λ is the wavelength, d_i is the lens-to-image distance, and D is the lens diameter. The image of two incoherently illuminated points separated by the Rayleigh separation is shown in Fig. 8-19 where the solid line is the image irradiance when both points are present. The dashed lines represent images of isolated points.

If the illumination of two points is mutually coherent, the amplitudes can interfere constructively or destructively, depending on the phase

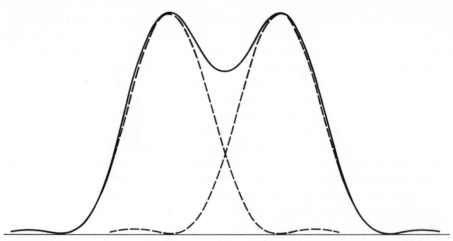

Figure 8-19   The solid line shows the central cross section of the image of two incoherently illuminated point objects separated by the Rayleigh distance. The dashed lines are irradiance distributions of each point image alone.

difference. If the two points are 90° out of phase, the image with coherent illumination is the same as with incoherent illumination. If the same two points are in phase, the amplitudes add, giving the distribution shown by the solid line of Fig. 8-20. Consequently, there is no dip between two peaks. Illumination with coherent light having 180° phase difference between the points produces an image irradiance with two peaks separated by more than the distance between the object points as shown by the dashed line. In the case of 180° phase difference, there is *always* a dip between the two image peaks, no matter how closely spaced the object points may be. Thus we must know the phase to properly interpret the image.

If we know something about the object, the so-called resolution limit can be exceeded, even with incoherent illumination. There are many approaches to exploit the knowledge of the object to achieve what is known as *superresolution* [8-9]. Although a full treatment of superresolution is not within the scope of this book, the following example illustrates one approach. Other approaches involve the reallocation of the available two-dimensional spatial frequency pass band of the system. If it is known that the object consists of either one or two nonresolvable points, then, referring to Fig. 8-19, we can infer that there is *always* a difference between the intensity distributions for a single-point and a two-point image. The decision between one or two points can be made if the S/N ratio is sufficient for accurate measurements of the intensity profile. Consequently, the S/N ratio, not the lens diameter, is the limiting factor [8-10].

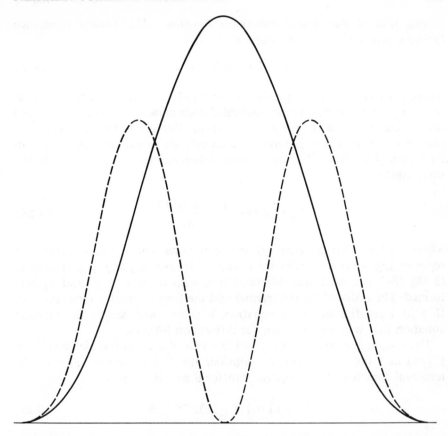

Figure 8-20 Central cross sections of the image irradiance of two coherently illuminated point objects separated by the Rayleigh distance. The solid line represents the image when the phase of the illumination of the two points is the same. The dashed line shows the change when there is a phase difference of 180°. The true locations of the points are near the crossings of the solid and dashed lines.

8.5 Partially Coherent Illumination

We have seen that a coherently illuminated imaging system is linear in amplitude and an incoherently illuminated system is linear in intensity. Systems employing partially coherent illumination can be shown to be linear in the mutual coherence function. That is, the mutual coherence of the wave is followed through the system rather than wave amplitude or intensity. It is also possible to obtain a transfer coefficient for the optical system which is a function of the coherence of the illumination.

Propagation of the mutual coherence function. The mutual coherence function was defined in Chapter 4 as

$$\Gamma_{12}(\tau) = \langle U(x_1, t+\tau) U^*(x_2, t) \rangle \qquad (8\text{-}57)$$

where U describes the complex wave amplitude, x_1 and x_2 are the points at which U is measured, τ is the time difference between measurements, and the angular brackets denote a time average. For simplicity, we show only the x coordinate, but in general the y coordinate would also appear. It can be shown [8-7, Ch. 3] that the mutual coherence function will satisfy the wave equation

$$\nabla_s^2 \, \Gamma_{12}(\tau) = \frac{1}{c^2} \frac{\delta^2 \, \Gamma_{12}(\tau)}{\delta \tau^2} \qquad (8\text{-}58)$$

where $s = 1, 2$. That is, there are two equations, one for each value of s representing the two coordinates x_1 and x_2. In showing that $\Gamma_{12}(\tau)$ satisfies (8-58), (8-57) is used and differentiation with respect to x_1 and x_2 performed. The order of the differential and the time integral is then reversed. If $\Gamma_{12}(\tau)$ satisfies the wave equation, it should also satisfy the imaging equation that was derived from the diffraction formula.

The imaging equation was derived in terms of a single wavelength. To fit $\Gamma_{12}(\tau)$ into the same form, we consider the Fourier components of the temporal function. The temporal transform pair is given by

$$\hat{\Gamma}_{12}(\nu) = \int \Gamma_{12}(\tau) e^{-i2\pi\nu\tau} \, d\tau \qquad (8\text{-}59)$$

$$\Gamma_{12}(\tau) = \int \hat{\Gamma}_{12}(\nu) e^{i2\pi\nu\tau} \, d\nu \qquad (8\text{-}60)$$

where the $\hat{\ }$ denotes a Fourier transform with respect to time and ν is the temporal frequency. We can follow each temporal frequency component of the mutual coherence function through the imaging system and then recombine the components. If we use $\hat{\Gamma}_{12}(\nu)$ in the diffraction formulas, we find [8-7, Ch. 7]

$$\hat{\Gamma}(x_1, x_2, \nu) = \int h(x_1, w_1, \nu) h^*(x_2, w_2, \nu) \, \hat{\Gamma}(w_1, w_2, \nu) \, dw_1 \, dw_2 \qquad (8\text{-}61)$$

where w is now the coordinate in the object plane, x is one image plane coordinate, and h is the impulse response of the imaging system. Unity magnification and a one-dimensional image are assumed. To follow the derivation of (8-61), it must be remembered that the subscripts refer to different points along the same coordinate, not the two coordinates of a plane, and (8-57) must be used. The expression in τ is obtained from (8-61) by use of the inverse transform (8-60). The resulting mutual coherence function in the image plane is

$$\Gamma(x_1, x_2, \tau) = \int \int \int \mathbf{h}(x_1 - w_1, \nu) \mathbf{h}^*(x_2 - w_2, \nu) \hat{\Gamma}(w_1, w_2, \nu) e^{i2\pi\nu\tau} d\nu \, dw_1 \, dw_2.$$

$$(8\text{-}62)$$

Spatial stationarity has been assumed in writing (8-62). This expression is easier to handle if the approximation $\Delta\nu \ll \bar{\nu}$ is made, where $\Delta\nu$ is the spectral width and $\bar{\nu}$ is the average temporal frequency. This is known as the quasimonochromatic assumption and is quite realistic. The use of this assumption allows (8-60) to be written as

$$\Gamma_{12}(\tau) \approx e^{i2\pi\bar{\nu}\tau} \int \hat{\Gamma}_{12}(\nu) d\nu = e^{i2\pi\bar{\nu}\tau} \Gamma_{12}(0), \qquad (8\text{-}63)$$

Consequently, from (8-59),

$$\hat{\Gamma}_{12}(\nu) = \Gamma_{12}(0)\delta(\nu - \bar{\nu}). \qquad (8\text{-}64)$$

The use of this relation in (8-62) gives

$$\Gamma_{12}(\tau) = e^{i2\pi\bar{\nu}\tau} \int \int \mathbf{h}(x_1 - w_1, \bar{\nu}) \mathbf{h}^*(x_2 - w_2, \bar{\nu}) \Gamma(w_1, w_2, 0) \, dw_1 \, dw_2. \qquad (8\text{-}65)$$

Consequently, we can find the mutual coherence in the image plane from the mutual coherence function in the object plane.

Let us now consider the two special cases of completely coherent and completely incoherent illumination and see how (8-65) reduces to the expressions previously obtained. In the case of coherent illumination, the mutual coherence function is

$$\Gamma(w_1, w_2, 0) = \mathbf{U}_o(w_1) \mathbf{U}_o^*(w_2). \qquad (8\text{-}66)$$

Expression (8-65) then becomes

$$\Gamma_{12}(0) = \int \int \mathbf{h}(x_1 - w_1) \mathbf{h}^*(x_2 - w_2) \mathbf{U}_o(w_1) \mathbf{U}_o^*(w_2) dw_1 \, dw_2 \qquad (8\text{-}67)$$

where the explicit variable $\bar{\nu}$ has been dropped. The intensity in the image plane is equal to the mutual intensity function for the image, so that $I_i = \Gamma_{11}(0)$ and

$$I_i(x_1) = \Gamma_{11}(0) = \int \int \mathbf{h}(x_1 - w_1) \mathbf{h}^*(x_1 - w_2) \mathbf{U}_o(w_1) \mathbf{U}_o^*(w_2) dw_1 \, dw_2.$$

$$(8\text{-}68)$$

Or, because the integrals are now separable,

$$I_i(x) = |\int \mathbf{h}(x-w)\mathbf{U}_o(w)\,dw|^2 = |\mathbf{U}_i(x)|^2, \tag{8-69}$$

and we see that the imaging process is linear in amplitude.

In the case of incoherent illumination,

$$\Gamma(w_1, w_2, 0) = I_o(w_1)\delta(w_1 - w_2), \tag{8-70}$$

which indicates that each point radiates independently of every other point. The correlation of complex wave amplitudes is zero for every case except autocorrelation. The use of (8-70) in (8-65) yields

$$\Gamma_{12}(0) = \int\int \mathbf{h}(x_1 - w_1)\mathbf{h}^*(x_2 - w_2)I_o(w_1)\delta(w_1 - w_2)\,dw_1\,dw_2$$

$$= \int \mathbf{h}(x_1 - w_1)\mathbf{h}^*(x_2 - w_1)I_o(w_1)\,dw_1. \tag{8-71}$$

The image intensity is then

$$I_i(x) = \Gamma_{11}(0) = \int |\mathbf{h}(x-w)|^2 I_o(w)\,dw, \tag{8-72}$$

which is a one-dimensional form of (8-36), the expression previously found for incoherent imaging.

We see that the mutual coherence function provides a variable in which the system is always linear. The results obtained for the two limiting cases of coherent and incoherent illumination agree with those obtained previously.

8.6 Aberrations and Apodization

A detailed study of aberrations is not appropriate here, but can be found in texts on optical system design. It is interesting to note, however, that any deviations of the lens characteristics from those of a perfect lens can be accounted for in the pupil function $\mathbf{P}$. Spherical aberration, coma, misfocusing, and so on, are represented by phase terms in $\mathbf{P}$ so that when the image description is obtained by convolution of $\tilde{\mathbf{P}}$ with the object distribution, $\tilde{\mathbf{P}}$ is not as desired, but is spread and/or distorted.

The effects of apodization can also be included in $\mathbf{P}$. If the lens transmissivity is reduced near the edges, for example, $\mathbf{P}$ is not uniform over the aperture, but has a tapered amplitude distribution. The transform of $\mathbf{P}$ then will have reduced intensity in the outer rings, sidelobes, or feet. Hence the term apodization, which means to cut off the feet. One application is to enable a dim object to be imaged even though it may be near a bright one. The Airy ring of the brighter object is reduced so that the dimmer one is not lost.

PROBLEMS

8-1. Two large lenses are used in a normally illuminated, coherent, double-lens imaging system. What is the diameter, a, of a stop in the transform plane that would give the same spatial frequency cutoff on-axis as would a lens of diameter D if used in a coherent, single-lens imaging system?

8-2. A coherent imaging system having a spatial frequency transfer function $\Pi(au)$ is used to image the distribution $1 + \cos[2\pi(x/4a)]$. We see from the transfer function and the spectrum of the distribution that the distribution will be imaged without alteration. However, the image can be written as

$$\frac{1}{a}\left[\operatorname{sinc}\left(\frac{x}{a}\right)\right] \otimes \left[1 + \cos\left(\frac{\pi x}{2a}\right)\right]$$

where the sinc function is the transform of the transfer function. This would indicate that the image is altered, being smoothed by the sinc. How do you resolve the apparent paradox? One approach is to evaluate the convolution integral. Think while integrating.

8-3. Show that the zero-frequency wave and one component of the angular spectrum pair, for a given frequency, overlap along one axis by an amount given by $(d_i D / d_o) - d_i \cot\theta$, where the variables are as shown in Fig. 8-3. Then show that, if overlap is required for only the zero-frequency wave and one off-axis wave (single-side band imaging), the cutoff frequency can be doubled when oblique illumination is used.

8-4. Show that the OTF of a square lens of width D and with no aberrations is given by $\Lambda(u/2u_c)\Lambda(v/2v_c)$ where Λ is the triangle function discussed in Chapter 2 and u_c and v_c are the cutoff spatial frequencies for the same lens when normally incident plane wave illumination is employed.

8-5. Compare coherent and incoherent imaging of a transparency having an amplitude transmittance of $t(x,y) = \frac{1}{2} + \frac{1}{2}\cos 2\pi x/\sigma$ and an intensity transmittance of $\tau(x,y) = [t(x,y)]^2$. Assume that the coherent spatial frequency cutoff lies between $1/\sigma < u_c < 2/\sigma$.

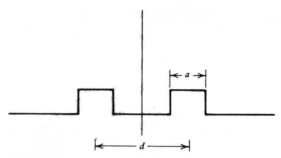

Figure P8-7

8-6. A perfect, single lens of diameter D is used to image an object having the amplitude distribution $1 + \cos 2\pi x / \sigma$. What image distance d_i will provide the largest spatial frequency band pass expressed in terms of the object spatial frequency? What is the image magnification under these circumstances?

8-7. A one-dimensional lens is apertured so that it has the amplitude transmittance shown in Fig. P8-7. Find the OTF for such a lens as a function of d. Sketch the OTF for the case where $d > 2a$. The result indicates that the images formed by apertures with different separations can be superimposed to get an image similar to that obtained with an unblocked large aperture [8-11].

8-8. Find the autocorrelation function of the amplitude distribution for the image formed using normally incident plane wave illumination of a moving diffuser before a transparent object. Show that if the autocorrelation of the phase distribution introduced by the diffuser is a peaked function with a width narrower than the impulse response of the system, the image intensity is described by an expression similar to that used in incoherent imaging [8-12, 8-13].

8-9. A one-dimensional perfect lens is modified so that it has an amplitude transmission decreasing linearly from unity in the center to zero at the edge. Find the OTF for such a lens. This is one form of apodization.

REFERENCES

8-1. E. Abbe, "Beiträge zur Theorie des Mikroskops und der Mikroskopischen Wahrnehmung," *Arch. Mikrosk. Anat.* **9**, 413–468 (1873).

8-2. John W. Strutt (Rayleigh), "On the Theory of Optical Images, with Special Reference to the Microscope," *Phil. Mag.* **42**, 167–195 (1896). Reprinted in *Scientific Papers by (John William Strutt) Lord Rayleigh* (Dover Press, New York 1964); Vol. IV: 1892–1901, p. 235.

8-3. Albert B. Porter, "On the Diffraction Theory of Microscopic Vision," *London, Edinburgh, Dublin Phil. Mag.* **11**, 154–166 (1906).

8-4. W. T. Cathey, "Derivation of the Scalar Diffraction Formulas by Use of the Angular Spectrum of Plane Waves," *J. Opt. Soc. Am.* **60**, 738 A (1970).

8-5. W. T. Cathey, "The Angular Spectrum and Imaging with Diffuse Illumination," *J. Opt. Soc. Am.* **60**, 1552 A (1970).

8-6. S. Herman, "Quasigeometric Approach to the Fourier Analysis of Imaging Lenses," *J. Opt. Soc. Am.* **61**, 1428–1429 (1971).

8-7. M. J. Beran and G. B. Parrent, Jr., *Theory of Partial Coherence* (Prentice-Hall, Inc., Englewood Cliffs, N. J., 1964).

8-8. W. T. Cathey, "An Angular Spectral Analysis of the Image of $|\cos \pi\alpha x|$," *J. Opt. Soc. Am.* **63**, 1300 A (1973).

8-9. See, for example, W. Lukosz, "Optical Systems with Resolving Powers Exceeding the Classical Limit," *J. Opt. Soc. Am.* **56**, 1463–1472 (1966); A. Bachl and W. Lukosz, "Experiments on Superresolution Imaging of a Reduced Object Field," *J. Opt. Soc. Am.* **57**, 163–169 (1967); A. W. Lohmann and D. P. Paris, "Superresolution for Non-birefringent Objects," *Appl. Opt.* **3**, 1037–1043 (1964); M. A. Grimm and A. W. Lohmann, "Superresolution Image for One-Dimensional Objects," *J. Opt. Soc. Am.* **56**, 1151–1156 (1966).

8-10. For example, G. Toraldo di Francia, "Resolving Power and Information," *J. Opt. Soc. Am.* **45**, 497–501 (1955); Neil J. Bershad, "Resolution, Optical-Channel Capacity and Information Theory," *J. Opt. Soc. Am.* **59**, 157–163 (1969).

8-11. G. W. Stroke, "Synthesis of Large-Aperture Optics by Successive Exposure of a Single Photographic Plate Through Successively Placed Small-Aperture Optics," *Opt. Commun.* **1**, 283–290 (1970).

8-12. C. J. Reinheimer and C. E. Wiswall, "An Intensity MFT Theory for Coherent Imaging," *Appl. Opt.* **8**, 947–949 (1969).

8-13. H. Arsenault and S. Lowenthal, "Partial Coherence in the Image of an Object Illuminated with Laser Light Through a Moving Diffuser," *Opt. Commun.* **1**, 451–453 (1970).

9

Classifications and Properties of Holograms

Many properties of holograms have been discussed in the natural development of the ideas of recording and retrieving information on a wave. Similarly, some classification has been done. Other aspects of holography, such as magnification, location of images, effects of finite resolution, and curvature of reference waves are treated here. In addition, further classification of the types of holograms is attempted although no standard notation has yet been accepted. Those types of holograms that have not yet been discussed are treated more completely. The effects of quantization of the exposure or phase as may occur in computer-generated holograms is considered after some techniques of generating holograms are presented.

The procedure followed in the section on holographic image formation, magnification, and image location is essentially that used in describing the imaging properties of a lens. In the case of a hologram, however, we have the option of changing wavelengths in the middle of the stream. That is, we can do one-half of the job—recording the hologram—at one wavelength and the other half—reconstructing the wavefront—at another.

9.1 Classification of Holograms

There is no universally agreed upon terminology in holography. Some terms are beginning to be preferred over others, however, and the confusion is being reduced somewhat. Some of the factors upon which classification of holograms is based are the type of moldulation used, the type of recording media, the geometry or the hologram recording configuration, the coherence of the hologram recording and/or illuminating waves, the type of information recorded, the wavelengths used, and the techniques

256

used in recording. Some of these factors have been discussed and are simply cataloged in this section. Others have not yet been discussed thoroughly, or they lead to special types of holograms. These are treated in the following sections.

Recording media and type of modulation. The operation of a volume recorder is treated in Chapters 3 and 6 and the results are summarized here. We have seen that holograms of the two types can have vastly different efficiencies and that the mechanism of wavefront reconstruction is not the same. The terminology commonly used is *thick*, or *volume*, *holograms* as opposed to *thin*, or *surface, holograms*. Both types can have variations in absorption, index of refraction, or both, to produce amplitude, phase, or complex modulation of the read-out wave. The terms formerly applied are amplitude and phase holograms, but because of possible confusion with another type of hologram sometimes referred to as a phase-only hologram, the terms absorption and dielectric are gaining acceptance.

The thin hologram must preserve and modulate *both* the amplitude and phase parameters of the wavefront by variations in the absorption (amplitude modulation) or dielectric constant (phase modulation) of the hologram. Amplitude modulation is described by

$$t(x,y) = |U(x,y)|^2 + |V(x,y)|^2 + U(x,y)V^*(x,y) + U^*(x,y)V(x,y) \quad (9\text{-}1)$$

whereas phase modulation is described by

$$t(x,y) = \exp\left\{ ip\left[|U(x,y)|^2 + |V(x,y)|^2 + U(x,y)V^*(x,y) \right. \right.$$
$$\left. \left. + U^*(x,y)V(x,y) \right] \right\}. \quad (9\text{-}2)$$

That is, one type of wave *modulation* is used to reconstruct waves with both phase and amplitude *information*. As shown in Chapter 3, the penalty paid is the reconstruction of two waves, the primary and the conjugate waves. The thick hologram, on the other hand, can use variations in the absorption or reflectivity (amplitude modulation) to directly impress the proper amplitude upon the reconstructed wave and variations in the dielectric constant (phase modulation) to directly impress the proper phase upon the wave. In this case, no secondary image need be generated. The amplitude transmissivity (or reflectivity) can be written as

$$t(x,y) = U(x,y)\exp[i\Psi(x,y)] \quad (9\text{-}3)$$

which requires separate modulation for the amplitude magnitude and the phase.

The adjectives that can be used to describe the recording medium and the type of modulation used are thick or thin and absorption or dielectric. If the terms amplitude or phase are used, the word modulation probably should be added, that is, phase modulation hologram, to avoid confusion with other types of holograms to be discussed.

Recording and reconstructing configurations. The recording configuration has a much greater effect on the performance of a hologram than might be expected. Recording configuration refers not only to path length differences, but also to the curvature of the reference wave, the location of the object, inclusion of lenses to image or Fourier transform, the location of the hologram, and source of the reference wave.

The path length problem arises because of the limited coherence length of the laser or other source used. As shown in Chapter 4, if the path from source to object to hologram and the path of the reference wave from source to hologram differ more than the coherence length of the source, no fringes are recorded. These paths can be simply measured and the restrictions obeyed. If the object is large, a technique of plotting contours of constant path difference may be helpful. This diagram, called a holo-diagram, is useful in placing the object with respect to the source, reference mirror, and hologram [9-1]. It is a family of ellipses with the source and the center of the hologram plate at the two foci. The reference mirror and the object are then placed on one ellipse to assure equal path lengths for the reference and object waves.

The effects of other variables in the configuration are more subtle. First, let us consider the relatively straightforward effects of object position. If the object is far enough from the hologram and/or small enough such that the hologram is in the far field of the object, a *Fraunhofer hologram* is recorded. The properties of this type of hologram are treated in a later section. Objects closer to the hologram or larger produce what have been called *Fresnel holograms*. The object can be placed close to the hologram, or a lens can be used to project an image onto the hologram to form a near image plane or an *image plane hologram*. Because these holograms have some rather interesting properties, a later section is also devoted to image plane holograms. The lens can be positioned to cast not an image, but a Fourier transform of the object onto the hologram. A *Fourier transform hologram* is then recorded. These are discussed further in the next section.

Rather than discussing the reference wave in terms of its curvature, let us specify the location of a point reference. Its distance from the hologram plate then becomes the radius of curvature. A plane wave is produced by a point reference at infinity and a converging wave by a virtual source on the opposite side of the hologram. Analysis of the effects of an arbitrary location of the point reference is rather involved and is given in Section

9.5. A name has been attached to only one geometry. The special case where the point reference is in the same plane as the object has been called a *lensless Fourier transform hologram* because of some similarities to a Fourier transform hologram [9.2]. Because a Fourier transform of the object distribution does not appear on the hologram, the term *quasi Fourier transform hologram* may be preferable [9-3]. This special case is treated in Section 9.4.

Size of reference.   If an extended reference rather than a point reference is used, the effect is to reduce the quality of the image. The term *extended reference* is taken to mean one that produces a reference wave which cannot be focused to a point. A uniform plane or spherical wave is not an extended reference.

An extended reference reduces the resolution of the image proportionally to the size of the reference. If a Fourier transform hologram is recorded, the object distribution is convolved with the reference distribution.

One form of hologram, the *local reference beam (LRB) hologram*, generates its own reference wave from the object distribution [9-4, 9-5]. This is done by splitting the wave from the object as shown in Fig. 9-1. A portion of the wave is imaged onto an iris which produces an extended reference source of variable size. If the object can be imaged small enough, the image itself can serve as a reference source for the wave passing through the rest of the system. One advantage of such a system is that, in making a hologram of a distant object, the coherence length of the source and the requirement to minimize the difference between the object and

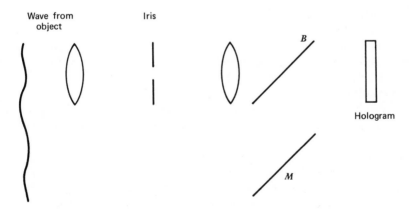

Figure 9-1 One method of recording a local reference beam hologram. *B*—beamsplitter, *M*—mirror.

reference wave path lengths create no problems. Another advantage is that object motion produces almost no problems with moving fringes because both the reference and the object waves have the same phase shifts. A disadvantage is that a strong reference wave is difficult to obtain.

A second form of hologram with an extended source is the *speckle reference hologram* [9-6]. In this case, a portion of the wave illuminating the object is focused to a point near the edge of the object. This focused beam serves as an extended reference source for recording the hologram. It has many of the advantages of the local reference beam hologram, but the angle of separation between the image and the on-axis light is small.

Degree of coherence. The first question to answer is, "Coherence of what?" There are three cases to be considered: coherence of the illuminating source, coherence of the read-out source, and the coherence between superimposed recordings on the hologram. If the illuminating source during recording is incoherent, the hologram is referred to as an *incoherent hologram*. The type of coherence considered is the *spatial* coherence of the source. The *read-out* wave for incoherent holograms is normally spatially coherent. The technique used in recording incoherent holograms is to make the wave from each point on the object interfere with itself.

Another technique used in making holograms with spatially incoherent illumination is a two step process called *integral photography* which results in a kind of pseudohologram. The technique is essentially an extrapolation of the stereo process except that multiple images are used rather than just two. An array of lenses is used to form separate images of a scene in the image plane of the array. Each lens sees the object from a different direction so that slightly different images are recorded. When the film is illuminated, the lenses project images of the object from different angles. If the film illumination is with a laser, and a reference wave is introduced, a hologram of the projected image can be recorded. This hologram produces an image similar to that seen in the multiple imaging process, rather than a normal holographic image. Caulfield and Lu have summarized much of the work in integral photography [9-7].

If the hologram is recorded with coherent illumination but the read-out beam is collimated white light, the hologram is called a *white light hologram*. This is the case when a thick hologram serves as an interference filter and hologram combination to reflect the wavelength with which the hologram was recorded. The principles of such holograms are treated in Chapters 3 and 6 and some of the properties are discussed in Section 9.9.

Another form of incoherent recording is the incoherent superposition on the same emulsion of coherently recorded holograms. These *multiple exposure holograms* are of interest when an image is to be synthesized from lines or spots. Their properties and limitations are considered in Section 9.8.

Wave parameters recorded. We have seen that both the amplitude and phase of a wave can be recorded and that the wave can be reconstructed with the proper amplitude and phase. The effects of recording only one of the parameters, amplitude or phase, are quite interesting. It is possible to record only the amplitude *or* phase *information* and partially reconstruct the wave using either amplitude or phase wavefront *modulation* (absorption or dielectric holograms). That is, amplitude modulation can be used to reconstruct only the phase of a wave—the amplitude of the reconstructed wave would be uniform. A hologram doing this would have an amplitude transmittance described by

$$t(x,y) = U_0^2 + V_0^2 + 2U_0V_0 \cos[2\zeta x + \Psi(x,y)]. \qquad (9\text{-}4)$$

The subscript zero denotes a constant value of the amplitude. Both a primary and a secondary wave are formed, but only the phase has the correct value. The amplitude is uniform, no matter what amplitude distribution the original wave had. Methods for discarding the amplitude information, reasons for doing so, and some of the effects are discussed in Section 9.11.

Holograms in which only the phase information is preserved could be termed phase holograms but that term has been used to apply to dielectric holograms in which phase modulation occurs. The label *phase-only holograms* has been used, but another term such as *phase information holograms* may be more descriptive and may help eliminate confusion. Strictly speaking, the process in which amplitude information is discarded is not holography because the wave is not recorded in its entirety. Gabor originally chose the term hologram because it indicates that *all* of the information is preserved. However, reference to (9-4) shows that the mathematical description is very similar to that used when the amplitude information is retained.

One implementation in which phase modulation is used to impress only the phase of a reconstructed wave has been called the kinoform [9-8]. The amplitude information is neglected in this case by not computing it; the kinoform is a computer-generated device. It is described further in Section 9.11.

The case in which the phase information is discarded, and only the amplitude of the original wave used, does not appear to be a practical implementation [9-9]. This type of hologram could be termed an *amplitude-only hologram* in analogy to the phase-only hologram, or, probably better, it should be called an *amplitude information hologram*.

Recording techniques. Quite often, the adjectives attached to the noun hologram relate to the techniques used in recording the hologram, the wavelength, or the type of energy in the wave. The wavelength is reflected in the name only when it is not within the band normally thought of as being used, that is, the visible region. Hence *microwave hologram* or *infrared hologram* is *recorded* using electromagnetic energy having centimeter or millimeter wavelengths or wavelengths in the IR. The read-out wave is usually visible. When it is not, other names apply. For example, a system recording a microwave distribution and then reconstructing it using microwaves is a microwave phased array or a retrodirective phased array. Similarly, we may record an *acoustical hologram*. These are discussed in Chapter 10. Related to techniques, we have, for example, *computer-generated holograms*, discussed in Section 9.12, and scanned holograms, discussed in Section 9.8. A *scanned hologram* or *scanned beam hologram* is one in which the reference is scanned on the hologram or the object is scanned with a beam rather than the entire object being illuminated at once.

Overlap of terms. It rapidly becomes obvious that specification of all parameters is not necessary and certainly not desirable. For example, *white light hologram* implies that a thick recording medium is used. However, if there is any possibility of confusion and an error would be important, the type of hologram should be specified.

9.2 Holographic Image Formation and Magnification

Before discussing further the various types of holograms and their operation, it is helpful to examine the holographic imaging process.

The procedure followed in this section is essentially that used in describing the imaging properties of a lens. In the case of a hologram, however, we have the option of changing wavelengths in the middle of the stream. That is, we can do one-half of the job—recording the hologram—at one wavelength and the other half—reconstructing the wavefront—at another. The size of the hologram can be changed before use, which has its effect. Another variable is the curvature of the reference and read-out wavefront. The option of varying the illumination wavefront was present in the case of imaging with a lens. Thus we shall see some similarities and some differences between a hologram and a lens.

Hologram recording. Assume that a hologram is to be recorded using the arrangement shown in Fig. 9-2. The wedge deflects a reference wave down toward the hologram, and the lens, which could have a positive or negative focal length, produces a reference wave with a spherical phasefront. The diffuser provides diffuse illumination of a transparent object. This arrangement allows us to analyze the imaging of diffusely illuminated or

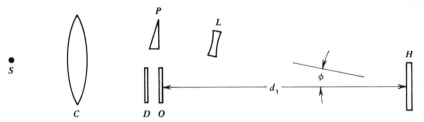

Figure 9-2 The recording of a hologram. *S*—point source, *C*—collimator, *P*—prism, *L*—lens to change curvature of reference wavefront, *D*—diffuser, *O*—object, *H*—hologram, d_1— distance from object to hologram, and ϕ—angle at which the reference wave is introduced.

reflecting objects. The imaging of a transparent object illuminated with a uniform wave of spherical phasefront can be treated by replacing the diffuser with a lens.

If the diffuser is close enough to the object to produce uniform illumination with variations in only the phase, the wave to the right of the object can be represented by

$$U_1(x_1,y_1) = e^{i\Psi(x_1,y_1)}t(x_1,y_1) \tag{9-5}$$

where $\Psi(x_1,y_1)$ represents, in this case, a random phase fluctuation and $t(x_1,y_1)$ is the amplitude transmittance of the object transparency. Use of the diffraction formula shows that, in the hologram plane,

$$U_2(x_2,y_2) = \frac{\exp(ik_1 d_1)}{i\lambda_1 d_1} h_1(x_2,y_2;d_1)$$

$$\times \int\int U_1(x_1,y_1)h_1(x_1,y_1;d_1)\exp\left[-ik_1\frac{(x_1 x_2+y_1 y_2)}{d_1}\right]dx_1 dy_1 \tag{9-6}$$

where the subscript on the **h** refers to the subscript on k or λ, that is,

$$h_1(x,y;d)\exp\left[ik_1\frac{(x^2+y^2)}{2d}\right]. \tag{9-7}$$

We assume that the hologram is formed using waves of wavelength λ_1.

The reference wave can be described by

$$V_2(x_2,y_2) = V_0\exp\left[ik_1\left(x_2\sin\phi+\frac{x_2^2+y_2^2}{2\rho_V}\right)\right] \tag{9-8}$$

where ϕ is the angle at which the wave is introduced and ρ_V is the radius of curvature of the wave. Notice that an off-axis point reference a distance ρ_V from the hologram would produce the same wave. If we now assume that the transmittance of the hologram is linearly dependent on the exposure, we obtain four terms in adding and squaring the object and reference waves. Let us follow only the primary wave, the one associated with the term

$$\mathbf{V}_0\mathbf{U}_2(x_2,y_2)\exp\left[-ik_1\left(x_2\sin\phi+\frac{x_2^2+y_2^2}{2\rho_V}\right)\right]. \qquad (9\text{-}9)$$

Hologram illumination. Now assume that the hologram is illuminated with a read-out wave described by

$$\mathbf{W}(x_2,y_2)=\mathbf{W}_0\exp\left[ik_2\left(x_2\sin\psi+\frac{x_2^2+y_2^2}{2\rho_W}\right)\right] \qquad (9\text{-}10)$$

where k_2 indicates that the illumination of the hologram is with wavelength λ_2, ψ is the angle between the direction of propagation of the read-out wave and the z axis, and ρ_W is the radius of curvature of the phasefront of the read-out wave. The distribution in the plane a distance d_2 from the hologram is described by

$$\mathbf{U}_3(x_3,y_3)=\mathbf{V}_0\mathbf{W}_0\frac{\exp(ik_2d_2)}{i\lambda_2d_2}\mathbf{h}_2(x_3,y_3;d_2)$$

$$\times\int\int\mathbf{U}_2(x_2,y_2)\mathbf{h}_1(x_2,y_2;-\rho_V)\exp[-ik_1x_2\sin\phi]$$

$$\times\mathbf{h}_2(x_2,y_2;\rho_W)\exp[ik_2x_2\sin\psi]\mathbf{h}_2(x_2,y_2;d_2)$$

$$\times\exp\left[-ik_2\frac{(x_2x_3+y_2y_3)}{d_2}\right]dx_2\,dy_2. \qquad (9\text{-}11)$$

Substitution of the expression for U_2 and combination of some of the terms yields

$$U_3(x_3,y_3) = -V_0 W_0 \frac{\exp[i(k_1 d_1 + k_2 d_2)]}{\lambda_1 \lambda_2 d_1 d_2} h_2(x_3,y_3;d_2)$$

$$\times \int \int \exp\left\{ i\pi(x_2^2+y_2^2)\left[\frac{1}{\lambda_2 d_2} + \frac{1}{\lambda_1 d_1} - \frac{1}{\lambda_1 \rho_V} + \frac{1}{\lambda_2 \rho_W} \right] \right\}$$

$$\times \exp\left[i2\pi x_2 \left(\frac{\sin\psi}{\lambda_2} - \frac{\sin\phi}{\lambda_1} \right) \right] \exp\left[\frac{-ik_2(x_2 x_3 + y_2 y_3)}{d_2} \right]$$

$$\times \int \int U_1(x_1,y_1) h_1(x_1,y_1;d_1)\,dx_1\,dy_1\,dx_2\,dy_2 \tag{9-12}$$

Following the procedure used in the single-lens imaging case, we reverse the order of integration to get

$$U_3(x_3,y_3) = \frac{V_0 W_0}{\lambda_1 \lambda_2 d_1 d_2} h_2(x_3,y_3;d_2) \int \int U_1(x_1,y_1) h_1(x_1,y_1;d_1)$$

$$\times \int \int \exp\left\{ i\pi(x_2^2+y_2^2)\left[\frac{1}{\lambda_1 d_1} + \frac{1}{\lambda_2 d_2} - \frac{1}{\lambda_1 \rho_V} + \frac{1}{\lambda_2 \rho_W} \right] \right\}$$

$$\times \exp\left[-i2\pi x_2 \left(\frac{\sin\phi}{\lambda_1} - \frac{\sin\psi}{\lambda_2} + \frac{x_1}{\lambda_1 d_1} + \frac{x_3}{\lambda_2 d_2} \right) \right]$$

$$\times \exp\left[-i2\pi y_2 \left(\frac{y_1}{\lambda_1 d_1} + \frac{y_3}{\lambda_2 d_2} \right) \right] dx_2\,dy_2\,dx_1\,dy_1, \tag{9-13}$$

where the minus sign and the exp $[i(k_1 d_1 + k_2 d_2)]$ have been dropped.

We can find a focusing condition that is similar to the focusing condition for the single lens. That is, we shall show that if

$$\frac{1}{\lambda_1 d_1} + \frac{1}{\lambda_2 d_2} - \frac{1}{\lambda_1 \rho_V} + \frac{1}{\lambda_2 \rho_W} = 0, \tag{9-14}$$

then (9-13) can be made to describe U_3 in terms of U_1 with no integrals involved. It is interesting to compare (9-14) with the imaging condition for a lens:

$$\frac{1}{d_1} + \frac{1}{d_2} - \frac{1}{f} = 0. \tag{9-15}$$

We shall return to the imaging condition, but first let us evaluate (9-13) when the condition (9-14) is imposed. When the quadratic term is missing in the integral over x_2 and y_2, the integral becomes the Fourier transform of unity. If the hologram size is taken into account, the integral becomes the Fourier transform of the pupil function. Because the Fourier transform of unity is a delta function, (9-13) can be written as

$$U_3(x_3, y_3) = \frac{V_0 W_0}{\lambda_1 \lambda_2 d_1 d_2} h_2(x_3, y_3; d_2)$$

$$\times \int \int U_1(x_1, y_1) h_1(x_1, y_1; d_1) \delta\left(\frac{\sin\phi}{\lambda_1} - \frac{\sin\psi}{\lambda_2} + \frac{x_1}{\lambda_1 d_1} + \frac{x_3}{\lambda_2 d_2} \right)$$

$$\times \delta\left(\frac{y_1}{\lambda_1 d_1} + \frac{y_3}{\lambda_2 d_2} \right) dx_1 \, dy_1. \tag{9-16}$$

The relation

$$\delta(ax) = \frac{1}{|a|} \delta(x) \tag{9-17}$$

can be used to write (9-16) as

$$U_3(x_3, y_3) = \frac{\lambda_1 d_1}{\lambda_2 d_2} V_0 W_0 h_2(x_3, y_3; d_2)$$

$$\times \int \int U_1(x_1, y_1) h_1(x_1, y_1; d_1) \delta\left(x_1 + \frac{\lambda_1 d_1}{\lambda_2 d_2} x_3 + d_1 \sin\phi - \frac{\lambda_1 d_1}{\lambda_2} \sin\psi \right)$$

$$\times \delta\left(y_1 + \frac{\lambda_1 d_1}{\lambda_2 d_2} y_3 \right) dx_1 \, dy_1. \tag{9-18}$$

Evaluation of this integral and defining the magnification in the x, y plane as

$$M = \frac{\lambda_2 d_2}{\lambda_1 d_1} \tag{9-19}$$

gives the result

$$U_3(x_3, y_3) = \frac{V_0 W_0}{M} h_2(x_3, y_3; d_2)$$

$$\times U_1\left(-\frac{x_3}{M} - d_1 \sin\phi + \frac{d_2 \sin\psi}{M}, -\frac{y_3}{M}\right)$$

$$\times h_1\left(-\frac{x_3}{M} - d_1 \sin\phi + \frac{d_2 \sin\psi}{M}, -\frac{y_3}{M}; d_1\right). \tag{9-20}$$

Before proceeding, let us compare this expression with (5-64) which describes single-lens imaging. Except for the λ's and the $V_0 W_0$, the factors before U_1 are the same. There is a phase error just as with single-lens imaging. The only difference in the U_1 is the displacement caused by the angles of the reference and read-out waves. We shall see that the primary image is usually a virtual image, in which case there is no inversion of the image. Remember that this equation applies to the primary image only. The secondary image is treated later.

Magnification and image location. It would appear from (9-19) that magnification of the image is linearly proportional to the ratio of the wavelengths used. This is not necessarily true, however, because the ratio d_2/d_1 is a function of wavelength. From the focusing condition (9-14) we find that

$$d_2 = \left(-\frac{1}{\rho_W} + \frac{\lambda_2}{\lambda_1 \rho_V} - \frac{\lambda_2}{\lambda_1 d_1}\right)^{-1}. \tag{9-21}$$

Consequently, to find the magnification we must use (9-19) and (9-21) together. If the preceding analysis were repeated for the secondary image, we would find that

$$d_2 = \left(-\frac{1}{\rho_W} - \frac{\lambda_2}{\lambda_1 \rho_V} + \frac{\lambda_2}{\lambda_1 d_1}\right)^{-1}. \tag{9-22}$$

Examination of (9-21) and (9-22) shows that *either* the primary image or the secondary image can be a real or a virtual image. If the variables are such that d_2 is negative, the image is virtual. If d_2 is positive, the image is real. In the case where the wavelengths are the same and the reference and read-out waves have the same form, the primary image is formed at $d_2 = -d_1$, that is, at a distance d_1 behind the hologram or on the opposite side of the viewer. The location of the secondary image depends on the values of ρ_V and d_1. Notice that for the magnitude of d_2 to be the same for both images simultaneously, either $\rho_V = \rho_W = \infty$, that is, plane reference and read-out waves, or $d_1 = \rho_V$. In the latter case, both images appear at the location of the point source producing the read-out wave.

The use of (9-19), (9-21), and (9-22) gives the expression for the transverse magnification, or magnification in the x,y plane, as

$$M = \left(-\frac{\lambda_1 d_1}{\lambda_2 \rho_W} \pm \frac{d_1}{\rho_V} \mp 1 \right)^{-1} \tag{9-23}$$

where the upper signs apply for the primary image and the lower signs apply for the secondary image. Again, if the wavelengths are the same and the reference and read-out wavefronts identical, the magnification of the primary image is -1. This negative number, coupled with the negative sign of the argument of U_1 in (9-20), indicates that the image is not inverted. For the magnification of the primary and secondary image to be the same, either $\rho_W = \rho_V = \infty$ or $d_1 = \rho_V$.

If, in addition to the variables considered above, the size of the hologram is changed between recording and reconstructing the wavefront, the magnification of the image is affected. Let the magnification of the hologram produced in copying the hologram be M_c. Such magnification (or demagnification) naturally occurs when the recording is made with an array of detectors and a longer wavelength illumination and the reconstruction is to be with a visible wavelength. In such cases a photographic reduction of the hologram is made. The result of changing the scale of the hologram is that the expression for magnification becomes

$$M = \frac{M_c}{\left(-\frac{\lambda_1}{\lambda_2} M_c^2 \frac{d_1}{\rho_W} \pm \frac{d_1}{\rho_V} \mp 1 \right)} \tag{9-24}$$

where the upper signs are for the primary image and the lower ones apply to the secondary image [9-10].

It is often more useful to express the magnification of the hologram in terms of angular magnification. This can be done by differentiating the angular size of the image, x_3/d_2, with respect to the angular size of the object, x_1/d_1. The result is

$$M_{\text{ang}} = -\frac{\lambda_2}{\lambda_1 M_c}. \tag{9-25}$$

The magnification in the longitudinal direction can be found by differentiation of the distance to the image point, d_2, with respect to the distance to the object point, d_1. This results in

$$M_{\text{long}} = -\frac{\lambda_1}{\lambda_2} M^2. \tag{9-26}$$

Equation (9-23) shows that the transverse magnification is not the linear function of the ratio of wavelengths implied by (9-19). In fact, if a plane read-out wave is used ($\rho_W = \infty$) the magnification does not vary with the ratio of wavelengths at all. The angular magnification is, however, a function of this ratio as seen in (9-25). So is the longitudinal magnification of (9-26). It is this effect that creates problems of large image distortions when the hologram is recorded and illuminated with radiation having greatly different wavelengths.

Aberrations.   The equations above reduce to those applicable to a lens if $\lambda_1 = \lambda_2$ and $M_c = 1$. Many of the techniques developed in conventional imaging system analysis can be used to analyze holographic images [9-11, 9-12]. Meier has calculated the third-order aberrations for holographic images, and discussed ways of minimizing them [9-10]. The analysis is not repeated here, but only the conclusions. If $\rho_V = \rho_W$, $\phi = \psi$, and the ratio of wavelengths is unity, there are no aberrations. There is no magnification either. For magnification without aberrations, $\rho_V = \rho_W = \infty$, and the hologram must be scaled according to the ratio of the wavelengths λ_1 and λ_2. This is often undesirable because of the possibilities of errors in the scaling process. If the point reference is in the same plane as the object ($\rho_v = d_1$) the spherical aberration is zero and *either* coma or astigmatism can be eliminated, but not both simultaneously. In this case, magnification is possible without scaling the hologram. Spherical aberration is normally introduced if $d_1 \neq \rho_V$. Coma is caused by a skewed reference wave and one method of compensation is to use a read-out wave with a corresponding amount of coma. The angle ψ must be related to ϕ by

$$\psi = -\frac{\lambda_2}{\lambda_1 M_c} \phi. \tag{9-27}$$

One image is then free of coma and the other has twice as much coma as with an on-axis read-out wave.

Armstrong has pointed out that the amount of aberration depends on the dimensions of the hologram [9-13]. If the linear dimension of the hologram is L, the spherical aberration varies as L^4, the coma as L^3, the astigmatism as L^2, and the distortion as L. Consequently, the spherical aberration is one of the first aberrations to be considered for correction. Armstrong also discussed and tabulated conditions for the removal of the distortions.

Aberrations off the optical axis were included in Champagne's study [9-14]. This study is of interest because holograms are often used in off-axis operations. A later study by Latta concentrated on a numerical analysis of holographic imagery [9-15]. This paper considered primarily four types of holograms: (*a*) in-line reference wave, (*b*) off-axis reference wave, (*c*) image near the hologram and (*d*) point reference in the plane of the object. It is shown that a considerable reduction in the total aberration can be achieved by balancing one aberration against another.

9.3   Fourier Transform Holograms

A Fourier transform hologram is, in essence, a hologram of a Fourier transform. In concept, it is the same as the complex spatial filter discussed in Chapter 7. One great difference is that the object can be a diffuse reflector or diffusely illuminated. In this section emphasis is placed on the distribution that appears on the hologram and how this affects the resolution and field of view of the image formed.

Recording.   The Fourier transform hologram is recorded in essentially the same way as a spatial filter. In the method shown in Fig. 9-3 the wave from the point source is collimated by the transforming lens to produce a plane reference wave. The distribution from the diffusely illuminated object is Fourier-transformed by the same lens. The wave from one point on the object is shown for use in the discussion of the distribution recorded on the hologram.

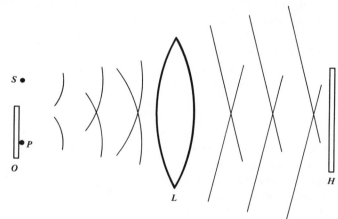

Figure 9-3 The recording of a Fourier transform hologram. *S*—point source reference, *O*—diffusely illuminated object, *P*—point on object, *L*—transforming lens, *H*—hologram.

A diffusely reflecting or diffusely illuminated object will not have a Fourier transform that is easily recognizable. The problem is that the quasi-random phase distribution causes a large spread in the transform plane. One way of treating the case of a diffusely illuminated transparent object is to use an angular spectral decomposition of the illuminating wave. That is, replace the diffuse illumination by its angular spectrum of plane waves. Each plane wave component will produce a Fourier transform of the object distribution in the transform plane. The distance off the optical axis will depend on the tilt of the illuminating plane wave. Consequently, there will be a smear of transforms, overlapping and with amplitudes and phases dependent on the amplitudes and phases of the illuminating plane waves. Another way of considering the effect is that the amplitude transmittance of the object is multiplied by a random phase distribution. In the transform plane, the transform of the object distribution is convolved with the transform of the random phase distribution, resulting in a smear of the object transform.

If we are not doing spatial filtering, a spread distribution in the transform plane is no problem. Its transform yields an image of the original, diffusely illuminated object. In fact, the diffuse illumination elemenates the overexposure of the film near the optical axis and removes the edge sharpening of the image.

Read-out. Figure 9-4 shows one read-out possibility. The point source that originally served as the reference source now is the read-out source. The second transforming lens produces two real images in the image plane, the secondary image being inverted with respect to the primary image. The second lens could be eliminated and the virtual images at infinity viewed directly. The magnification of the images is determined as if the system were a two-lens imaging system with the exception that the wavelength can change in the transform plane. This is equivalent to changing the scale of the transform.

Figure 9-4 Read-out of a Fourier transform hologram producing two real images in the image plane. S—point source, L—transforming lenses, H—hologram, I—image plane.

Distribution on the hologram.   Let us first investigate the distribution due to a single object point and then consider the effects of multiple points. Figure 9-3 shows the waves from a reference point and one point on the object. The two plane waves produced by the lens interfere to form fringes that are recorded by the hologram. The period of the fringes can be seen to be a function of the separation of the reference and object points in the input plane. The farther the object point is from the reference point, the closer the spacing of the fringes.

When this fringe system is illuminated, as shown in Fig. 9-4, three plane waves are formed which are focused by the second lens to produce primary and secondary image points along with a point formed along the axis of the read-out wave. If the primary plane wave is viewed before entering the second lens, it appears to have come from a point at infinity.

This approach of representing the object as a number of points accurately describes the images but does not predict the the true distribution associated with the $|\mathbf{U}|^2$ term. If the hologram material is assumed to be in the linear region of its $\mathbf{t} - \mathcal{E}$ curve, the effects of all the points can be superimposed to obtain the waves off the axis of the read-out wave. This is because the image wave is linearly related to the object wave even though the hologram material responds to energy and is hence a nonlinear device in amplitude. On the axis of the read-out wave, the effects of interference between the various object points will appear.

Thus far we have looked at one object point and said that we may do the same thing for each object point and add the results. Let us consider further the justification for this. Assume that the object produces, in the hologram plane, the distribution described by $\mathbf{U}(x,y)$. If the object is thought of as being a number of points, $\mathbf{U}(x,y)$ can be written as the sum of plane waves,

$$\mathbf{U}(x,y) = \sum_m \mathbf{U}_m \exp(i\alpha_m x), \qquad (9\text{-}28)$$

where α_m is the propagation constant of the mth plane wave component. The intensity distribution of the hologram is then

$$\left| \sum_m \mathbf{U}_m \exp(i\alpha_m x) \right|^2 + \mathbf{V}_0^2$$

$$+ \mathbf{V}_0 \exp(i\zeta x) \sum_m \mathbf{U}_m \exp(-i\alpha_m x)$$

$$+ \mathbf{V}_0 \exp(-i\zeta x) \sum_n \mathbf{U}_m \exp(i\alpha_m x), \qquad (9\text{-}29)$$

where $\mathbf{V}_0 \exp(i\zeta x)$ is the reference wave. If the waves from the object points are thought of as being recorded separately, the distribution on the hologram is described by

$$\sum_m |U_m|^2 + V_0^2 + V_0 \exp(i\zeta x) \sum_m U_m \exp(-i\alpha_m x)$$

$$+ V_0 \exp(-i\zeta x) \sum_m U_m \exp(i\alpha_m x). \qquad (9\text{-}30)$$

The two representations of the hologram distribution (9-29) and (9-30) differ only in the first term. The terms describing the primary and secondary waves are identical and are linear in U_m.

Resolution required. We can see from the discussion above that to record an image of the entire object, the resolution of the recorder must be sufficient to record the fringes formed by the reference wave and the wave from the object point farthest from the reference point. This would imply that the object should be placed as closely as possible to the point reference. But, if the image is to be viewed directly, it should appear at a relatively large angle from the read-out wave.

A more fundamental limit is the spread of the central distribution. The point object does not show the effect, but the first term of (9-29) contains the square of the object distribution. The transform provided by the second lens produces an autocorrelation of the image in the central distribution. Consequently, the central distribution is twice as wide as the image. To avoid overlap, the closest edge of the image must be at least the width of the image from the axis of the read-out wave. Figure 9-5 shows the amplitude distribution in the image plane for a rectangular image with uniform brightness.

From the relation in the image plane, we find that the closest object point must be at least one object width from the reference point. Consequently, the resolution is at least double that predicted if the spread of the central distribution is neglected.

Effect of finite resolution. We have considered the resolution needed to achieve a good holographic image. The question of the effect of not

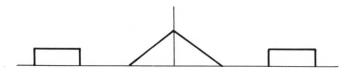

Figure 9-5 Amplitude distribution in the image plane of a Fourier transform hologram. The image is of a uniformly bright strip.

providing that resolution now arises. We can see from the preceding discussion that the resolution limit of the film or other recorder sets a limit upon the size of the image, or the field of view. The modulation transfer function (MTF) of the recorder can be superimposed upon the distribution in the object or image plane to indicate the effect on the field of view. Of course, the coordinate of the MTF must be changed from cycles per unit length to units of length by considering the relation between fringe period and object point location. The MTF is centered on the reference point. An example is shown in Fig. 9-6 where the MTF is superimposed on the object distribution to show where the image becomes dim and eventually vanishes. The image *amplitude* distribution is obtained by multiplying the MTF and the object distribution. Because the image irradiance is proportional to the intensity of the wave, the *brightness* of the image is affected by the square of the MTF. Object O_3 does not appear at all in the image.

The effect of the resolution of the film on the resolution of the image is only a secondary one. The problem is essentially that of how close two fringes can be in frequency and still be separated. This is primarily a function of the *size* of the hologram or how many cycles are on the hologram.

9.4 Quasi Fourier Transform Holograms

Two names have been applied to the type of hologram discussed in this section—*lensless Fourier transform hologram* and *quasi Fourier transform*

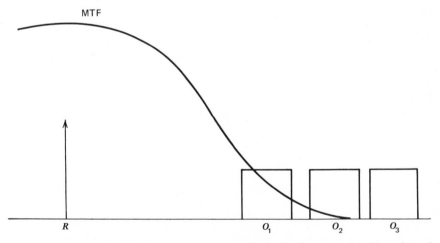

Figure 9-6 Recorder MTF replotted on plot of amplitude distribution in the object plane. R —reference point, O_1—object 1, O_2—object 2, O_3—object 3.

hologram [9-2, 9-3]. The adjective quasi, meaning in a certain sense or degree, appears to be more exact because the Fourier transform of the object distribution does not appear on the hologram. There are, however, many similarities to the Fourier transform hologram.

Recording. The arrangement often used in recording a quasi Fourier transform (QFT) hologram is shown in Fig. 9-7. It is essentially the same as Fig. 9-3 without the lens. The wave from the reference point has the same curvature at the hologram as the wave from the object point. Consequently, when the sum of the two waves is squared, the phase term vanishes and the fringes are *the same as fringes formed with interfering plane waves*.

Distribution on the hologram. Even though the fringes on the QFT hologram are the same as on a Fourier transform hologram of a point object, the distribution is not the same for a more general object. The reason for this can be traced to the quadratic phase term on the object wave. Obviously, the Fourier transform of the object distribution cannot appear on the hologram. However, the same conclusions can be made concerning the effect of the resolution of the hologram on the image as were made for the Fourier transform hologram.

Read-out. Illumination of the hologram can be provided by the reference point or with a plane wave. If the same point source is used to provide a read-out wave, the read-out wave is identical to the reference wave and a virtual image of the object appears in the location of the object just as with any configuration of reference wave. If a plane read-out wave is used, the fringes on the hologram give rise to plane image waves that must be Fourier-transformed with a lens to obtain image points. Both primary and secondary images are real images appearing in the back focal plane of the

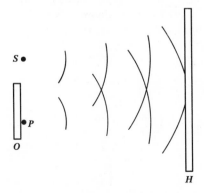

Figure 9-7   The recording of a quasi Fourier transform hologram. *S*—point source reference, *O*—object, *P*—point on object, *H*—hologram.

Fourier-transforming lens. This is another similarity to the Fourier transform hologram. In fact, the intensity distribution in the image (irradiance) is the same as if a Fourier transform hologram had been used.

Analysis of quasi Fourier transform holograms It is interesting to see exactly what the distribution is in the image plane of a QFT hologram. Another question relates to the read-out configuration in which a lens is used to Fourier transform the hologram distribution. If we Fourier transform the hologram distribution to get the image, but the hologram distribution is not a transform of the object distribution, of what *is* it the Fourier transform?

If the distance from the object to the hologram of Fig. 9-7 is d_1, the object wave at the hologram, $U_2(x_2,y_2)$, is related to the object distribution U_1 by

$$U_2(x_2,y_2) = U_1(x_2,y_2) \otimes h(x_2,y_2; d_1), \qquad (9\text{-}31)$$

where it is assumed that the object is on the z axis. If the reference point is offset by a, the reference wave at the hologram is $h(x_2 - a, y_2; d_1)$. Assuming a linear $t-\mathcal{E}$ curve, the amplitude transmittance of the hologram is given by

$$t(x,y) = |h(x_2 - a, y_2; d_1)|^2 + |U_1(x_2,y_2) \otimes h(x_2,y_2; d_1)|^2$$

$$+ h(x_2 - a, y_2; d_1)[U_1(x_2,y_2) \otimes h(x_2,y_2; d_1)]^*$$

$$+ h(x_2 - a, y_2; -d_1)[U_1(x_2,y_2) \otimes h(x_2,y_2; d_1)]. \quad (9\text{-}32)$$

If the hologram is illuminated with a uniform plane wave, we must Fourier transform the distribution to get the image.

Let us consider only the primary wave denoted by U_p. The offset a does not influence the result except to provide an offset for the image. Dropping the a makes the manipulations somewhat simpler. The Fourier transformation of the last term of (9-32) yields

$$U_p(x_3,y_3) = \mathcal{F}\left\{ h(x_2,y_2; -d_1) \int\int U_1(w,s)h(x_2 - w, y_2 - s; d_1)\, dw\, ds \right\}$$

$$= \int\int \exp\left[\frac{-ik(x_2^2 + y_2^2)}{2d_1}\right] \exp\left[\frac{-i2\pi}{\lambda f}(x_3 x_2 + y_3 y_2)\right]$$

$$\times \int\int U_1(w,s)\exp\left\{ \frac{ik}{2d_1}\left[(x_2 - w)^2 + (y_2 - s)^2\right]\right\} dw\, ds\, dx_2\, dy_2.$$

$$(9\text{-}33)$$

Expansion of the quadratice term in $x_2 - w$ gives a quadratic exponential term that cancels the first quadratic exponential term. In addition, we can reverse the order of integration to yield

$$U_p(x_3, y_3) = \int\int\int\int U_1(w, s) \exp\left[\frac{ik}{2d_1}(w^2 + s^2)\right] \exp\left[\frac{-i2\pi}{\lambda d_1}(x_2 w + y_2 s)\right]$$

$$\times \, dw \, ds \exp\left[\frac{-i2\pi}{\lambda f}(x_3 x_2 + y_3 y_2)\right] dx_2 \, dy_2. \tag{9-34}$$

Letting $U_1 \exp[\,]$ be replaced by U_1', we find that the inner integral is a Fourier transform yielding

$$U_p(x_3, y_3) = \lambda^2 d_1^2 \int\int \tilde{U}_1'\left(\frac{x_2}{\lambda d_1}, \frac{y_2}{\lambda d_1}\right)$$

$$\times \exp\left[\frac{-i2\pi d_1}{f}\left(\frac{x_2 x_3}{\lambda d_1} + \frac{y_2 y_3}{\lambda d_1}\right)\right] d\left(\frac{x_2}{\lambda d_1}\right) d\left(\frac{y_2}{\lambda d_1}\right) \tag{9-35}$$

where the tilde represents the Fourier transform. The second transform, obtained in evaluating (9-35), gives

$$U_p(x_3, y_3) = \lambda^2 d_1^2 U_1'\left(\frac{-d_1 x_3}{f}, \frac{-d_1 y_3}{f}\right) \tag{9-36}$$

so that we do obtain an image of the object, but there is a phase error of

$$\exp\left[\frac{ik}{2d_1}\left(\frac{d_1^2 x_3^2}{f^2} + \frac{d_1^2 y_3^2}{f^2}\right)\right] = \exp\left[\frac{ikd_1}{2f^2}(x_3^2 + y_3^2)\right]. \tag{9-37}$$

The phase error does not appear in the case of the Fourier transform hologram, but if the intensity of the wave is the only parameter of interest, there is no difference and the image irradiance is the same in both cases.

We can now see that the QFT hologram can be considered as a Fourier transform of the object distribution multiplied by a quadratic phase factor. If $f = d_1$, giving unity magnification to the image, we see that the phase term multiplying the object distribution is

$$\exp\left[\frac{ik}{2d_1}(x_1^2 + y_1^2)\right]. \tag{9-38}$$

Consequently, the distribution on the hologram is the Fourier transform of the object distribution multiplied by the phase factor of (9-38).

9.5 Effects of the Locations of a Point Reference and the Finite Resolution of the Recorder

The preceding sections show that the resolution of the hologram affects the field of view of the image but not its resolution. This is true only for the Fourier transform hologram and the quasi Fourier transform hologram. If a plane wave reference is used, the resolution, *not* the field of view of the image, is reduced as the resolution of the hologram is lowered. In other cases, both resolution *and* field of view are affected.

Point reference at infinity.   A point reference at infinity provides a uniform plane wave at the hologram plane. In this case, it is easier to consider the object distribution in terms of its angular spectrum of plane waves rather than a set of point sources. Figure 9-8 shows the plane reference wave, the object, and a few of the angular spectral components of the object wave.

We can again use superposition in working with the primary and secondary waves. The plane reference wave interferes with one plane wave

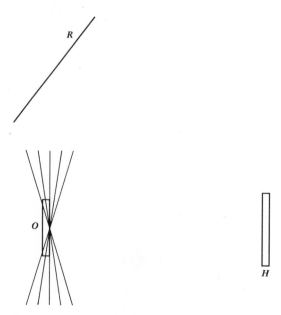

Figure 9-8   The recording of a hologram with a uniform plane reference wave. *R*—reference, *O*—object and portion of angular spectrum of plane waves, *H*—hologram.

of the angular spectrum to give linear fringes on the hologram. The fringe period in this case relates to the angular frequency, or tilt, of the plane wave from the object spectrum. The effect of a finite resolution in the hologram is that of an off-axis spatial filter. The filter is off-axis because of the tilt of the reference wave. From Fig. 9-8 we see that the higher-frequency plane wave is closer to the angle of the reference wave, hence produces the lower-frequency fringes. The absolute value spectral plot show in Fig. 9-9 illustrates this.

The object spectrum should not overlap the reference spatial frequency because object frequencies equal increments above and below the reference spatial frequency result in fringes of the same period being recorded on the hologram. That is, the image spectrum would be the same as the object spectrum folded about the reference frequency. Figure 9-8 shows that this means the tilt of the reference wave must be greater than the tilt of the plane wave components of the object.

More careful consideration indicates that the spatial frequency of the reference wave must be at least three times the highest frequency component of the object. Interference between the reference wave and an object plane wave component produces a fringe that, when illuminated, causes waves to appear at angles equal to the difference between the angles of the reference and object plane waves. This results in an image spectrum that is the same as the object spectrum but reversed and repeated about the reference spatial frequency. The interference of the object plane waves with each other causes the spectrum of the on-axis image plane distribution to be twice as wide as the object spectrum. Interference between the highest positive and lowest negative plane wave components of the object causes a fringe that produces waves at twice the angle of the interfering waves. In this case, taking the difference between the angles results in a greater, not smaller number because one angle is negative. Figure 9-10 shows the magnitude of the spectrum in the image plane.

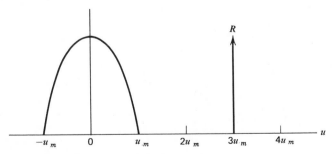

Figure 9-9 The object spectrum with maximum spatial frequency u_m and single spatial frequency plane wave reference R.

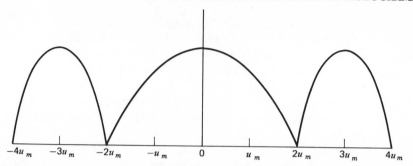

Figure 9-10 Spectra of the on-axis distribution and the primary and secondary images. The spectra of the images are shifted to centers at the spatial frequency of the reference.

We can see from Fig. 9-10 that the resolution of the hologram recording medium must be at least four times the highest-frequency component of the object. If high-resolution images are to be obtained, high-resolution holographic material must be used. The effect of the MTF of the recorder is shown in Fig. 9-11 where the MTF is superimposed on the spectral distribution of Fig. 9-9. The MTF is centered at the reference frequency because all fringes producing the primary and secondary images are formed by interference with the reference wave. This also shows why the resolution of the recorder must be at least four times the highest spatial frequency component of the object. The MTF of the film is seen to act as an off-axis spatial filter.

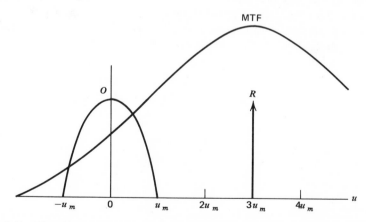

Figure 9-11 The object spectrum O and reference wave spectrum R with the MTF of the recorder shown on the same plot centered at the reference frequency.

Our work with lenses has shown that a mask across the lens can produce results similar to those of a spatial filter. The same is true here. The effect of the MTF of the recorder is the same as that of placing over the hologram a mask with transmittance

$$t(x,y) = \left| \mathbf{H}\left(\frac{x}{\lambda d_1} - u_r, \frac{y}{\lambda d_1} \right) \right| \qquad (9\text{-}39)$$

where u_r is the spatial frequency of the reference and $|\mathbf{H}|$ is the MTF of the recorder. Consequently, we find that a plane wave reference or reference point at infinity produces the effect of a mask over the hologram which reduces the effective aperture, hence the resolution of the image. In the case of a QFT hologram, the mask appeared on the object.

Point reference at arbitrary location. We have seen that the resolution of the hologram can have vastly different effects on the image depending upon the form of the reference wave. If the point reference is at an arbitrary location, the mask having an amplitude transmittance proportional to the MTF of the recorder appears to be in the plane of the point reference [9-16, 9-17, 9-18]. The coordinates in the plane of the reference are related to the spatial frequency coordinate of the MTF plot by

$$x = \rho_V \lambda u \qquad (9\text{-}40)$$

where ρ_V is the distance from the point reference to the hologram, λ is the wavelength, and u is the spatial frequency coordinate of the MTF. The effect of the MTF is then found by projecting the MTF through the object point onto the hologram.

Figure 9-12 shows the means of projecting the MTF when the point reference lies between the object and infinity. The projections for two points on the MTF are shown for object point O_1. A line is drawn through O_1 and a point on the hologram to the abscissa of the MTF curve about the reference point. The value of the MTF at the point of intersection on the abscissa is plotted on the corresponding point on the hologram. This gives a plot of the visibility of the fringe on that portion of the hologram due to the interference between the waves from the reference point and the object point. Because the visibility is related to the efficiency of the hologram in diffracting energy to the primary wave, this is also the form of a plot of the efficiency of each portion of the hologram. If a similar process is followed for point O_2, we obtain the dashed line projected on the hologram. Not only is the effective size of the hologram reduced, reducing the resolution, but the fringes related to object point O_2 are of low

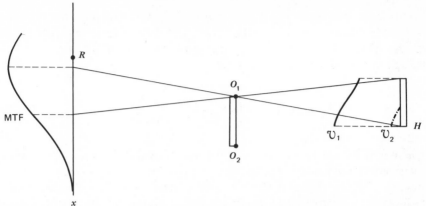

Figure 9-12   The effect of the recorder MTF on a hologram with a general point reference. R —reference point, O_1—object point 1, O_2—object point 2, $\mho_1$—visibility variation across hologram for fringe associated with object point 1; $\mho_2$—visibility variation across hologram for fringe associated with object point 2, H—hologram.

visibility, causing a reduction in the brightness of that portion of the image. An object point much below O_2 would not produce recorded fringes at all. Consequently, *both* the field of view and the resolution of the image are affected.

We can see from Fig. 9-12 that if the object is in the same plane as the reference point, projection of the MTF curve indicates that the fringes have the same visibility over the entire hologram for a given object point. Consequently, the resolution is not reduced because of a reduction in the effective aperture of the hologram. The fringes caused by different object points do, however, have different visibilities, and the field of view can be reduced.

If the point reference moves to infinity, the plot of the MTF in the plane of the point reference becomes infinite in extent. Its angular size is

$$\frac{x}{\rho_V} = \lambda u. \qquad (9\text{-}41)$$

Because of its infinite size and distance, the same variation in visibility appears over the hologram for all object points. There is no restriction on the field of view. However, because the angular size of the MTF plot is finite, projections onto the hologram, through any object point, give a reduction in the effective aperture of the hologram.

Space-spatial frequency bandwidth product.   We have seen that the resolution of the hologram and the effective size of a hologram are related by the location of the point reference. We also saw that the size of the hologram affects the ability to discriminate between the spatial frequency of two fringe patterns in the Fourier transform and QFT holograms. That is, the size affects the image resolution. The effect of hologram size can also be inferred by analogy to the effect of lens size.

It is important to note that it is the *product* of resolution and size that matters. That is, the number of resolution elements per millimeter is not necessarily important, but the *total* number of resolution elements is. Put another way, the number of samples, not the number of samples per millimeter, is the controlling number. If the analogy is made to temporal systems, the space-spatial frequency bandwidth product is the parameter of interest. Many studies have been made of the effects of resolution, hologram size, and configuration upon the number of resolvable image points that can be reconstructed [9-19, 9-20, 9-21].

9.6 Image Plane Holograms

An image plane hologram can be formed by placing the object close to the hologram, using a lens or lenses to form an image of the object on the hologram, or using a regular hologram to form a real image in the second hologram plane. The easiest is to use a lens. The image, however, is then pseudoscopic. That is, it is inverted in all three dimensions and the parallax effects are reversed. The image of an object that appears to be in front vanishes when the image of an object that appears to be in back is lined up behind it. The recording of an image plane hologram using the real image of another hologram removes that effect because the real image of a hologram is also pseudoscopic. The recording of the wave from a pseudoscopic image (the real image of the first hologram) in a process that produces a pseudoscopic image of a normal object will cause a normal, orthoscopic image to be formed. This, however, requires the making of two holograms.

Figure 9-13 illustrates the recording of an image plane hologram. Note that not all of the image is in the plane of the hologram. The portions not in the plane are seen as being very much out of focus in diffuse illumination. However, upon illumination with light that is closer to plane wave illumination (sunlight will do), those portions of the image out of the plane of the hologram can be seen.

One of the advantages of an image plane hologram is that the read-out source can be a white light or wide spectral range source. The effects of the spectral width of the source are described in the next section. A difference between a Fresnel zone hologram and an image plane hologram is the localization of the information storage in the image plane hologram. The

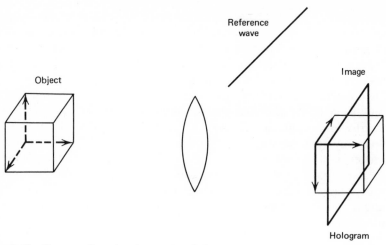

Figure 9-13   The recording of an image plane hologram.

information is not stored over the entire hologram which would provide large dynamic range images and scratch or noise resistance.

Effects of spectral width of the read-out wave.   Much of this discussion follows the paper by Brandt in which he describes many of the properties of image plane holograms [9-22]. The effect of more than one wavelength on the image can be obtained by considering the hologram as a generalized grating. The location of an image point is a function of the angle between the reference and object waves when recorded, the angle of the read-out wave, and the wavelength. The image location x_i is given by

$$x_i = d_i \lambda_2 \frac{\sin(\chi + \phi)}{\lambda_1} \tag{9-42}$$

where d_i is the distance to the image, λ_1 is the recording wavelength, λ_2 is the reconstructing wavelength, χ is the angle between the z axis and the direction of propagation of the object wave, and ϕ is the angle between the z axis and the direction of propagation of the reference wave. The spread of the image point, Δx_i, is given by

$$\Delta x_i = \frac{d_i \Delta \lambda_2 \sin(\chi + \phi)}{\lambda_1}. \tag{9-43}$$

We see that the image points closest to the hologram are spread the least because d_i is small.

Another effect reduces the anticipated smearing of the image when it is viewed through a small aperture such as the pupil of the eye. The effects of the dispersion of the hologram are limited, in fact, by the finite aperture. The angular blur $\Delta x_i / d_V$ seen is given by

$$\frac{\Delta x_i}{d_V} = \frac{d_i D}{d_V} \tag{9-44}$$

where d_V is the distance to the viewer and D is the diameter of the viewing optical system [9-22].

Effect of extended read-out source.   It can be shown [9-22] that the spread in the image point as a function of the extent of the read-out source is given by

$$\Delta x_i = \frac{\lambda_1 M_c^2 d_o}{\lambda_2 \rho_W} \Delta S \tag{9-45}$$

where ΔS is the extent of the read-out source and the other variables are as defined in Section 9.2. The blur of the image goes to zero as the object-hologram distance d_o goes to zero. In the case of $d_o = 0$, we are, in effect, simply illuminating a transparency with the image recorded in the plane of the transparency. Similar results are obtained for the case of an extended reference source. It is, of course, the cases where $d_o \neq 0$ that are of interest in holography. Otherwise, we have a photograph with no depth information.

Local reference beam image plane holograms.   Because of the high tolerance of image plane holograms for extended reference sources, they are often recorded using a locally generated reference wave. Figure 9-14 shows one implementation. The light from the object serves as a reference wave for the image formed with the lens. A lens could be used in the reference path to produce an image nearer the hologram that would serve as an extended reference source for the image on the hologram [9-22]. This image reference source could be reduced in size by using an iris as in Fig. 9-1. The local reference beam image plane (LRBIP) hologram has the same advantages as the LRB hologram. There is no problem with object vibration or short coherence length illumination because the object and reference waves are derived from the same source.

Figure 9-15 shows the image obtained when a LRBIP hologram is illuminated with white light. The recording technique used was that illustrated in Fig. 9-14. Portions of this image show up almost white with little evidence of the dispersion normally seen. The reason is that the wave from an image point is recorded with waves from a number of reference

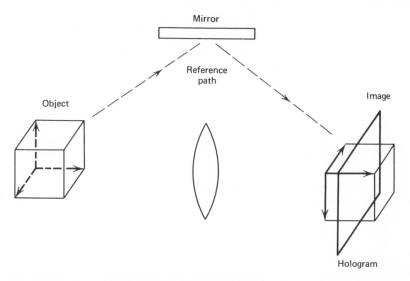

Figure 9-14 The recording of a form of local reference beam image plane hologram.

Figure 9-15 Image obtained when a LRBIP hologram is illuminated with white light [9-22].

286

points. This produces a number of recorded gratings of different periods which diffract different wavelengths to the image point. When added, these waves produce white light.

Optimum reference-to-object wave intensity ratio. Holograms recorded in the Fresnel region of the object have the highest diffraction efficiency when the object and reference waves have equal brightness. Because of the nonlinearities often encountered for a low reference-to-object wave intensity ratio K, large values of K are often used. See Chapter 6 for a complete treatment of nonlinearities that produce multiple images and reduce image contrast. Image plane holograms appear not to have the same problems with nonlinearities because of the localized nature of the recording. Cross modulation can occur only over a very small region.

The contrast of the image of a diffuse object recorded on a Fresnel region hologram is dependent on the size of the hologram. This is because the entire hologram can diffract light to a single image point. In the case of the image plane hologram, however, image contrast is closely related to the film properties. This is the result of the use of only a small region of the hologram to record the information concerning the image point.

Another difference in the images formed with image plane holograms is that they can be negatives over part or all of the image. This is best explained by reference to Fig. 9-16 which shows the diffraction efficiency of image plane holograms as a function of the object wave exposure for three different reference beam exposures. The curves were obtained from diffraction efficiency versus amplitude transmittance curves taking into account the amplitude transmittance versus exposure curve [9-22]. If the reference exposure is such that the entire range of object wave exposure falls on the portion of the curve to the left of the peak, the image is positive. If the entire range is on the linear region of the curve, the image intensity is directly proportional to object intensity. If the range of object wave exposures is to the right of the peak, an increase in object exposure results in a decrease in the diffraction efficiency of the hologram and a decrease in the image intensity. Consequently, brighter object points appear dimmer in the image, and a negative image is obtained. If the average exposure is near the peak of the diffraction efficiency curve, both brighter and dimmer object points will appear as dimmer image points. For best results, then, the exposure should be controlled so that the range of exposures is over the linear region with a positive slope.

9.7 Fraunhofer Holograms

The term Fraunhofer hologram applies to holograms recorded in the Fraunhofer region or far field of the *individual elements* of an object. It is especially applied to an object that consists of a number of small particles.

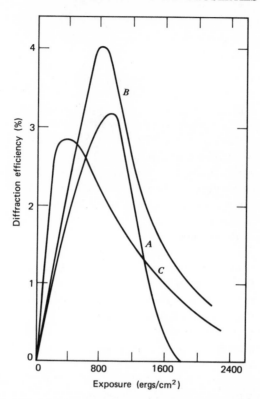

Figure 9-16 Diffraction efficiency of image plane holograms as a function of object wave exposure for reference beam exposures: A—200 ergs/cm^2, B—650 ergs/cm^2, and C—1600 ergs/cm^2 [9-22].

The hologram is farther than a^2/λ from the object, where a is the diameter of the largest particle in the object field. If this restriction is obeyed, an on-axis reference wave can be used with little difficulty caused by the secondary image. This is the chief advantage of Fraunhofer holograms.

One of the first applications of holography, other than in displays, was to measure the sizes and distribution of particles [9-23]. The advantage of using holography is that the image can be scanned in depth and the particle sizes measured under magnification. If conventional photographs were taken, they would be in focus in only a small region. A series of photographs focused in different planes would allow the distribution to change between photographs. By recording a sequence of holograms, particle velocity as well as size and distribution can be determined.

An on-axis reference is provided by the undiffracted wave passing through the group of particles. Both the primary and secondary images then appear on the axis with the undiffracted part of the read-out wave. The unwanted image causes no problems, because when the real image is in focus, for example, the virtual image of the particles is far away. Hence the energy from the point images is spread over a large area, simply reducing contrast [9-24].

9.8 Multiply Exposed Holograms

Multiply exposed holograms are those on which a number of separate holograms are superimposed. That is, one object wave is recorded and then a different object wave is recorded on the same plate. Strictly speaking, the cases discussed in Chapter 6 where different image waves can be formed by illumination of a volume hologram at different angles are multiply exposed holograms. However, the term is normally applied to cases where the waves individually recorded are reconstructed such that all images appear simultaneously. For example, a number of holograms of point objects can be superimposed by separate recordings on the same photographic plate. Upon reconstruction, all the point images combine to form a composite image.

A multiply exposed hologram consisting of many holograms of point objects are useful in the synthesis of images [9-25]. One example is the synthesis of a three-dimensional display for a vector cardiogram [9-26, 9-27]. In this case, a point object is moved to represent the tip of the vector representing the electric field in the vicinity of the heart. Multiple exposures are made at equal time intervals throughout the cycle. Consequently, the distance between the points is a measure of the rate at which the vector is changing length or direction.

Point objects. Most of the analysis of multiply exposed holograms have assumed point objects because they are most common in experiment. The mathematical description is similar to (9-30), which described the effect of assuming that the object is a number of separate point objects. The difference is that the summation must be taken over the separate reference waves also and the waves being added are usually spherical object waves and plane reference waves. If the phase due to the direction of the wave is incorporated into the phase of the complex amplitude, the description of the distribution on the hologram is

$$\sum_{m} \left\{ |U_m|^2 + |V_m|^2 + U_m V_m^* + U_m^* V_m \right\}. \tag{9-46}$$

The difference between this expression and (9-29) is that there is no on-axis distribution caused by mutual interference of the object waves. In this case, the object waves do not exist simultaneously. Consequently, there is a narrower on-axis distribution.

The most important problem is the buildup of the uniform exposure of the hologram. This causes saturation of the recording medium, limiting the number of object waves that can be recovered. If this background exposure or bias point and fringe visibility are the only factors considered, it can be shown that the optimum reference-to-object wave intensity ratio is $1:1$ for sequential exposures of point objects and $N^2:1$ for simultaneous exposure of each of N point objects [9-28]. This neglects considerations of S/N ratio.

In recording multiply exposed holograms, the exposure of each hologram must be reduced so that the total exposure is within the range of the recorder. If many exposures are necessary, the exposure of a single object point becomes so low that if only one exposure were made, *no* image wave would be reconstructed. However, when the other exposures are added, the film is biased off the toe of the H & D curve and *all* images waves are reconstructed. Such biasing can also be provided with uniform illumination either before or after recording the hologram [9-29].

We might expect that the first exposures of the hologram would result in dimmer images than the later exposures. The converse appears to be true. In an experiment using Agfa 8E70 and four different developers, the first of 12 exposures produced an image four to eight times brighter than the twelfth exposure [9-30]. The postulate is that this effect arises from the latent image-forming process in the emulsion.

Scanned holograms: scanned object illumination.   In some cases the entire object may not be illuminated at the same time, but a scanned beam may be used to provide illumination. A form of multiple recording results because holograms of different portions of the object are recorded sequentially.

Scanned holograms: scanned reference beam.   The entire object can be illuminated and a reference beam scanned over the hologram. This has application in holographic television systems because the resolution required of the television camera can be reduced by a factor of 4 [9-31]. A complete hologram is not formed at the transmitting end of the system. The reference wave has a different frequency than the object wave. The resulting intensity distribution in the hologram plane (on the face of the vidicon) is

$$I = U^2(x,y) + V_0^2 + 2V_0 U(x,y) \cos[\Delta\omega t + 2\zeta x + \Psi(x,y) - \Phi]. \quad (9\text{-}47)$$

Consequently, the signals related to the first two terms can be removed by band-pass filtering of the video signal.

Rather than scanning the intensity distribution with an electron beam, a photodetector such as a photomultiplier can be used if the reference beam of a different frequency is scanned. The scanner need not have the resolution necessary to resolve the interference fringes because they are no longer needed to separate the terms of (9-47). The separation is done electronically with the temporal band-pass filter. Let the amplitude of the reference beam be written as a moving spot, $|V| = \delta(x - s_x t, y - s_y t)$ where s_x and s_y are the horizontal and vertical scan velocities. Use of this in (9-47) and integration over the area of the detector yields

$$I(t) = 2U(s_x t, s_y t) \cos\left[(\Delta\omega + 2\zeta v_x)t + \Psi(s_x t, s_y t) - \Phi \right]. \qquad (9\text{-}48)$$

Notice that *either* $\Delta\omega$ or $2\zeta v_x$ can provide the necessary offset needed for separation of the terms. The spatial carrier can be set to zero (i.e., $\zeta = 0$) and $\Delta\omega$ chosen to provide the separation. Now, the aperture of the scanning beam need resolve only the spatial frequencies present in U, not at least four times the highest spatial frequency in U as it would if the fringes were needed for image separation.

Another reason for scanning the reference beam is to reduce the effects of vibration [9-32]. If the energy of the reference wave can be increased by the ratio of the area of the hologram to the area of the scanned reference wave, the exposure time of the scanned element is reduced accordingly. Object motion during the time of the exposure of the whole hologram will cause an image blur because different portions of the hologram will form an image at different locations, but the image can be recorded.

9.9   Color and White Light Holograms

These holograms are recorded using coherent illumination and read out using white light. The white light read-out illumination is usually collimated unless the hologram is also an image plane hologram, in which case diffuse illumination may be permissible. If the recording of the hologram is made with a single wavelenght, the image will either be single color or white, depending on the techniques used. Multiple wavelength recording can produce full color images.

Volume white light holograms.   Volume or thick holograms have been discussed in Chapters 3 and 6. The planes in the emulsion act as spectral filters, reflecting only those wavelengths near the wavelength with which the hologram was recorded. This assumes no shrinkage of the emulsion and illumination at or near the Bragg angle. Otherwise, the reflected wavelength will not be the same as the recorded one. These holograms are normally used in the reflection mode [9-33, 9-34]. That is, the reference wave is introduced such that the reflecting surfaces are parallel to the surface of the hologram.

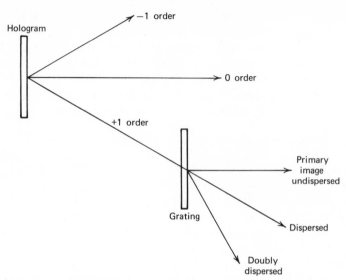

Figure 9-17   Dispersion of a hologram compensated by a grating [9-34].

Surface white light holograms.   If a surface or thin hologram is illuminated with white light, there is no mechanism for removing the unwanted wavelengths unless a separate color filter is used. Unless the hologram is an image plane hologram, the dispersion due to the grating effect causes many images of different color to be formed with small displacements.

One technique to compensate for the dispersion of the hologram is to use a grating to reverse the dispersion [9-35, 9-36]. The dispersion is compensated because the second diffraction is in the opposite direction. See Fig. 9-17. The positions of the hologram and grating can be interchanged. If this is done, a blazed grating with most energy going into the first order can be used.

If the hologram and grating are closely spaced, the zero order of the first diffracting element and the undispersed primary image resulting from two dispersions lie along the same axis. The zero order can be removed by placing a venetian blind structure between the two elements with the openings parallel to the direction of the first order of the first element. This permits only the undispersed primary image to be seen [9-36].

Volume color hologram.   The volume color hologram is essentially the same as the volume white light hologram, but more than one wavelength is used to record the image. For example, if three wavelengths are used to record the hologram, three superimposed holograms are recorded, one for each wavelength. It is shown in Chapter 6 that the reflection hologram will reflect only a narrow range of wavelengths, depending on the thickness of the recorder. If there is no shrinkage of the hologram and the read-out wave is introduced at the same angle as the reference wave, the band of reflected wavelengths is centered at the recording wavelength. If illumination occurs outside of this band of wavelengths, no energy is reflected. Consequently, if the three wavelengths used are separated more than the band of reflected wavelengths of each color, a good color image is formed with no crosstalk.

Crosstalk in a color hologram occurs when more than one of the superimposed holograms forms an image with the same wavelength illumination. The image formed when a hologram is recorded with one wavelength and illuminated at a second wavelength is not in registration with the other color images. It is displaced an amount proportional to the wavelength differences.

Crosstalk is not normally a problem in volume reflection color holograms because the large Bragg angle reduces the spectral spread of the reflected illumination. See Chapter 6. Emulsion shrinkage in photographic holograms is, however, a great problem, because changes in the spacing of the reflecting surfaces change the colors of the reflected waves. Techniques to prevent shrinkage are described in Appendix 3.

Volume transmission holograms have fewer problems due to shrinkage, but the small Bragg angle allows a larger range of wavelengths to be diffracted into the image. Quite often, transmission holograms are illuminated by the same laser wavelengths with which they were formed. This is possible because shrinkage of the emulsion does not affect the spacing of diffracting surfaces normal to the hologram surface. The broader range of acceptable wavelengths further reduces the effect of any shrinkage.

Surface color holograms.   Surface color holograms are formed when the transmission holograms are recorded with fringe spacings large with respect to the emulsion thickness. Because no color selection exists, illumination must be with the same wavelengths used in recording the hologram. In this case, all recorded holograms form images at all wavelengths of the illumination. For example, holograms recorded and illuminated with wavelengths λ_1 and λ_2 will form four images. Each hologram will form an image with both wavelengths of illumination. Very careful image registration techniques must be followed. Alternatively, a composite hologram can be made with a sort of checkerboard array of different color reference waves. If the array of different color references is used, the different wavelength holograms are formed in different sections of the array [9-37]. In forming the image, the hologram must be placed in registration with the array of different color read-out waves.

9.10   Incoherent Holograms

The term *incoherent hologram* is generally reserved for holograms *recorded* with incoherent illumination. The illumination of the hologram can be with either coherent or incoherent light. There are three general types of incoherent holograms. In one, the hologram formation is the second step in a two-step process in which the first step is the recording of an array of photographs at different locations. In another approach to incoherent holography, fringes are formed on a hologram which preserve the information about the object. The fringes are not formed in the same manner as in coherent holography, but are produced by interference of the light from one image point with itself. That is, two images of the same object are produced and the light from corresponding image points can interfere. A third approach is to modify the methods for recording coherent holograms so that they can be used with incoherent illumination. This includes achromatic fringe systems.

Achromatic fringe systems.   Achromatic fringe systems produce two interfering beams that have the same phase difference for a broad band of wavelengths [9-38–9-40]. One technique for doing this is shown in Fig. 9-18. The grating in plane P_1 is imaged into plane P_4 where the hologram is recorded. The image of the grating produces the necessary fringes. The diffracted waves are focused in the back focal plane of the lens where any two can be selected for use in forming the hologram. Figure 9-18 shows the selection of the $+1$ and -1 orders of the grating. The object is placed in one beam in plane P_3. The two beams are combined in what would ordinarily be the image plane for the grating, P_4. In this case, however, one beam has been spatially modulated by the object. Consequently, the fringes show the same form as in other methods of recording.

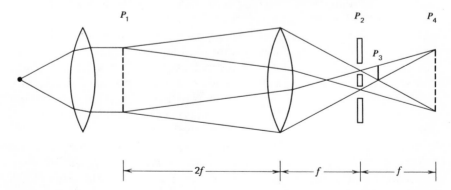

Figure 9-18   An achromatic fringe system using a grating. The grating in plane P_1 is imaged in plane P_4 [9-38].

In practice, it is better to use the zero order to illuminate the object. This provides undispersed illumination of the object. Either the $+1$ or -1 order can be used as a reference wave. Dispersal of the reference wave between the grating and its image does not matter. The colors are all registered in the image of the grating where the hologram is recorded.

Twin image technique. There are a number of implementations of the twin image technique. All of them use some procedure to split the wave from the object into two parts and form two images. The light from the corresponding "twin" points on the two images interfere to produce fringes or a zone plate. Light from each point on the image interferes only with light from its twin, not with light from any other point because light waves from non twin points in the two images are mutually incoherent. The result is similar to that obtained with superimposed or multiply exposed holograms. Thus the background exposure on the hologram increases proportionately with the number of resolvable points on the object. Each point is recorded as if with a separate exposure because of the lack of interaction with the light from other points.

A buildup of the bias exposure produces a reduction in the number of exposures that can be superimposed. As an example, Fig. 9-19 shows the reduction in quality of the reconstructed image as the number of object points is increased. Figure 9-19a shows the object with 450 resolvable points. The image (b) is of the entire object. The number of resolvable points was reduced to approximately 250 in recording the image (c). The image (d) was produced using a hologram that was exposed to only about 90 points. The images were all normalized with respect to the background noise. We can see that the S/N ratio decreases rapidly with an increase in the number of points. A continuous tone object and the image made from an incoherent hologram are shown in Fig. 9-20.

A twin image technique can be used to record Fourier transform holograms [9-41]. Because the transform is of the intensity, not the amplitude transmittance, the *amplitude* of the image wave is proportional to the object *radiance*, and the image irradiance is proportional to the *square* of the object radiance. If the object has binary transmittance, this presents no problem.

A concise summary of some techniques used to record incoherent holograms and a list of references is given by Kozma and Massey [9-42]. Two techniques of recording incoherent holograms are discussed here: a modified Linnik interferometer and a triangle interferometer. Both produce twin images with interference between waves from twin points.

The Linnik interferometer is shown in Fig. 9-21. The focal lengths f_1 and f_2 are not equal. Consequently, the two images formed appear at different distances from the hologram. Waves from twin points then have different

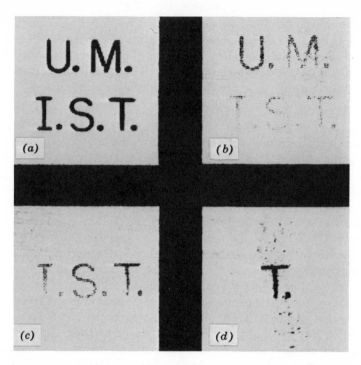

Figure 9-19 The effect of increasing the number of object points on the brightness of the image formed with an incoherent hologram. (*a*) The object, (*b*) image from an incoherent hologram of the entire object, (*c*) and (*d*) images where only parts of the object contributed to the hologram [9-39].

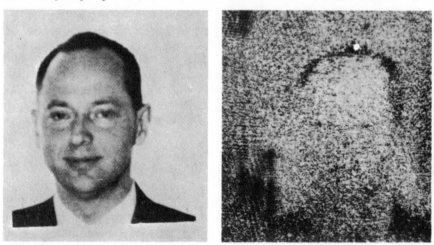

Figure 9-20 A continuous tone object (left) and the image (right) from an incoherent hologram [9-42].

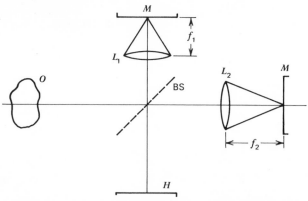

Figure 9-21 A modified Linnik interferometer for recording incoherent holograms. L_1 and L_2—lenses of focal lengths f_1 and f_2, M—mirrors, O—object, H—hologram plane, BS—beam splitter.

radii of curvature. When these waves interfere, a zone plate is formed with its center aligned with the center of the object point forming it. Consequently, zone plates appear over the entire hologram corresponding to the locations of the object points. An offset frequency can be provided by using half-plane stops in the mirror planes. These have the effect of introducing phase shifts on the image points that show up in the zone plates.

A triangle interferometer can also be used to record holograms [9-43, 9-44]. Figure 9-22 shows that two lenses of different focal length are again used but both lenses enter into the imaging. The lenses are placed so that their front focal planes coincide with the object plane and their back focal planes coincide in a plane some distance from the hologram. Waves going through the system clockwise produce images with a magnification of $-f_1/f_2$, and waves going through counterclockwise produce images with a magnification of $-f_2/f_1$. This difference in magnification produces zone plates on the hologram in much the same manner as the Lennik interferometer. The difference is that, in the case of the triangle interferometer, the object is placed off axis during the recording and the centers of the zone plates blocked during reconstruction [9-44]. This is done to provide separation of the images.

Holographic sterograms. One two-step process for making holograms starts with a series of photographs of a scene [9-45]. Each photograph is made with natural illumination and at equally spaced positions about an object or along a line. These photographs are then used to make the hologram. The advantage of such a process is that holograms can be made

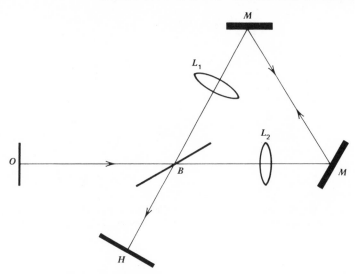

Figure 9-22 A triangle interferometer for use in recording incoherent holograms. *O*—object, *B*—beamsplitter, *M*—mirror, *H*—hologram, L_1—lens of focal length f_1, and L_2—lens of focal length f_2 [9-43].

that show three-dimensional views of large objects with natural illumination. The view is not exactly the same as if a hologram had been made of the scene directly, but parallax is preserved. The result is similar to viewing a stereo pair of images, but the glasses are not required and many views of the object, not just two, are used.

Figure 9-23 shows one technique used in recording holographic stereograms. A series of ordinary photographs of a three-dimensional object are taken using reversal film. The resulting two-dimensional positive transparencies are sequentially illuminated by a coherent source and projected onto a translucent screen as shown in Fig. 9-23 . The hologram is exposed through a movable vertical slit in a mask in front of the photographic plate. One vertical strip hologram is then recorded for each transparency. Vertical slits are used if parallax is recorded only in the horizontal direction. If an array of photographs were made, with different views taken vertically, an array of holograms would provide vertical parallax as well.

The slit width is chosen to be smaller than or equal to the aperture of the eye so that changes in image parallax occur smoothly. Jumps in the image as the viewer moves are thus avoided. If the slits are too small, the horizontal resolution of the image is degraded.

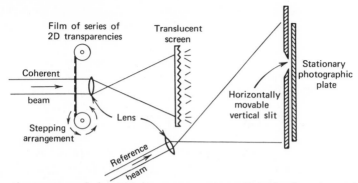

Figure 9-23 Arrangement for conversion of a series of two-dimensional transparencies into a composite hologram of as many contiguous vertical strip holograms [9-45].

Integral photography. *Integral photography* is the term used to describe a technique of three-dimensional imaging using an array of lenses. The technique can be used to produce three-dimensional images directly, or a hologram can be formed in a two-step process. Burckhardt has derived the optimum size for the lenslet of the array and shows that the resolution needed to record an integral photograph is proportional to the fourth power of the linear resolution in the image [9-46].

The recording of the necessary array of photographs is illustrated in Fig. 9-24. Each object point is imaged onto the photographic plate behind each lens. Illumination of the object can be with natural light. After the exposed

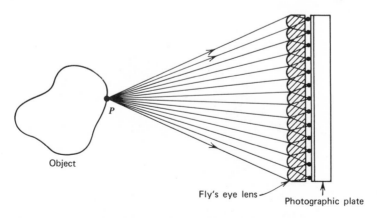

Figure 9-24 Recording an integral photograph [9-46].

plate is developed in reversal, the plate is illuminated as shown in Fig. 9-25. Waves from each small transparency propagate from different directions to be imaged to points at P. Because the waves converge from different directions, the image has depth, requiring the viewer to refocus when viewing different regions of the image.

The image obtained in this procedure is pseudoscopic, but an orthoscopic image can be obtained by a two-step process, in which the recorded images are inverted. Some two-step processes result in images that are limited in field of view or repeat as the angle of viewing increases. One method that avoids these problems is described in reference [9-47].

Holograms of the real image can be made if the illumination of Fig. 9-25 is coherent and a reference wave introduced [9-48]. The hologram is then viewed, giving either a real or virtual image, depending on the direction of illumination.

9.11 Phase Information Holograms and Kinoforms

A *phase information hologram* is one in which the amplitude information contained in a wavefront is disregarded and only the phase information is retained. The name phase information hologram implies nothing about the type of modulation used. Either phase or amplitude modulation can be used. Amplitude modulation of a wave impressing only the phase information onto the read-out wave would be provided by a transparency with transmittance

$$\mathbf{t}(x,y) = U_0^2 + V_0^2 + 2U_0V_0\cos[2\zeta x + \Psi(x,y)]. \qquad (9\text{-}49)$$

Phase modulation to impress the same phase information onto the wave could be described by

$$\mathbf{t}(x,y) = \exp\left(ip\left\{U_0^2 + V_0^2 + 2U_0V_0\cos[2\zeta x + \Psi(x,y)]\right\}\right). \qquad (9\text{-}50)$$

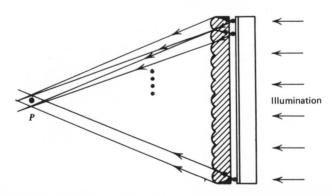

Illumination

Figure 9-25 Formation of the real image using an integral photograph [9-46].

Because only one parameter is recorded and impressed onto the read-out wave, the spatial subcarrier $\cos(2\zeta x)$ is not needed. The phase of the read-out wave can be directly modulated using a transparency with a transmittance given by

$$\mathbf{t}(x,y) = \exp[ip\Psi(x,y)]. \tag{9-51}$$

To reconstruct the same wave as recorded (without amplitude variations), $p = 1$. This is the principle used in the kinoform. Kermisch has performed an analysis of the effect of discarding the amplitude information of a wavefront reflected from a diffusely reflecting object [9-49]. His results show that approximately 78% of the radiance of an image formed with a phase information hologram constructs an exact image of the diffusely reflecting object. The remaining 22% consists of convolutions of the object distribution. The image irradiance is given by

$$I_i = K\left[I_0 + \tfrac{1}{8}I_0 \otimes (I_0 \star I_0) \right.$$

$$\left. + \frac{3}{64}I_0 \otimes (I_0 \star I_0) \otimes (I_0 \star I_0) + \cdots \right] \tag{9-52}$$

where K is a constant and I_0 is the normalized object radiance [9-49]. The degree of degradation of an image depends on the form of the convolutions and correlations of the object distribution.

Three means of discarding amplitude information are considered: hard-limiting or clipping the hologram to produce a binary hologram, recording only the phase as in acoustical holograms, and computer-generated holograms where the amplitude information can readily be dropped.

Hard limited holograms. Kozma and Kelly found that the amplitude information need not be preserved in making matched spatial filters [9-50]. If the amplitude spectrum of the signal is near uniform, the decrease in the S/N ratio in the output is small. A hard-limited hologram would have an amplitude transmittance of

$$\mathbf{t}(x,y) = U_0 + V_0 + 2U_0V_0 \frac{\cos[2\zeta x + \Psi(x,y)]}{|\cos[2\zeta x + \Psi(x,y)]|}. \tag{9-53}$$

This describes a hologram of completely opaque and transparent lines. The information is preserved in the *position* modulation of the lines.

Phase information holograms. This term is often reserved for holograms in which the amplitude information is never recorded, as opposed to hard-clipped holograms in which it may have originally appeared but is then discarded. Microwave images have been formed in which the amplitude information was neglected, but recent emphasis has been on the recording of phase information holograms using acoustical energy [9-51, 9-52].

Figure 9-26 shows one means of recording a phase information acoustical hologram. Because acoustical detectors are linear detectors, the reference wave can be added after detection. The electronic controls provide a constant amplitude signal to the CRT which is photographed to produce the hologram. The photograph is demagnified before illumination with visible wavelength light. Figure 9-27 shows the image obtained after the conjugate image is removed by spatial filtering. This image compares favorable with others obtained from acoustical holograms. The low resolution is largely due to the longer wavelength of illumination.

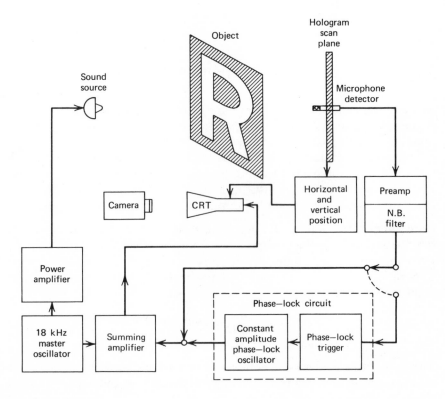

Figure 9-26 Block diagram showing the equipment used to record a phase information acoustical hologram [9-52].

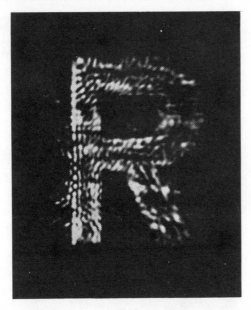

Figure 9-27 The optical image formed when a demagnified phase information acoustical hologram is illuminated [9-52].

Kinoforms. The kinoform is a phase modulator in which the phase information is directly impressed on the read-out wave. The amplitude information is ignored [9-8]. In the generation of a kinoform, the phase distribution of the wave a given distance from an object distribution is calculated using a digital computer. This information is used to produce a bleached photograph which has a phase variation proportional to $\Psi(x,y)$. Because of the possible large range of $\Psi(x,y)$, the calculated phase is reduced modulo 2π. The phase distribution is then plotted on a 32-level gray scale plotter with a possible resolution of 12000×8000. The plot is photoreduced to approximately $12 \times 8\,\text{cm}$ and bleached. The exposure, development, and bleaching processes must be carefully controlled to preserve the phase distribution. That is, light passing through a region of phase $\Psi = 0$ must be retarded one wavelength with respect to light incident on a phase $\Psi = 2\pi$ region. This is known as phase matching.

Figure 9-28 shows the effects of mismatching the phase of a kinoform. These effects were described analytically by Kermish [9-49]. The two effects are the appearance of a bright spot in the center of the image and the production of two additional images on either side of the bright spot. Figure 9-28 shows a two-dimensional image, but three-dimensional images are possible.

Figure 9-28 Photographs of images produced by kinoforms in which (*a*) the exposure time was too low, producing too small a phase modulation constant, (*b*) near optimum conditions for phase matching, and (*c*) overexposure, producing too large a phase modulation constant [9-8]. (Copyright 1969 by International Business Machines Corp.)

The amplitude and phase can be controlled if the kinoform is recorded on color film [9-53].

9.12 Computer-Generated Holograms

A digital computer is often used to calculate the hologram distribution that will produce a desired image. This is sometimes useful in matched filtering, three-dimensional displays of computer data, and testing of optical surfaces against a computed master. In a few cases, a computer-generated reconstruction of an image is useful, such as in studies of the effects of holographic materials [9-54].

In generating the hologram, an exact analogy to the optically recorded hologram can be made. The distribution in the hologram plane is calculated, the reference wave added, the sum squared, and the result plotted on a gray scale plotter. This output is then photoreduced to provide a transparency for visible wave reconstruction. In generating a hologram by computer, however, some additional flexibility is possible. For example, the extent and angular spectrum of the on-axis distribution can be reduced by producing a transparency with transmittance

$$t(x,y) = K + 2V_0 U(x,y) \cos[2\zeta x - \Psi(x,y)] \qquad (9\text{-}54)$$

where K is a constant. This cannot be done with an optically recorded hologram, but is easily produced by computer [9-55]. The kinoform is another example of the additional varieties of techniques possible with a computer.

Two other techniques for generating holograms with a computer are discussed in more detail below. A good survey of techniques and applications is given by Huang [9-56].

Aperture size and location modification. Rather than recording continuous fringes with variations in the transmittance of the hologram, small apertures can be used. The hologram is then either transparent or opaque, and the transmittance varies locally by variations in the size of the aperture. The phase is coded by displacements in the aperture from the regular array. In the earlier work the displacement was determined by the phase at the sample point, but in later work an extrapolation was used to position the aperture according to the phase at the aperture locations rather than according to the phase at the sampling points [9-57]. This removed much of the phase distortion. An example of a computer-generated binary hologram is shown in Fig. 9-29 and an image formed with such a hologram is shown in Fig. 9-30.

In computer generation of a hologram, the transmittance level must be

Figure 9-29   Computer-generated binary hologram with 16,000 apertures [9-57].

digitized and a finite number of samples taken in the hologram plane. Given a fixed number of bits per hologram, a decision must be made how to optimally allocate them between sampling and quantization. It is possible to make the allocation to minimize the mean square errors in the image [9-58]. Three types of images have been treated: a known image, a stochastic image, and a sampled stochastic image. Without knowing the type of image, conclusions concerning the optimum procedure are difficult to make. It does appear, however, that in making a computer-generated Fourier transform hologram of a complex object, the quantization should be coarse and more bits allocated to samples. This is generally true for objects with a wide spatial frequency spectrum.

Multiple aperture representation.   Another method of constructing a hologram by computer is to code the complex-valued sample using four positive real samples [9-59]. That is, any complex number can be written as

$$\mathbf{g} = g_1 - g_3 + ig_2 - ig_4 \qquad (9\text{-}55)$$

where g_1, g_2, g_3, and g_4 are real and positive. Only two of the numbers need be used. This can be implemented in a hologram as shown in Fig. 9-31. When illuminated with a uniform plane wave, an image wave is produced in the $\cos^{-1}\lambda/d$ direction, where d is the total separation of the four apertures. The path delay of $\lambda/4$ between each aperture produces a phase shift of $\pi/2$ between apertures giving the signs of $+1$, $+i$, -1, and $-i$ to

Figure 9-30   Image from a binary hologram showing resolution, gray tones, and noise from a simulated diffuser [9-57].

the wave amplitudes from the four apertures. In the case of a single aperture per sample point, the path difference in the direction of the image is modified by moving the aperture. In the four aperture per sample case, the apertures do not move, but the wave amplitudes at each are modified.

Four apertures are not needed. Three apertures can be used if the phase difference between the waves from the three apertures is $2\pi/3$ [9-60]. This is easily seen by drawing three vectors separated by $2\pi/3$ radians and noting that any complex number can be represented by using two of them. The advantages are that the required memory size and plotter resolution are reduced. The multiple-aperture approach then requires three times more resolution in the direction of the image than a single-aperture technique.

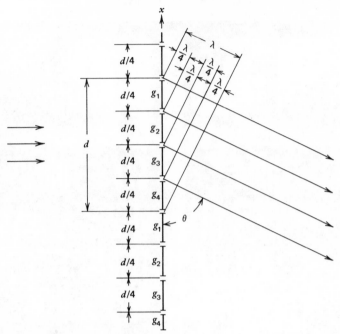

Figure 9-31 The use of four apertures to obtain the proper complex wave amplitude at a sample location [9-56].

9.13 Effects of Sampling and Quantization

Sampling of holograms occurs whenever a discrete number of detectors or scans are used in recording the hologram or when the display is discrete. Examples are microwave, acoustical, or computer-generated holograms. Quantization can occur in such cases as hard clipped holograms already treated, but most often occurs in computer-generated holograms. The effects of quantization are similar to those of nonlinearities of the detector just as the effects of sampling are similar to those due to finite resolution.

In some cases, digital communication theory should be useful in the analysis of the effects of digitizing the information. For example, the technique of varying the aperture width and position in computer-generated holograms is analogous to pulse width modulation and pulse position modulation. The multiple-aperture technique is analogous to pulse amplitude modulation.

Sampling. Many effects of sampling a hologram distribution have been treated and are not discussed here. The possibility of nonuniformly placed samples has been examined and may in some cases prove useful [9-61,

9-62]. The sample locations can be chosen by knowing the most probable location of the information. For example, in a Fourier transform hologram more samples may be needed about the origin. Monte Carlo techniques have also been used in the selection of the sampling points [9-61].

In addition to the replication of images due to sampled hologram distributions, finite-size samples cause spatial frequency distortion of the images [9-63].

Quantization.   The quantization of the signal level of a hologram is a form of nonlinearity that produces results similar to other nonlinearities. False images can occur if the object is discrete, and a reduction in contrast is the result if the object is diffusely illuminated or diffusely reflecting.

An extreme form of quantization is the binary representation of the entire hologram distribution [9-64]. Another common form of quantization is to pick only one level for the amplitude information and quantize the phase finer. This is the approach in the kinoform. Phase quantization has been examined for the case of the Fourier transform of the original distribution [9-65 through 9-67]. When the transform of a distribution is quantized in phase, the inverse transform yields

$$g_q(x) = \sum_{m=-\infty}^{\infty} \mathrm{sinc}\left[m + \left(\frac{1}{N}\right)\right] g_m(x) \qquad (9\text{-}56)$$

where $g_q(x)$ is the retrieved image after quantization of the phase of its transform, N is the number of quantization levels, and $g_m(x)$ is

$$g_m(x) = \int_{-\infty}^{\infty} |G(u)| \exp[i(Nm+1)\Psi(u)]$$

$$\times \exp(i2\pi ux)\, du. \qquad (9\text{-}57)$$

Equation (9-56) shows that the image is replicated, and if the object made up of a single small spot, the phase of $g_m(x)$ causes the extra images to be displaced in angle from the primary image. If, however, the object has a random phase, the higher orders produce overlappings of the image which reduce contrast and cause distortions of the primary image.

PROBLEMS

9-1.   Incorporate the effect of the hologram size into the description of the hologram imaging process and show that the image obtained is the image distribution from an infinitely large hologram convolved with the Fourier transform of the pupil function of the hologram.

9-2. Derive (9-22).

9-3. Incorporate the hologram scale change M_c into the derivation of (9-20) and derive (9-24).

9-4. Perform the necessary differentiation to obtain M_{ang} and M_{long} of (9-25) and (9-26).

9-5. A Fourier transform hologram is to be made using laser illumination at 632.8 nm, the geometry of Fig. 9-3, a lens with a focal length of 10 cm, and the two types of film listed below with their resolution in lines/mm:

> Kodak SO-243: 300 lines/mm
> Agfa 14C75: 1500 lines/mm

Find the radius of the circle centered at the reference point within which images will be produced of object points. Assume that the amplitude of the reference wave is at least twice the amplitude of object wave.

9-6. Repeat problem 9-5 for a quasi Fourier transform hologram with the object and point source reference 10 cm from the hologram.

9-7. Write the convolution in the last term of (9-32) in integral form and modify to show that the expression describes the Fourier transform of an object distribution which is multiplied by a quadratic phase factor.

9-8. Obtain a plot of the variation of the visibility of fringes on a hologram when the point reference is closer to the hologram than the object point. Use an arbitrary recorder MTF.

9-9. In using the Linnik interferometer to record an incoherent hologram, an offset spatial frequency is provided by half-plane stops over the mirrors in the focal planes of the lenses. (See Fig. 9-21.) Consider a point object and show how this procedure produces a phase shift multiplying the point image.

REFERENCES

9-1. N. Abramson, "The Holo-Diagram: A Practical Device for Making and Evaluating Holograms," *Appl. Opt.* **8**, 1235–1240 (1969).

9-2. G. W. Stroke, "Lensless Fourier-Transform Method for Optical Holography," *Appl. Phys. Lett.* **6**, 201–203 (1965).

9-3. J. T. Winthrop and C. R. Worthington, "Fresnel-Transform Representatation of Holograms and Hologram Classification," *J. Opt. Soc. Am.* **56**, 1362–1368 (1966).

9-4. W. T. Cathey, "Local Reference Beam Generation for Holography," U.S. Patent 3,415,587, December 10, 1968, filed December 9, 1965.

9-5. H. J. Caulfield et. al., "Local Reference Beam Generation in Holography," *Proc. IEEE* **55**, 1758 (1967).

9-6. J. P. Waters, "Object Motion Compensation by Speckle Reference Beam Holography," *Appl. Opt.* 11, 630–636 (1972).

9-7. H. J. Caulfield and S. Lu, *The Applications of Holography* (Wiley-Interscience, New York, 1970), Ch. 8.

9-8. L. B. Lesam, P. M. Hirsch, and J. A. Jordan, Jr., "The Kinoform: A New Wavefront Reconstruction Device," *IBM J. Res. Dev.* 13, 150–155 (1969).

9-9. J. Powers, J. Landry, and G. Wade, "Computed Reconstructions from Phase-Only and Amplitude-Only Holograms," in *Acoustical Holography,* Vol. 2, A. F. Metherell, ed. (Plenum Press, New York, 1971).

9-10. R. W. Meier, "Magnification and Third-Order Aberrations in Holography," *J. Opt. Soc. Am.* 55, 987–992 (1965).

9-11. R. W. Meier, "Cardinal Points and the Novel Imaging Properties of a Holographic System," *J. Opt. Soc. Am.* 56, 219–223 (1966).

9-12. R. W. Meier, "Optical Properties of Holographic Images," *J. Opt. Soc. Am.* 57, 895–900 (1967).

9-13. J. A. Armstrong, "Fresnel Holograms: Their Imaging Properties and Aberrations," *IBM J. Res. Dev.* 9, 171–178 (1965).

9-14. E. B. Champagne, "Nonparaxial Imaging, Magnification, and Aberration Properties in Holography," *J. Opt. Soc. Am.* 57, 51–55 (1967).

9-15. J. N. Latta, "Computer Based Analysis of Hologram Imagery and Aberrations," *Appl. Opt.* 10, 599–618 (1971).

9-16. R. F. van Ligten, "Influence of Photographic Film on Wavefront Reconstruction. I. Plane Wavefronts," *J. Opt. Soc. Am.* 56, 1–9 (1966).

9-17. R. F. van Ligten, "Influence of Photographic Film on Wavefront Reconstruction. II. 'Cylindrical' Wavefronts," *J. Opt. Soc. Am.* 56, 1009–1014 (1966).

9-18. J. C. Urbach and R. W. Meier, "Properties and Limitations of Hologram Recording Materials," *Appl. Opt.* 8, 2269–2281 (1969).

9-19. A. Macovski, "Hologram Information Capacity" *J. Opt. Soc. Am.* 60, 21–29 (1970).

9-20. A. Kozma and J. S. Zelenka, "Effect of Film Resolution and Size in Holography," *J. Opt. Soc. Am.* 60, 34–43 (1970).

9-21. J. T. Winthrop, "Structural-Information Storage in Holograms," *IBM J. Res. Dev.* 14, 501–508 (1970).

9-22. G. B. Brandt, "Image Plane Holography," *Appl. Opt.* 8, 1421–1429 (1969).

9-23. B. J. Thompson, "A New Method of Measuring Particle Size by Diffraction Techniques," *Jap. J. Appl. Phys.* 4, Supplement 1, 302–307 (1965).

9-24. J. B. DeVelis, G. B. Parrent, and B. J. Thompson, "Image Reconstruction with Fraunhofer Holograms," *J. Opt. Soc. Am.* 56, 423–427 (1966).

9-25. G. W. Stroke, F. H. Westervelt, and R. G. Zech, "Holographic Synthesis of Computer Generated Holograms," *Proc. IEEE* 55, 109–111 (1967).

9-26. W. T. Cathey, D. Shander, and R. C. Bigelow, "Three-Dimensional Display of a Vector Cardiogram," in *Biomedical Sciences Instrumentation*, Vol. 7, R. J. Gowen, C. E. Tucker, and J. K. Aikawa, eds. (Instrument Society of America, Pittsburgh, 1970), pp. 30–33.

9-27. D. Shander and T. M. Duran, "A New Three-Dimensional Display of the Spatial Vectorcardiogram Employing Holography," *Chest* 59, 438–440 (1971).

9-28. M. Lang, G. Goldman, and P. Graf, "A Contribution to the Comparison of Single Exposure and Multiple Exposure Storage Holograms," *Appl. Opt.* 10, 168–173 (1971).

9-29. H. J. Caulfield, "Biasing for Single-Exposure and Multiple-Exposure Holography," *J. Opt. Soc. Am.* **58**, 1003–1004 (1968).

9-30. N. Nishida and M. Sakaguchi, "Improvement of Nonuniformity of the Reconstructed Beam Intensity from a Multiple-Exposure Hologram," *Appl. Opt.* **10**, 439 (1971).

9-31. L. H. Enloe, W. C. Jakes, Jr., and C. B. Rubinstein, "Hologram Heterodyne Scanners," *Bell Syst. Tech. J.* **47**, 1875–1882 (1968).

9-32. J. C. Palais and I. C. Vella, "Some Aspects of Scanned Reference Beam Holography," *Appl. Opt.* **11**, 481 (1972).

9-33. G. W. Stroke and A. E. Labeyrie, "White-Light Reconstruction of Holographic Images Using the Lippmann-Bragg Diffraction Effect," *Phys. Lett.* **20**, 368–370 (1966).

9-34. J. Upatnieks, J. Marks, and R. J. Fedorowicz, "Color Holograms for White Light Reconstruction," *Appl. Phys. Lett.* **8**, 286–287 (1966).

9-35. D. J. DeBetetto, "White-Light Viewing of Surface Holograms by Simple Dispersion Compensation," *Appl. Phys. Lett.* **9**, 417–418 (1966).

9-36. C. B. Burckhardt, "Display of Holograms in White Light," *Bell Syst. Tech. J.* **45**, 1841–1844 (1966).

9-37. R. J. Collier, C. B. Burckhardt, and L. H. Lin, *Optical Holography* (Academic Press, New York, 1971), Ch. 17.

9-38. E. N. Leith and J. Upatnieks, "Holography with Achromatic-Fringe Systems," *J. Opt. Soc. Am.* **57**, 975–980 (1967).

9-39. R. H. Katyl, "Compensating Optical Systems. Part 1: Broadband Holographic Reconstruction, Part 2: Generation of Holograms with Broadband Light, Part 3: Achromatic Fourier Transformation," *Appl. Opt.* **11**, 1241–1260 (1972).

9-40. E. N. Leith and B. J. Chang, "Space-Invariant Holography with Quasi-Coherent Light," *Appl. Opt.* **12**, 1957-1963 (1973).

9-41. J. Braat and S. Lowenthal, "Short-Exposure Spatially Incoherent Holography with a Plane-Wave Illumination," *J. Opt. Soc. Am.* **63**, 388–390 (1973).

9-42. A. Kozma and N. Massey, "Bias Level Reduction of Incoherent Holograms," *Appl. Opt.* **8**, 393–397 (1969).

9-43. G. Cochran, "New Method of Making Fresnel Transforms with Incoherent Light," *J. Opt. Soc. Am.* **56**, 1513–1517 (1966).

9-44. P. J. Peters, "Incoherent Holograms with Mercury Light Source," *Appl. Phys. Lett.* **8**, 209–210 (1966).

9-45. D. J. DeBitetto, "Holographic Panaramic Stereograms Synthesized from White Light Recordings," *Appl. Opt.* **8**, 1740–1741 (1969).

9-46. C. B. Burckhardt, "Optimum Parameters and Resolution Limitation of Integral Photography," *J. Opt. Soc. Am.* **58**, 71–76 (1968).

9-47. R. L. deMontebello, "Integral Photography," U.S. Patent 3,503,315, March 1970 (filed December 1966).

9-48. R. V. Pole, "3-D Imagery and Holograms of Objects Illuminated in White Light," *Appl. Phys. Lett.* **10**, 20–22 (1967).

9-49. D. Kermisch, "Image Reconstruction from Phase Information Only," *J. Opt. Soc. Am.* **60**, 15–17 (1970).

9-50. A. Kozma and D. L. Kelly, "Spatial Filtering for Detection of Signals Submerged in Noise," *Appl. Opt.* **4**, 387–392 (1965).

9-51. B. A. Sichelstiel, W. M. Waters, and T. A. Wild, "Self-Focusing Array Research Model," *IEEE Trans.* **AP-12**, 150–154 (1964).

9-52. L. Larmore, H. M. A. El-Sum, and A. F. Metherell, "Acoustical Holograms Using Phase Information Only," *Appl. Opt.* **8**, 1533–1536 (1969).

9-53. D. C. Chu, J. R. Fienup, and J. W. Goodman, "Multiemulsion On-Axis Computer Generated Hologram," *Appl. Opt.* **12**, 1386–1388 (1973).

9-54. W. H. Carter and A. A. Dougal, "Field Range and Resolution in Holography," *J. Opt. Soc. Am.* **56**, 1754–1759 (1966).

9-55. J. J. Burch, "A Computer Algorithm for the Synthesis of Spatial Frequency Filters," *Proc. IEEE* **55**, 599–601 (1967).

9-56. T. S. Huang, "Digital Holography," *Proc. IEEE* **59**, 1335–1346 (1971).

9-57. B. R. Brown and A. W. Lohmann, "Computer-Generated Binary Holograms," *IBM J. Res. Dev.* **13**, 160–168 (1969).

9-58. R. A. Gabel and B. Liu, "Minimization of Reconstruction Errors with Computer Generated Binary Holograms," *Appl. Opt.* **9**, 1180–1191 (1970).

9-59. W. H. Lee, "Sampled Fourier Transform Hologram by Computer," *Appl. Opt.* **9**, 639–643 (1970).

9-60. C. B. Burckhardt, "A Simplification of Lee's Method of Generating Holograms by Computer," *Appl. Opt.* **9**, 1949 (1970).

9-61. H. J. Caulfield, "Wavefront Sampling in Holography," *Proc. IEEE* **57**, 2082–2083 (1969).

9-62. H. J. Caulfield, "Correction of Image Distortion Arising from Nonuniform Sampling," *Proc. IEEE* **58**, 819 (1970).

9-63. W. T. Cathey, "The Effect of Finite Sampling in Holography," *Optik* **27**, 317–326 (1968).

9-64. A. R. Sass, "Binary-Intensity Holograms," *J. Opt. Soc. Am.* **61**, 910–915 (1971).

9-65. J. W. Goodman and A. M. Silvestri, "Some Effects of Fourier-Domain Phase Quantization," *IBM J. Res. Dev.* **14**, 478–484 (1970).

9-66. W. J. Dallas, "Phase Quantization—A Compact Derivation," *Appl. Opt.* **10**, 673 (1971).

9-67. W. J. Dallas and A. W. Lohmann, "Phase Quantization in Holograms—Depth Effects," *Appl. Opt.* **11**, 192–194 (1972).

10

Applications of
Holography

Many applications of holography have been discussed and illustrated in the process of examining the different types of holograms, and the division between techniques and applications is not always clear. Nevertheless, an attempt has been made to place in this final chapter some of the more important applications of holography. They are not ordered by importance, but more or less by ease of presentation and predominance of use.

Displays are treated first and then the related topic of data storage. The recording of a hologram of moving objects is seen to be possible with a pulsed laser. Both continuous wave and pulsed lasers have been used in various types of interferometry. Several examples of the applications of holographic interferometry are given. Another useful application is holographic microscopy where a greater depth of field is possible than with conventional microscopy. Microwave and acoustical holograms are examples of recording the hologram at a long wavelength and illuminating it at a shorter wavelength. Many of these applications are used in biomedical fields. The last three topics are somewhat related though varying in their feasibility: imaging through random media, compensation for atmospheric effects, and bandwidth reduction.

10.1 Displays

The use of holograms as displays is limited primarily by the imagination of the user. This does not mean that one is not limited by the state of the art, but that the uses possible within that limit have not been exhausted. Many of the displays, such as commercial ones, are "useful," but the more interesting of them have been made by artists who are not constrained by

314

cost effectiveness or the type of recording.

Some of the types of holograms used in displays are treated in the sections that follow. In general, the read-out wave need not be provided by a laser. The resolution requirements are lax enough that a filtered mercury source is adequate. One advantage of the mercury source is that there is no speckle in the image. Another is that there is no worry about laser·safety when the display is accessible to the public.

Cylindrical holograms. This is an interesting format for a display because the image can be viewed over a 360° angle [10-1]. One way of making such a hologram is shown in Fig. 10-1. A portion of the light from the microscope objective illuminates the cylinder of film directly, providing the reference wave. Another portion falls on the object, providing the object wave. Illumination of the resulting hologram with the same source produces an image that can be viewed from any position about the hologram. Photographs of the image produced by such a hologram are shown in Fig. 10-2.

White-light, image plane holograms. This type of hologram is used quite often in displays. The principles of the white-light image plane hologram are given in Chapter 9 and are not repeated here. The wavelength selection done by the reflection hologram allows illumination with a white-light source, and the fact that the image is in the plane of the hologram allows the formation of an acceptable image even though the spread of wavelengths is not narrow and the illuminating beam is not collimated. The hologram can be illuminated by a flashlight, spotlight, or sunlight to give an image suitable for display purposes.

Computer generated holographic displays. Any computer-generated hologram can, of course, be used for display purposes, but one possible application is its use to display computer data. For example, the perspective display of graphs and shapes could be made with holograms [10-2].

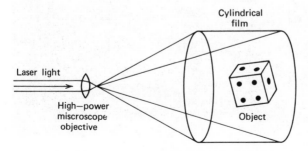

Figure 10-1 Configuration for making a 360° hologram [10-1].

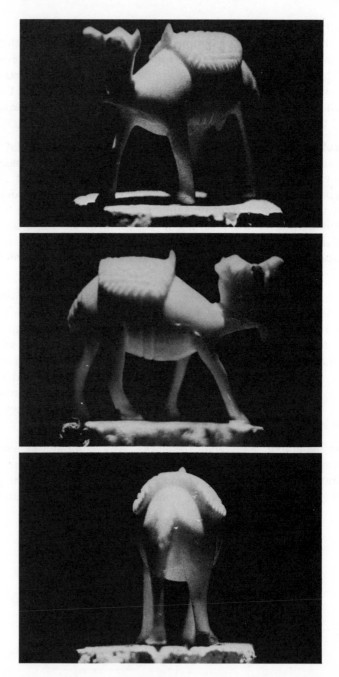

Figure 10-2 Photographs of various views of an image produced with a cylindrical hologram. [Courtesy T. H. Jeong, Lake Forest College.]

316

The different views can be calculated and used in making a hologram as with the holographic panoramic stereograms described in Chapter 9. From that point, all storage is in the hologram and the computer memory is free to be used elsewhere. Or the display can be stored, not calculated each time, thereby conserving machine time.

Holographic motion pictures.   Holographic motion pictures and television are two tantalizing topics. Television requires bandwidth reduction to be practical and is treated in the last section. The primary difficulty with movies is obtaining a large-screen image. If the image is viewable only through or in line with a hologram, the viewing angle is restricted considerably. A practical projection scheme is needed. Another problem area is that of obtaining a good color image.

The motion picture film can be framed as with conventional techniques to produce sequentially recorded holograms [10-3]. These are then viewed with a system sequentially illuminating each hologram. A continuous drive can be used if the holograms are narrow vertically and full width horizontally [10-4]. The vertical parallax is eliminated, but the result is a smoothly changing image as the holograms are pulled through the illumination. Another means of allowing continuous hologram motion without image jitter is to use Fourier transform or quasi Fourier transform holograms.

Copying holograms.   There are many methods of copying a hologram. One of the simplest is illustrated in Fig. 10-3. The illuminating wave passes through the film on which the copy is to be made and illuminates the original reflecting hologram. The result is that the copy film is illuminated on one side by a uniform reference wave and on the other by the image wave of the original hologram. In this case, the copy is not a duplicate of the original hologram but is capable of producing the same image. This copying technique is not restricted to reflection holograms but is appropriate whenever the emulsion is thick compared to the fringe spacing on the original hologram causing only one image to be formed. If this is not so, twin images are obtained [10-5].

In the case of phase holograms, an embossing technique has been successfully used. RCA's HoloTape is made by using a nickel master to emboss onto plastic a strip of phase holograms. This provides high-quality, inexpensive copies.

10.2   Information Storage and Retrieval

The primary difference between the requirements for data storage and displays is image resolution. The resolution of the display need be no more than the human eye can measure, and even that degree of resolution is not

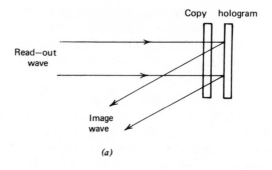

(a)

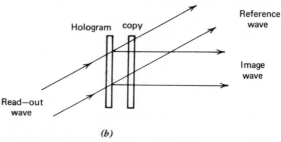

(b)

Figure 10-3 Method of making a copy of (a) reflection hologram, and (b) transmission hologram where only one reconstructed wave falls on the copy.

needed. In the case of information storage, however, high-resolution, high-quality images are necessary. It is desirable to have the highest possible data density at the lowest cost, but maintaining noise resistance and an error free output. The Fourier transform hologram, or variations of it, provides the best recording format. Means are then sought to reduce the speckle associated with the images. In the implementation, consideration must be given to the type of storage needed—read only, read-write—and to the time needed to write and read.

Data density. The large data-storing capacity of holograms was recognized early in the development of holography. First emphasis was given to storage of many holograms in a volume recorder [10-6]. The holograms are separated by the Bragg effect as described in Chapter 6. More work has been done using an array of small holograms that are addressed by a movable read-out beam. Figure 10-4a shows a 32×32 array of holograms having 1.2-mm diameters [10-7]. Each hologram can form a real image, such as shown in Fig. 10-4b. This image falls onto an array of photodetectors to read out the presence or absence of a bright spot. One such detector

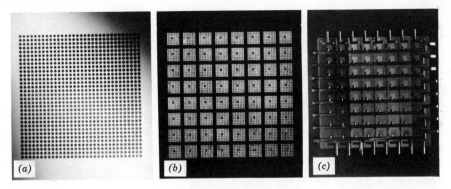

Figure 10-4 (*a*) A 32×32 array of holograms; (*b*) image typical of that available from any one hologram; (*c*) array of phototransistors with each element of the array containing 64 phototransistors corresponding to the possible locations of data spots [10-7].

array is shown in Fig. 10-4*c*. Systems of this type can be made with an array of 10^4 holograms each producing patterns having 10^4 bits. Consequently, the array of holograms can store 10^8 bits of information.

The addressing of the holograms is done with an acoustooptic deflector capable of random switching within 0.5 μsec. This is done, of course, using a page-organized memory. Problems of cross-talk, storage density, and optical efficiency of holographic memories have been treated by Hill [10-8]. The design relationships needed to achieve maximum capacity of a block oriented random access memory were worked out by Vander Lugt [10-9].

Considerable work is being done on a page composer, or the means of getting the information into the memory. Some possibilities are the use of Pockel's effect devices or band edge absorption in Cds [10-10, 10-11].

Operating mode. The more practical uses of holographic memory use read-only mode of operation. The hologram is recorded on a photographic emulsion or embossed plastic and is not changed. Some techniques, such as the thermoplastic holograms described in Chapter 6, can be used to make read-write memories. The cycle time, however, is very long compared to cycle times of semiconductor memories. Consequently, holographic memories for conventional computer operations must await better materials. Holographic memory can be practical for cases where large amounts of permanent data must be stored, such as table look-up, or where infrequent changes are needed, such as customer credit data or payroll information.

There are at least two other types of data storage that are read-only. One is the storage of television programs on a plastic strip of holograms [10-12], and the other is the recording of printed material by holographic ultrafiche [10-13]. The storage of 400 pages/1 in.^{2} has been demonstrated.

Types of holograms used. Most types of holograms provide resistance to loss of data by damage to the storage plate. Dust, scratches, and imperfections in the recording medium have little effect on the holographic image, whereas a normal microfilm image would suffer. To avoid the problems of alignment, a Fourier transform, quasi Fourier transform, or near Fourier transform hologram is used. A near Fourier transform hologram is one in which the object is in the front focal plane of a lens but the recording is made after the back focal plane at some distance d. A uniform plane wave is used as a reference. All of these forms of holograms have the property that translation of the hologram does not cause the image to move. The positioning requirements on the hologram or the read-out beam are thus relaxed considerably.

The shift theorem shows that a lateral change in the position of a Fourier transform hologram produces only a phase shift in the image. The discussion of quasi Fourier transform holograms in Chapter 9 indicates that if they are illuminated in the same fashion, the same effect occurs. A similar property can be shown to be possessed by a near Fourier transform hologram [10-12].

Diffusers for Fourier transform holograms. One reason for wanting to record a hologram in the Fourier transform plane of the data is that the minimum space-spatial frequency bandwidth is required. A disadvantage is that the bright zero-frequency region of the transform saturates the recording medium. To reduce this effect, a diffuser can be used with the data mask (object) to spread the energy in the transform plane. If a simple ground glass diffuser is placed behind a data mask, the resulting image may have speckles across a single data spot, increasing the probability of error. There should be no speckle, hence no random phase variations across a data point. The high energy concentration about the zero-frequency region of the hologram can be avoided by providing uniform phase across a data point but random phase between data points. One approach is to use a phase mask that produces 180° phase shift to half of the data spots chosen at random [10-14]. The phase over a single data spot is uniform. Rather than selecting 0 or 180° phase shift, four levels can be used, which reduces the rms intensity fluctuations on the hologram and in the image by $\sqrt{2}$ [10-15]. It has been shown that four levels of phase quantization is nearly as good as a continuous random variation of phase between data spots [10-16]. Deterministic diffusers can be designed to give a nearly uniform amplitude distribution in the transform plane [10-17].

Redundancy. If the information to be stored is not in the form of binary data, phase masks are not as practical, although one could visualize systems in which the size of the data spot was related to the desired image resolution. An alternative approach is to pass the object wave through a grating, so that in the Fourier transform plane an array of holograms is produced. If the hologram is recorded out of the transform plane, the problem of saturation of the zero-frequency region is eliminated. Figure 10-5 shows an embossed plastic tape having approximately 20 sub-holograms for each composite hologram Even though the subholograms are displaced laterally, the images fall in the same place because the object was placed in the front focal plane of the transforming lens when the near Fourier transform holograms were recorded. The redundancy reduces the sensitivity to dust and scratches [10-18, 10-19]. Figure 10-6 illustrates the reduction in noise when nine subholograms are used.

Figure 10-5 Photograph of an embossed holographic tape. The illumination is with white light so that the subholograms are easier to see [10-12].

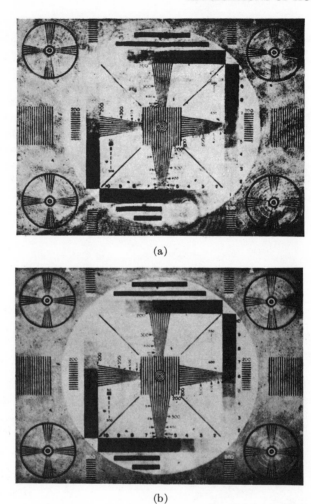

(a)

(b)

Figure 10-6 Comparison of images when (a) one 2×2 mm hologram is used and (b) when nine holograms are used [10-12].

10.3 Pulsed Laser Holograms

The short exposure time associated with high-power pulsed lasers makes possible the recording of holograms of a variety of objects that could not be considered otherwise. The short coherence length associated with most pulsed lasers make special matching techniques necessary [10-20, 10-21], but lasers with special mode control can now produce coherence lengths of several meters. The use of Q switches and mode controls provides coherent pulses of durations shorter than 40 nsec. Consequently, stability of the object ceases to be a problem, and holograms of rapidly moving objects can be recorded. Photographic emulsions become less sensitive at short exposure times, however, and the total exposure must be increased. The total exposure for the Agfa emulsions must be increased by a factor of 2 to 4 [10-22]. The exposure of Kodak 649-F must be increased by a factor of 10 to 100 [10-23]. In addition, 649-F is a factor of 10 less sensitive at the ruby wavelength, 694.3 nm that at the HeNe wavelength, 632.8 nm. The sensitivity of 649-F to short exposures can be increased by increasing the development time to 8 mins.

Pulsed lasers. A ruby laser is normally used in pulsed laser holography because it supplies high-energy pulses in the visible region. One arrangement for using the laser is shown in Fig. 10-7. A mode-controlled and Q-switched ruby oscillator is used to provide the reference wave and to drive the ruby amplifier [10-24]. The illumination must be through a diffuser if live objects are used. The extended diffuse source reduces the energy density on the retina below the level at which eye damage can occur. Care must be taken to ensure that no specular reflections from the reference wave can be seen in the region of the object. Figure 10-8 is an example of an image of a moving object.

Typical uses. In addition to holograms of people and bursting light bulbs, pulsed laser holograms have been used to study the distribution of particle sizes in a moving medium, and in holographic interferometry. Measurements of fuel mixtures and fuel droplet sizes have been made in both rocket and diesel engines. The uses of pulsed laser in vibration analysis and interferometry are discussed in the next section.

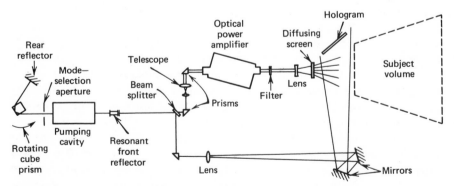

Figure 10-7 One setup for pulsed laser holography. The oscillator consists of a rear reflector, a rotating Q switch, mode-selecting aperture, ruby pump, and a front reflector. The amplifier consists of a beam expanding telescope and ruby pump [10-24].

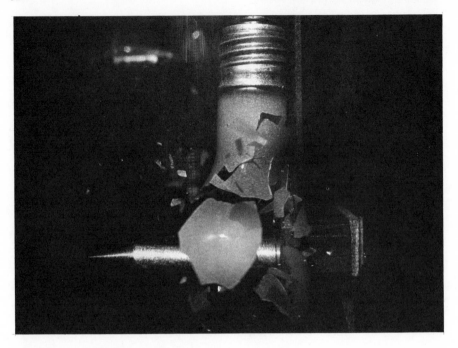

Figure 10-8   Holographic image of light bulb bursting [10-24].

10.4   Holographic Interferometry

One unique feature of holographic interferometry is that the wave that serves as a standard of comparison can be stored and used at a later time. In classical interferometers, such as the Mach-Zender, the wave in one arm is the standard to which the wave in the other is compared, and both must be present at the same time. Another feature is the ability to compare waves that are reflected from a diffusely reflecting object at two different times. Classical interferometers can be used only with transparent objects or specular reflectors. These two properties of holographic interferometry make it possible to record the wave from an object at one time, apply stress to the object, and, by comparing the stored wave with the new wave from the object, determine the strain induced.

As with classical interferometry, the information desired must be extracted from the patterns of fringes that form when the two waves interfere. The fringes can be either *live fringes*, which means that the wave from the object interferes with the stored wave in real time, or *frozen fringes*, which arise when two holograms are recorded on the same plate—one with the object without stress and the other with stress.

Other types of interferometry can also be done holographically. In addition, fringes can be formed that show contours of constant distance from some reference point or contours with vibration patterns.

These topics are briefly treated here. For more detail, the reader is referred to more specialized treatments [10-25, 10-26, 10-27] or to the references cited later.

Experimental arrangement. A simple configuration that can be used is illustrated in Fig. 10-9. This is essentially the same as is used in making holograms for any purpose. Care must be taken that no motion of the apparatus occurs between exposures if a double-exposure, frozen-fringe hologram is made. Or, if a live fringe study is being conducted, the hologram must either be developed in place or replaced in exactly the same position as recorded. If these precautions are not taken, fringes occur because of the shift between the two hologram positions. These fringes may be misinterpreted as being due to motions of the object.

In using the system of Fig. 10-9 to view live fringes, the reference wave from the mirror is used to illuminate the hologram. This causes a virtual image of the object to appear in the same location as the object. The wave producing the virtual image can be denoted as wave number one in the figure. At the same time, the object is illuminated and the wave from the object, wave 2, passes through the hologram to be viewed. If these waves differ, interference fringes occur that are related to the object (or hologram) motion after the hologram was recorded. If the hologram was double exposed, both waves 1 and 2 represent recorded virtual images.

The frozen-fringe method is most often used with rapidly changing events, and a pulsed laser is used to record the hologram. If the spacing between two pulses of a laser is variable, changes in the object can be observed for different time intervals. A classical example of frozen-fringe

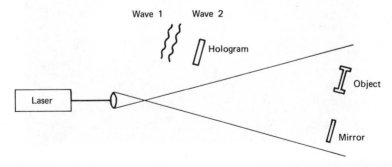

Figure 10-9 Simple holographic interferometer for observing changes in an object.

holographic interferometry is shown in Fig. 10-10. The first pulse of the laser caused the recording of a wave through space with no disturbance. The second pulse was triggered by the passage of the bullet. The backlighting provided allows not only a silhouette of the bullet to be seen, but also the effects due to changes in the density of the medium through which the bullet is passing. The shock waves can be seen in three dimensions.

Another advantage holographic interferometry has over conventional interferometry is that the optics need not be near perfect. For example, the windows of a wind tunnel need not be optically flat to a fraction of a wavelength because *both* waves go through the same window and any

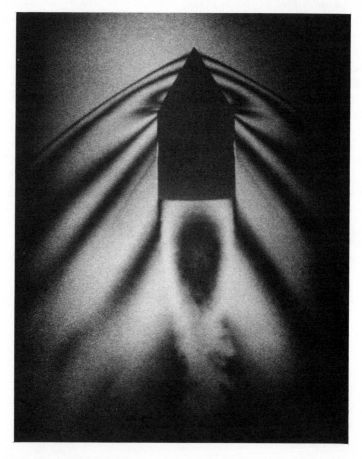

Figure 10-10 Photograph of the images obtained in a doubly exposed hologram of a bullet. The first exposure, before the bullet arrived, provided the comparison wave [10-20].

aberrations affect both waves in the same manner. In the case of a Mach-Zender interferometer, however, the two waves go along separate optical paths. The easiest way to make separate paths equal optically is to have very good (and very expensive) optical components.

If the changes to be investigated are slow, or if real-time observation is desired, live-fringe interferometry may be preferable. Figure 10-11 is a photograph of fringes produced when the object is stressed [10-28].

Quantitative analysis. The fringes seen in Fig. 10-11 indicate where strain occurs but it is sometimes not easy to determine the type of motion associated with it. For example, object translation, rotation, bending, or twisting can all cause fringes to appear [10-29]. One of the parameters of interest is where the fringes appear in space. They may or may not be in the same plane as the object, but a *surface of localization* where the fringes have maximum contrast can be found. The fringes are localized when a change in viewing angle does not cause a change in the fringe position. This occurs if the viewer focuses on the surface of localization. A shift in the position of the fringe is caused by a change in the phase difference between the two interfering waves. Consequently, the surface of localization can be found by writing an expression for the phase difference, differentiating with respect to the viewing angle, and setting the result equal to zero. This gives the conditions for no change in phase difference, hence no fringe motion.

Let us consider the fringes produced by different types of motion and then examine the means of determining the type of motion from the fringe patterns. The approach followed is similar to that used by Sollid [10-30] and Aleksandrov and Bonch-Bruevich [10-31]. Figure 10-12 shows the paths of the light illuminating and scattered from a point on a surface displaced only by translation. If we assume plane wave illumination, the difference in phase δ_p between the two paths is

$$\delta_p = k[\Delta x(\cos\theta_i + \cos\theta_s) - \Delta y(\sin\theta_i + \sin\theta_s)], \qquad (10\text{-}1)$$

where Δx and Δy are the incremental movements along the x and y axes and θ_i and θ_s are measured from the average of the surface profile. Differentiation of δ_p with respect to θ_s shows that the condition for no change in δ_p is

$$\theta_s = \tan^{-1}\left(-\frac{\Delta y}{\Delta x}\right). \qquad (10\text{-}2)$$

The exact value of θ_s is unimportant. The important aspect is that it does not depend on the location of the point on the object. That is, there exists

Figure 10-11 Photograph of fringes produced when a bar is stressed [10-28].

328

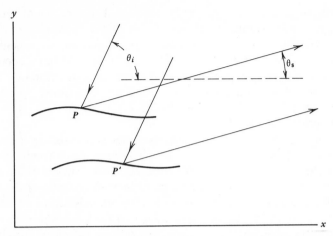

Figure 10-12 The angles of illumination θ_i and scattering θ_s from a point P displaced by translation.

an angle at which *all* points contribute energy at the same phase. At another angle, all points will contribute energy with some different phase relation. Consequently, to see the fringes the angular distribution must be converted into a spatial one with a lens. Alternatively, the viewer could be at infinity. The fringes in this case are parallel and localized at infinity if the illumination is with a plane wave. Note that (10-1) relates the fringe period not only to the displacement of the object, but also to the angles of illumination and viewing.

If the object encounters only rotation about an axis, in the surface, a similar analysis shows that the fringes are localized at

$$d = -\frac{x \cos\theta_s \sin^2\theta_s}{\sin\theta_i + \sin\theta_s} \tag{10-3}$$

where d is measured from the surface of the object [10-25, 10-31]. From this we see that if the viewing angle is normal to the surface, then $\theta_s = \pi/2, d = 0$, and the fringes are normalized on the surface. In general, the surface of localization of the fringes does not correspond to the surface of the object, but it will intersect the object at $x = 0$. The phase difference between the two paths to the scattering point is

$$\delta_p \approx -kx\varphi[\sin\theta_i + \sin\theta_s] \tag{10-4}$$

where φ is the angle through which the object is rotated. Consequently, if it is known that rotation is the only motion, the illumination and viewing angles can be measured, and the angle of rotation determined from the fringe period.

Combinations of the two rigid body motions discussed above produce fringes that are quite similar to those of one of the motions alone. This presents a problem in determining what type of motion is present. The problem becomes much more complicated when deformations occur. In-plane surface strain can be measured using techniques described by Ennos [10-32]. Motions of a point on an object can be determined by multiple observations on a single hologram [10-25, 10-30, 10-31] or by making multiple holograms [10-30, 10-32].

If a single hologram is used, the viewing device is focused on a point on the object and the viewing angle changed. The number of fringes passing the point as the angle is changed are counted. Three such measurements give sufficient data to determine the magnitude of the motion of the point. A fourth measurement is needed to determine the direction of motion. Two problems arise in this technique. One is that the hologram may be too small to permit movement far enough to count a sufficient number of fringes. Another difficulty is that if the fringes are not localized near the object, both the object and the fringes cannot be in focus simultaneously. The size of the viewing aperture can be reduced to increase the depth of field, but if taken too far, the size of the speckle approaches the size of the fringes, and the fringes become difficult to count.

The multiple hologram approach is similar, but the fringes cannot be counted in going from one viewing point to another. This technique depends on finding a common reference such as zero-order fringes. Zero-order fringes are those that remain fixed relative to the object surface even when the viewer moves. If an object point can be found that has zero-order fringes, the fringes between that point and the point of interest can be counted to determine the *order* of the fringe at the point of interest. This is done at one viewing angle through one hologram. The same reference point with zero-order fringes is then found looking through a second hologram. The order of the point of interest can now be found at this new viewing angle. The difference between the orders is the number of fringes that would have been counted if the two holograms had been part of one very large hologram and the fringes counted as in the single hologram approach. The same number of measurements must be made in either case, but in this technique, an object point having zero-order fringes must be found.

Qualitative analysis. In many cases, a precise measurement of object motion is not needed. For example, if, in stressing a material, fringes appear clustered near a region, it is clear that the region has undergone greater strain than the rest of the material. Interruption of the fringes can indicate a crack in the object. This is one way in which holographic interferometry can be useful in materials testing.

Holographic test pieces. In constructing optical components such as lenses and mirrors, a standard *test piece* is used to check the accuracy of the surface of the new component. In making the check, the surface to be tested and the test surface are brought into contact so that an interferometric check can be made. This results in eventual damage to the expensive test piece. A noncontacting test can be conducted by making a hologram of the surface of the test piece [10-33]. The hologram is repositioned by moving it until the fringes, caused by interference between the light from the virtual image and light from the surface of the test piece, vanish. The test piece is then removed and the object to be tested put in its place. The differences between the holographically formed image of the test piece and the object are shown in the interference fringes formed. This technique works only with objects having smooth surfaces. If the surfaces are diffuse reflectors, the surface detail of each piece is different, even if the overall shape is the same. No observable fringes are formed using two different diffuse surfaces.

The construction of a test plate in the first place can be very expensive, particularly if the surface is not spherical. The test plate need not be made, however, if a computer-generated hologram is used. The methods of deviated phase and multiple apertures discussed in Chapter 9 have both been used in making holograms to check aspheric optical elements [10-34].

Vibration analysis. One of the early applications of holographic interferometry was that of vibration analysis. Here the object can be vibrating during the exposure of the hologram. If the exposure is long compared to the period of vibration, a hologram is recorded showing fringes across the object. This occurs because the object pauses at the ends of its path of vibration long enough for a hologram to be recorded of the two extreme positions. Consequently, fringes are seen that show the contours of the object position. The light falling on the hologram while the object is in motion contributes to the background exposure of the hologram. An analysis of this type of interferometry was given by Powell and Stetson [10-35].

Similar results are obtained if a live-fringe holographic system is used and the object then set in vibration. The observer does the time averaging which eliminates all the fringes except those produced at the two ends of the vibration path. In this case, the reference position is the rest position of the object. One advantage is that the fringes can be watched as they change with the modes of vibration. Figure 10-13 shows three modes of vibration of a diaphragm as detected with this method.

Both of the techniques can be modified by using stroboscopic illumination. If it is used during the recording of the hologram, the hologram may be exposed only by light from the object in the extreme positions. This

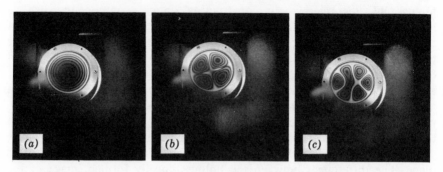

Figure 10-13 Three modes of vibration seen with holographic interferometry. [Courtesy W. G. Gottenberg, Mechanical Engineering Department, University of Colorado.]

produces less background exposure and fringes of higher visibility. Other types of modulation of the reference or object wave can be helpful in detecting small-amplitude vibrations or determining the phase of the vibrations [10-36].

Contour generation. Fringes can be formed on the image of an object that are contours of constant range or depth. There are many ways in which contours can be formed [10-37 through 10-41]. Two holograms can be recorded on the same plate. The object can be illuminated with two wavelengths simultaneously; illuminated with only one wavelength but from two directions; or with one wavelength but with a medium of different index surrounding the object. Alternatively, the hologram can be recorded with one wavelength, replaced, and it and the object illuminated with a second wavelength. The interference between the wave from the image (magnified because of the shift in wavelength) and the wave from the object causes contours to appear on the image. Similar results are obtained if the hologram is recorded while the object is in a medium of one refractive index, and the hologram and the object then illuminated while the object is in another medium.

Let us first consider the technique where the objects are illuminated with point sources in different locations. The phase of the wave from the object is a function of the phase of the illuminating wave. If a double-exposure hologram is made with the illuminating source moved between exposures, the resulting image irradiance has sinusoidal variations with a period dependent on the movement of the source. For the fringes to be contours of constant height, the point sources must be at infinity. That is, the object illumination should be with plane waves. Also, the line of sight should be *perpendicular* to the bisector of the angle between the directions of propagation of the two illuminating waves [10-38]. This means that the hologram must be made looking at the side of the object (assuming the *front* is toward the illumination). The result is that some regions of the object, as seen from the hologram, are not illuminated. The distance between contours is then

$$\sigma = \frac{\lambda}{2\sin(\Delta\theta_i/2)} \qquad (10\text{-}5)$$

where $\Delta\theta_i$ is the angle between the propagation vectors of the two illumination waves.

When one wavelength is used to illuminate the object and a second used in the read-out of the hologram, the image is displaced from the object because of the grating effect of the hologram. To compensate, the angle of the read-out wave can be modified so that the image is superimposed on the object. Lateral magnification can be eliminated by using plane reference and read-out waves. The longitudinal magnification is a function of wavelength, however, so fringes are formed when the image and object waves interfere. If the illumination and viewing angles are the same, the period of the fringes is [10-38]

$$\sigma = \frac{\lambda_1 \lambda_2}{2(\lambda_1 - \lambda_2)} \qquad (10\text{-}6)$$

Similar results are obtained if the hologram is simply double exposed, once for each wavelength. Or, both wavelengths can be used simultaneously. The contours can then be thought of as arising because of the periodicity in the amplitude of the sum of the signals from the source. As the path length difference between the object and reference waves increases, the amplitude, hence the visibility of the interference pattern recorded on the hologram, varies sinusoidally. That is, those points that cause path differences such that the amplitude is zero are not recorded on the hologram. Because these points all lie at multiples of the same distance from the illuminating source, dark contour bands appear on the image where the corresponding waves are not reconstructed. This is illustrated in Fig. 10-14, which is a photograph of a holographic image recorded using a two-frequency pulsed ruby laser [10-42]. The period of the contour fringes is

$$\sigma = \frac{\lambda_1 \lambda_2}{2(\lambda_1 - \lambda_2)\cos[(\theta_i - \theta_s)/2]} \qquad (10\text{-}7)$$

If the index of refraction of the medium surrounding the object is changed between exposures of a hologram, the optical distance (measured

Figure 10-14 Photograph of a holographic image recorded using a two-frequency pulsed ruby laser. The 23-mm contour spacing is equal to the optical thickness of the resonant reflector in the ruby laser [10-42].

in wavelengths) to different portions of the object changes. This produces a phase shift similar to that obtained by changing the wavelength of the illumination [10-39, 10-41].

Desensitized hologram interferometry. One problem with holographic interferometry is that it is too sensitive for some applications. Deformations or motions of the object on the order of a few wavelengths are sufficient to produce fringes. If large motions are involved, the fringes become too closely spaced to be easily seen or interpreted. The use of a longer wavelength will give larger spacing if suitable sources and detectors are available. In some cases, a desensitization of the measurement using visible wavelengths is possible.

The desensitization technique uses a hologram giving contours on the image, then stressing or moving the object before making a second hologram to show the new contours [10-43]. The moiré pattern that appears is the same as a holographic interferogram made with the equivalent wavelength

$$\lambda_{eq} = \frac{\lambda_1 \lambda_2}{\lambda_1 - \lambda_2}. \tag{10-8}$$

Speckle interferometry. Much information can be found from *photographs* of objects if coherent illumination is used and the speckle in the image is used to advantage [10-44–10-46]. One technique involves double exposing a photograph and Fourier transforming the result to get quantitative measurements. The other is to note the visibility of the speckle of a vibrating object.

When a double exposed photograph is made of an object that has been displaced in the plane of the object, the speckle pattern of the image is simply displaced proportionally to the object displacement. The result is a random array of exposed regions on the transparency coupled with the same array displaced a small distance. When this distribution is Fourier transformed optically, fringes appear which are proportional to the separation of the exposed regions due to speckle (see Problem 2-24). Because the speckle motion is related to the object motion, the object motion is determined by the fringe spacing in the transform plane.

It is also possible to measure the tilts of an object [10-47, 10-48]. If the tilt is small, the tilt of the wave in the object plane produces only a shift in position in the Fourier transform plane. Hence if the Fourier transform is recorded before and after the tilting by double exposing an emulsion, two identical, speckly Fourier transforms are recorded with a small separation between them. Illumination of this recording and Fourier transformation gives fringes proportional to the angle of tilt of the object between the two exposures.

Vibrations out of the plane of the object can be detected by adding an in-line reference wave to interfere with the wave from the diffusely reflecting object [10-44]. When the object moves, the relative phase between the speckles of the image and the reference wave changes, causing flickering of the speckles. The nodes of vibration have nonflickering speckles and clearly stand out when a time exposure is made of the image-reference combination. The clarity of the display can be increased by recording the image with a video detector, using a temporal frequency filter to remove the varying signal due to the changing speckle, and displaying the image on a television receiver [10-49]. It is also possible to detect the amplitude of the vibration and to measure very small motions with variations of this technique [10-50, 10-51].

10.5 Microscopy

There are two aspects of holographic microscopy. One is to use holograms to achieve magnification, and the other is to use the hologram to record a wave either before or after magnification is provided with a conventional microscope. Or, magnification can occur both before and after the hologram is recorded.

Holographic Magnification. Gabor's goal was to record a hologram with electron waves and illuminate it with visible light, circumventing some of the problems with electron lenses and obtaining magnification with the change in wavelengths. Even though this was not accomplished, the early work in magnification with holograms stressed the magnification possible in the holographic process. Some very good images having high magnifications (X 120) were obtained [10-52]. However, if the recording and reconstruction of the wave is to be done in the visible region, there are few reasons not to use a conventional microscope. Holograms have the same aberrations as lenses.

Holographic microscopy. It is not clear to which processes the term holographic microscopy should refer. It has been applied to almost any procedure involving a hologram and a microscope. In some applications, it is convenient to record a hologram and later magnify the real or virtual image. In others, the magnification comes first. In either case, the attraction of the hologram is the large depth of field available. In using a conventional microscope, a series of photographs could be taken with the microscope focused at different planes. If the object is in motion, however, the state of the object may have changed before the series of photographs is completed. If a hologram is made, the investigation of the various planes of the object can be made at leisure.

One case in which it is convenient to form a hologram and then examine the real image with a microscope is in studying the growth of gas bubbles in a liquid under pressure [10-53]. The system is to image the capillaries in a hamster's cheek pouch over a 3-mm field. A resolution better than 3 μm was needed to see the 7 μm red blood cells and to see the gas bubbles grow. Because of the flow rate, 10-μsec recordings were made at 20 frames/sec. A microscope stage and collecting lens were inside the pressure chamber and a variable speed 16-mm camera was outside. The object illumination beam entered one chamber window, then passed through the object, collecting lens, and second window. The reference beam bypassed the chamber in a modified Mach-Zender arrangement. The resolution achieved was better than 200 lines/mm.

One-micrometer resolution has been obtained in an image of neurons by magnifying the image before recording the hologram [10-54]. The arrangement used is similar to that of Fig. 10-15. The 1-μm resolution was available throughout a scanning depth of more than 40 μm. That is, objects could be brought into focus in one plane, then by moving the eyepiece, a plane 40 μm deeper in the object could be focused with as much resolution. Similar experiments [10-55] have been conducted using the modified interference microscope shown in Fig. 10-16.

10.6 Microwave Holography

Microwave holography can be traced back to work in optical processing of radar data. The process can be explained in terms of recording a hologram with microwaves and illuminating it with visible waves. Many microwave processes such as synthetic aperture radar and frequency-modulated pulse compression can be described using a holographic viewpoint. This can lead to new insights and new methods of signal processing [10-56]. Later work used microwave energy to both record and illuminate the hologram. The reconstruction of the wave can be done by either illuminating a hologram or properly controlling an array of radiators. When visible waves are used to illuminate the hologram, attention must be given to the image-distorting effects of the large change in wavelength. This problem is treated last.

Synthetic aperture antenna processing. The resolution of an antenna is proportional to its size in wavelengths. In some cases—on an airplane, for

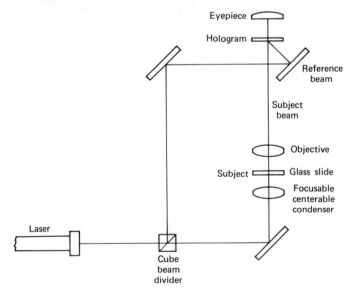

Figure 10-15 Schematic of a holographic microscope [10-55].

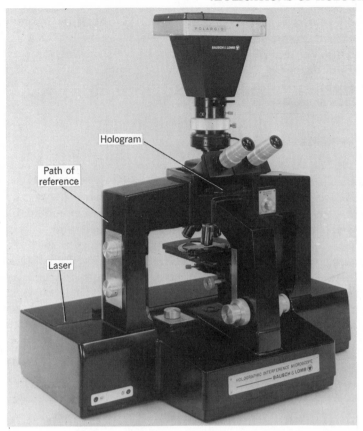

Figure 10-16 Photograph of a holographic microscope. [Courtesy K. Snow; Bausch and Lomb.]

example—a large antenna cannot be used. The high resolution needed for mapping the ground can be obtained with a side-looking radar. In this case, the radiation is directed down and to the side of the aircraft as shown in Fig. 10-17. Resolution in the r_0-direction can be obtained by measuring the time of arrival of short pulses or by other range-measurement techniques [10-57]. The means of obtaining resolution in the x direction is best described in terms of holography [10-58].

As the antenna is moved, a record of the wave from the object is made by photographing the face of a cathode ray tube. This is equivalent to having many antennas along the path of flight. If a coherent reference signal is provided, both the amplitude and the phase of the received signal can be preserved. The signal from a point object as shown in Fig. 10-18 is recorded with the reference to produce a one-dimensional zone plate.

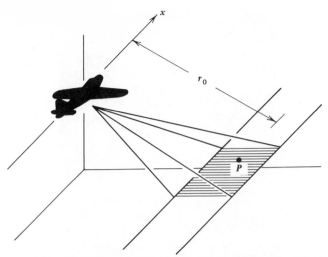

Figure 10-17 Geometry of the synthetic antenna radar system [10-58].

Some of these zone plates can be seen in Fig. 10-19, which is a recording of the raw data from a side-looking radar.

The recording of Fig. 10-19 is a hologram in only the azimuth direction. In the range dimension, it is an image. Consequently, the two axes must be treated differently. Figure 10-20 shows the optical system used to image the record in the range dimension and the virtual image in the azimuth direction. The conical lens is necessary because the azimuth images would otherwise fall on a tilted plane. Alternatively, a tilted plane can be used in recording the image [10-59]. The focal positions of the azimuth images depend on the range of the object. We can see from Fig. 10-18 that the focus of the one-dimensional zone plate produced by a point is determined by the range to the point object. The output of the imaging system is shown in Fig. 10-21.

Figure 10-18 Recording positions of the antenna and the spherical wave from a point target *P*.

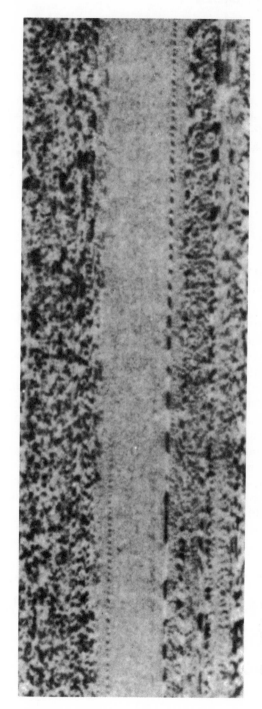

Figure 10-19 Raw data from a side-looking radar. The vertical dimension is range and the horizontal is the azimuth dimension [10-58].

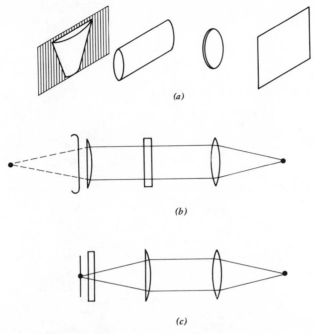

(a)

(b)

(c)

Figure 10-20 (a) The lens system used to bring the range and azimuth focal planes into coincidence. (b) Top view showing the conical and spherical lenses reimaging the virtual image of the signal record in the azimuth dimension. (c) Side view showing the cylindrical and spherical lenses reimaging the signal record in the range dimension [10-58].

This explanation was not used in the original work, but the close analogy between a coherent radar system and a coherent optical system was recognized [10-60].

Let us now look at the analysis of this type of imaging. Let the transmitted radar signal be $\cos \omega_r t$. The return signal is then $a \cos[\omega_r(t - 2r/c)]$ where r is the range from the aircraft to the target, c is the velocity of light, and a is dependent on target reflectivity and distance. If we use the usual approximation for r of

$$r \approx r_0 + \tfrac{1}{2}\left[\frac{(x - x_0)^2}{r_0} \right], \qquad (10\text{-}9)$$

we obtain

$$s(t) = a \cos\left[\omega_r t - \frac{4\pi r_0}{\lambda_r} - \frac{2\pi (x - x_0)^2}{\lambda_r r_0} \right] \qquad (10\text{-}10)$$

Figure 10-21 Output of the imaging system [10-56].

where $\lambda_r = 2\pi c/\omega_r$ is the radar wavelength, x is the position of the aircraft along the flight path, and x_0 is the target position in the direction of the flight path. This signal is mixed with an oscillator to shift the frequency down to an intermediate frequency. The signal is then sent to a cathode ray tube (CRT), for example, which exposes a strip of moving film. The CRT scan is in the range direction, and the film moves in the azimuth direction. The recorded signal can be represented by

$$s(x) = s_b + a\cos\left[\omega_x x - \frac{2\pi(x - x_0/w)^2 w^2}{\lambda_r r_0}\right] \qquad (10\text{-}11)$$

where s_b is a film bias term, ω_x hides the relation between the radar frequency, the intermediate frequency, and the velocity of the strip of film, and w is a scaling factor between the aircraft coordinate and the film coordinate. The symbol x now stands for the coordinate along the strip of film. The constant phase term $4\pi r_0/\lambda_r$ was dropped.

If this recording is a transparency illuminated by a light wave $\cos\omega_L t$, we have

$$s(x)\cos\omega_L t = s_b \cos\omega_L t$$

$$+ \tfrac{1}{2}a\cos\left[\omega_L t + \omega_x x + \frac{2\pi}{\lambda_r}\frac{(x - x_0/w)^2 w^2}{r_0}\right]$$

$$+ \tfrac{1}{2}a\cos\left[\omega_L t - \omega_x x - \frac{2\pi}{\lambda_r}\frac{(x - x_0/w)^2 w^2}{r_0}\right]. \qquad (10\text{-}12)$$

The third term, which can be written as

$$\tfrac{1}{2}a\cos\left[\omega_L t - \omega_x x - \frac{\pi}{\lambda_L}\frac{(x - x_0/w)^2}{[(\lambda_r/2w^2\lambda_L)r_0]}\right], \qquad (10\text{-}13)$$

represents a cylindrical wave from a line source at at a distance

$$d = \left(\frac{1}{2w^2}\right)\left(\frac{\lambda_r}{\lambda_L}\right)r_0 \qquad (10\text{-}14)$$

from the signal record. The second term produces a real image of the same source. That is, the signal record produces real and virtual images at that distance. Equation (10-14) shows the variation in focus of the image with range which must be corrected by the conical lens or the tilted plane.

Another interesting feature of this type of radar is that a broad illuminating beam, consistent with a small antenna, is necessary for high resolution. This is in contrast to the usual situation where a narrow beam is necessary to illuminate and image only a small region at a time. The broad beam of the side-looking system is better because the length of the synthetic antenna that can be generated is the distance the antenna can travel while illuminating a target. This distance is the width of the beam. If all elements of the synthetic array are not equidistant to the target, the resolution is limited to the change in range unless additional processing is provided [10-61].

Adaptive arrays. If a microwave hologram is considered to be a device that records and reconstructs the same microwave, then an array of antennas can fit the description. An array of 64 apertures was used in 1963 to demonstrate that the array could act as a retrodirective array to focus back on a point or to form a microwave image of a reflector in the form of an R [10-62]. That is, a real image was formed of either a point object or of an R. The resolution was quite good for an 8×8 array. Only the phase was controlled. The amplitude from each radiating element was the same. Figure 10-22 shows the microwave image of the R.

Microwave recording and microwave reconstruction. If an array of radiators is not used to detect and directly reconstruct the waves, conventional holographic techniques can be used with minor modifications. A reference wave can be introduced in space as with visible wave holograms so that the distribution on the recorder is the square of the sum of the amplitude of the reference and object waves. The recorder could still use an array of detectors or a scanning detector, but other techniques are available. The dependence of the developing speed of Polaroid film on temperature can be employed to record a microwave field. The field distribution heats a uniformly preexposed Polaroid packet and records a hologram because of the variations in developing speed. This can then be used to make, for example, a photoetched copy on a copper-coated circuit board. The result is an off-axis or Leith and Upatnieks hologram which can be illuminated with microwaves to form a microwave image [10-63].

Microwave recording and visible wave illumination. If a visible wave reconstruction of the wave is desired, the microwave recording must be converted into a transparency for visible waves. One of the most common techniques is also the oldest [10-64]. The output from a scanning detector intensity modulates a CRT, while the beam is deflected to track the position of the microwave detector. A reduced-scale transparency is made from a photograph of the CRT display. This results in a rather small hologram, but, if a reference signal is used so that the reconstructed

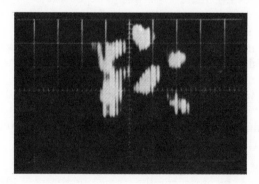

Figure 10-22   Microwave image of R obtained with an adaptive array [10-62].

primary and conjugate waves can be separated, the fringes must be closely spaced with respect to the read-out wavelength.

The Polaroid technique can be used to record microwave holograms to be photoreduced. The use of a liquid crystal is another temperature-sensitive method of detecting the microwave energy. The change in color of the sheet of liquid crystals is photographed through a color filter to yield the transparency to be photoreduced [10-65]. The problem with the temperature-sensitive methods is their insensitivity. For example, the output of the microwave source should be measured in watts, even for a relatively small object. A scanning heterodyne detector can be orders of magnitude more sensitive, but it has lower resolution or requires a lot of scan time. Photochromic materials have also been used [10-66].

An example of the quality of images available in microwave holography is given in Fig. 10-23. The illumination was from a 4.3-mm source and the wave reflected from the object was heterodyne-detected with a single detector spirally scanned over a 75-cm diameter circle [10-67]. The amplitude information was discarded in making this hologram.

One application of microwave holography is to study the source itself. By recording a hologram of the pattern from a radiating aperture, a visible image of the field in the plane of the aperture, in the Fresnel region, and in the far field of the radiation, can be obtained [10-68].

Microwave holographic interferometry has been demonstrated, but the speckle in the image of a diffusely reflecting object is very large [10-69]. This is one disadvantage of using a longer wavelength. The technique may prove useful, however, in observing deformations of large structures [10-70].

Distortions due to wavelength change. It is shown in Chapter 9 that the magnification in general depends on the ratio of wavelengths used. The equations governing the lateral and longitudinal magnification are

$$M_{\text{lat}} = \frac{M_c}{-\dfrac{\lambda_1}{\lambda_2}M_c^2\dfrac{1}{\rho_W} \pm \dfrac{d_1}{\rho_V} \pm 1} \qquad (10\text{-}15)$$

and

$$M_{\text{long}} = -\frac{\lambda_1}{\lambda_2}M_{\text{lat}}^2, \qquad (10\text{-}16)$$

where M_c is the magnification of the hologram produced by copying, ρ_V is the radius of curvature of the reference wave, and ρ_W is the radius of curvature of the read-out wave.

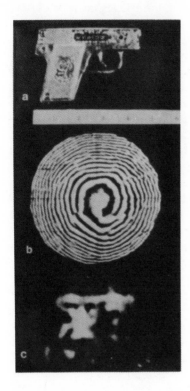

Figure 10-23 (a) Toy pistol object; (b) hologram; (c) image of object inside a handbag [10-67].

We can see from (10-16) that only when

$$M_{\text{lat}} = \frac{\lambda_2}{\lambda_1} \qquad (10\text{-}17)$$

is $M_{\text{long}} = M_{\text{lat}}$. If the longitudinal maginification is not equal to the lateral magnification, the image is distorted. The requirement is often simplified by assuming

$$\rho_V = \rho_W = \infty, \qquad (10\text{-}18)$$

which reduces the condition for no distortion to

$$M_c = \frac{\lambda_2}{\lambda_1}. \qquad (10\text{-}19)$$

This means that the microwave hologram must be demagnified by the ratio of the visible to microwave wavelengths.

In many cases this type of distortion of the image is not important. One common case is the imaging of a two-dimensional object. Even when the distortion is important, the demagnification of the hologram by a factor of 10^3 to 10^4 leaves a very small hologram indeed. The resulting image is also quite small. See (10-17).

Because the side-looking radar processor images directly in one dimension (range) and uses a form of microwave hologram in the other, the effects of these distortions must be corrected [10-59].

Several schemes for ameliorating the distortion have been proposed, but perhaps the most practical for displays (and, by hindsight, the most obvious) is not to worry about the image as formed, but as perceived by the viewer. It can be shown that by using two prisms, one in front of each eye, the apparent distance, hence the apparent depth-width ratio, can be changed. By proper choice of the wedge of the prisms, much of the perceived distortions can be eliminated [10-71].

10.7 Acoustical Holography

There has been much more work done in acoustical holography than in microwave holography. Part of the reason is interest in ultrasonic diagnostics in medicine and in ultrasonic testing of materials. Acoustical reconstruction of waves is quite rare, however. In almost all cases, the recording is with acoustical waves, but the reconstruction uses visible light. We consider the detectors, the techniques used, and some applications. For more information, the reader should refer to the review articles and the series of symposium proceedings [10-72 through 10-76].

Liquid surface system. One of the easiest ways to form an acoustical hologram is to let the object and reference waves interfere to produce ripples on the surface of a liquid [10-72, 10-77]. The ripples are then illuminated with a laser as shown in Fig. 10-24. The surface deformations produce a phase modulation hologram, and the point reference in the plane of the object causes it to be of the quasi Fourier transform type. The greatest advantage of this technique is that the output is available in real time.

Alternative reference systems. Acoustical holograms can have a reference signal in the form of a local oscillator. If the frequencies of the reference and object waves are not the same, an intermediate frequency signal results that has the amplitude and phase information of the original object wave. To get a linear relation between the input wave and the output signal then

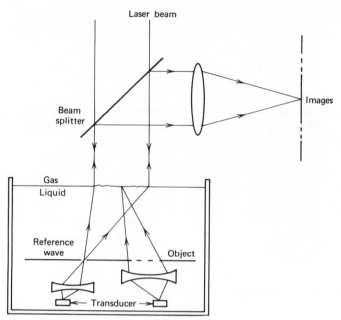

Figure 10-24 A liquid surface holography system generating a quasi Fourier transform hologram at the liquid surface [10-72].

requires a heterodyne operation. *Linear* acoustical detectors are available, however, that make the addition of an electronic reference signal possible. Figure 10-25a shows the conventional image of a bolt and Fig. 10-25b is the image made from an ultrasonic hologram [10-78]. The bolt is 1.27 cm in diameter and has five threads to the centimeter. The 5-MHz ultrasonic wave had a 0.3-mm wavelength in water. The hologram was reduced one-tenth and illuminated with a 632.8-nm wave to obtain the image of Fig. 10-25b. Range gating was used to eliminate reflections from other structures.

Many techniques used in the microwave region are applicable in acoustical holography as well. A scanning microphone, for example, has been used to modulate the brightness of a scanning light that was photographed to form the hologram [10-79]. Similarly, liquid crystals, photochromics, and Polaroid film have been used in recording acoustical holograms. Acoustic cameras, designed for conventional acoustical imaging, can also be used to record holograms [10-80].

Materials that record the acoustical pattern directly, such as liquid crystals, are rather insensitive, requiring high energy levels (approximately 1 Watt/cm^2). Electronic systems, on the other hand, can detect signals as low in power as 10^{-11} W/cm^2.

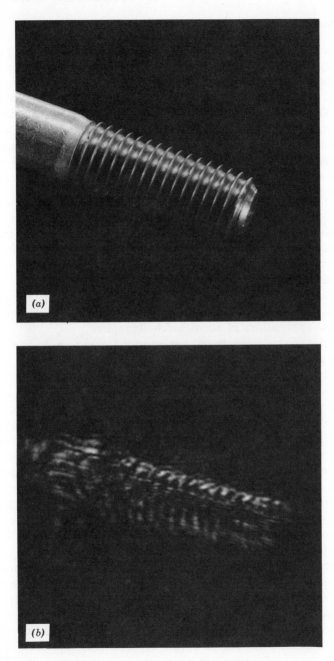

Figure 10-25 (*a*) Conventional photograph of a 1.27 cm-diameter bolt. (*b*) image from an ultrasonic hologram [10-78].

10.8 Biomedical Applications

In this section we apply to biology and medicine some of the techniques previously discussed. They can be generally divided into two groups: information gathering and information display. Some cases, such as conventional holography and acoustical holography, can be placed in both groups. Others, such as x-ray holography and holographic microscopy, pertain more to information gathering. Spatial filtering and image enhancement relate primarily to information display. Holograms can be used to display information gathered by conventional x-ray and electronic data-gathering techniques. A summary of other applications is given in reference [10-81].

Information gathering. One of the earlier suggestions for applying holography to medicine was to use it to obtain an image of the eye [10-82]. The primary advantage is the large depth of field possible. Much of the work in holographic microscopy has used biological samples because of the usefulness of the large depth of field.

Acoustical holography can be used to form images of internal organs that are transparent to x-rays. This is probably one of the more interesting biomedical applications of acoustical holography.

X-ray holograms have been demonstrated using an in-line hologram recorded with 0.114-nm x-rays [10-83]. The greatest limitation is the lack of sufficiently coherent x-rays to produce an off-axis reference wave.

Another approach is to use what is equivalent to an incoherent x-ray or γ-ray hologram. In conventional γ-ray imaging, the γ-ray source must be small to get high-resolution images because the image is essentially a γ-ray shadow of the object. As the size of the source is reduced, however, the available radiation is reduced, and the exposure time increased. A large source can be used if it is used in conjunction with a spatial modulator that produces a coded image to be recorded. One such modulator is a zone plate mask placed between the object and the γ-ray source [10-84]. Alternatively, the source itself can be coded as in the case where the x-ray emitting material of an anode was made in the shape of an off-axis zone plate [10-85].

The imaging characteristics of both systems are similar and are explained in terms of point objects. As shown in Fig. 10-26, the source is imaged through each point in the object onto the film. Illumination of the film recording with spatially coherent light produces point images because of the focusing effect of the zone plates. The use of an off-axis zone plate pattern causes the other foci to appear off-axis where they can be blocked by the iris. A general object can be imaged by covering it with a fine mesh to break the object into points.

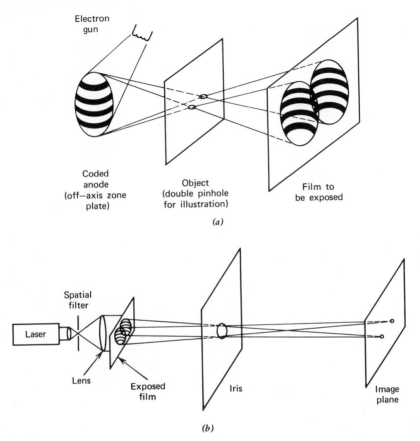

Figure 10-26 (*a*) The x-ray recording system; (*b*) the visible wave reconstruction system [10-85].

The similarity of this and incoherent holography can be seen in the building up of the background exposure and the reduction in image S/N ratio as the number of incoherently recorded zone plates is increased. Nevertheless, a larger, more intense source can be used if the source aperture is coded, giving 2 orders of magnitude decrease in the exposure time [10-84].

Information display. Optical processing techniques can be used to perform image enhancement or to produce special displays. Many of the uses in processing medical data are shown in reference [10-86]. Integral photography and holography have been used to produce holographic displays of x-ray images [10-87].

Electronically recorded electrocardiograms can be made along three orthogonal coordinates to produce vectorcardiograms that show the locus of the vector representing the electrical potential across the heart. Graphs of the head of the vector are normally made to show the three views of the path that forms a loop. Stereo techniques can be used to obtain a three-dimensional view, but they are limited in field of view and versatility. A holographic display can be made by multiply exposing a hologram with the end of an illuminated optical fiber serving as a point object. Repeated exposures are made, one for each recorded x-y-z value of the electrical vector. The resulting image gives a three-dimensional view of the locus of the top of the vector [10-88, 10-89].

10.9   Imaging Through Random Phase Media

The preservation of phase information is the property of holography that allows the retrieval of image information in spite of random phase variations of a wave caused by a nonuniform medium. There are two basic approaches in compensating for random phase effects. One is to pass the conjugate wave back through the same medium to form a real image of the object in the same location as the object [10-90, 10-91]. The other is to cause the object and reference waves to pass through the same phase-distorting medium so that the effect of the medium is canceled when the square law recording takes place [10-92]. In addition, speckle inter-ferometry can be done through a random medium. Both approaches and applications are discussed.

Aberration compensation.   The process of imaging back through a random medium is a form of aberration compensation. The general procedure is illustrated in Fig. 10-27. The object wave propagates through the phase-distorting medium to the hologram. Let the wave immediately to the left of the distorting medium be represented by $U''(x,y)e^{i\Psi''(x,y)}$. Assume that the medium modifies only the phase of the wave and is relatively thin. The wave just to the right of the distorting medium can then be described by

$$U'(x,y)e^{i\Psi'(x,y)} = U''(x,y)e^{i\Psi''(x,y)}e^{i\Psi_d(x,y)} \qquad (10\text{-}20)$$

where $\Psi_d(x,y)$ is the phase shift caused by the distorting medium. The object wave at the hologram is given by $U(x,y)e^{i\Psi(x,y)}$ where the functions U and Ψ could be calculated from knowledge of U'', Ψ'', and Ψ_d using the wave equation. This is not necessary, however.

If, after recording and processing, the hologram is replaced and il-luminated by a plane wave antiparallel to the original reference wave, the conjugate image wave travels toward the object. For this process to work,

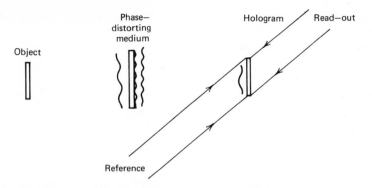

Figure 10-27 Holographic compensation for aberrating medium.

the read-out wave must be the conjugate of the reference wave, and this is easiest to arrange if the two waves are plane. The wave that arrives just to the right of the distorting medium can be described by $U'(x,y)e^{-i\Psi'(x,y)}$. The conjugation reflects the shift in the direction of propagation. The shapes of the two waves are identical. When the waves travel to the left through the medium, the phase of the wave just to the left of the medium is given by

$$U'(x,y)e^{-i\Psi'(x,y)}e^{i\Psi_d(x,y)} = U''(x,y)e^{-i\Psi''(x,y)}e^{-i\Psi_d(x,y)}e^{i\Psi_d(x,y)}$$

$$= U''(x,y)e^{-i\Psi''(x,y)}, \tag{10-21}$$

which continues to propagate to the left, forming a real image in the object plane.

The description above assumes that there is perfect alignment and no noise introduced. If the read-out wave is not the conjugate of the reference wave, if the hologram is too small to give adequate resolution, or if the hologram or the phase distorting medium is moved, the noise in the image rapidly increases. Any of the above mentioned factors causes the phase distortion not to be canceled completely.

The compensation process works just as well for any number of distorting media or for a distorter extending from the object to the hologram. This can be seen by repeating the argument above for propagation from one thin distorting medium to the next.

The first demonstrations of the cancellation of phase aberrations used random phase distorting media such as shower glass or a diffuser. An application of aberration compensation is to consider an imperfect imaging element such as a lens in the place of the distorting medium [10-93, 10-94]. If the object used in the recording is a point, the hologram-lens combination can produce an image system with a sharper impulse response than the low-quality lens alone.

Compensation for the effects of a random medium has been demonstrated where the illuminating wave and the object wave both go through the medium [10-95]. This is the case when one attempts to image through a random medium such as the atmosphere. The possibility of compensating for atmospheric phase fluctuations in real time has been demonstrated [10-96]. The hologram was really a two-element array operating at 10.6 μm, but the feasibility of short wavelength adaptive arrays was shown.

Seeing through random media. In many cases an image of the object in the region of the object is worthless; what is needed is an image of the object on the opposite side of the phase-distorting medium. This can be provided by causing the reference and object waves to pass through the same phase distortions. Consider the object and reference points as placed in Fig. 10-28. The requirement that the object and reference waves pass through the same portions of the phase-distorting medium is adequately satisfied if

$$\frac{\delta}{z_0} < \frac{\Delta_d}{z_d} \tag{10-22}$$

where δ is the separation of the reference and object points, z_0 is the distance from the two points to the hologram, z_d is the distance from the phase distorting medium to the hologram, and Δ_d is the average linear dimension over which the phase of the wavefront emerging from the medium is constant to within $\lambda/8$.

From the expression (10-22) we see that if the phase distorting medium is close to the hologram, or if the distortions are small, the object can be farther from the reference and the two waves encounter the same phase distortions. If so, the distribution on the hologram is described by

$$\left| U(x,y) e^{i\Psi(x,y)} e^{i\Psi_d(x,y)} + V(x,y) e^{i\Phi(x,y)} e^{i\Psi_d(x,y)} \right|$$

$$= U^2(x,y) + V^2(x,y) + U(x,y)V(x,y)e^{i\Psi(x,y)}e^{-i\Phi(x,y)}$$

$$+ U(x,y)V(x,y)e^{-i\Psi(x,y)}e^{i\Phi(x,y)}. \tag{10-23}$$

This is exactly the same as that which would be recorded if the distorting medium were absent. We can see, however, that a more accurate description would require $\Psi_d(x-a, y-b)$ in one wave. If the variations in phase

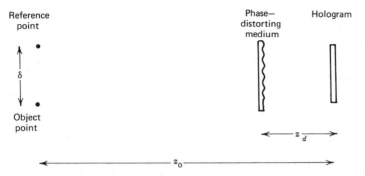

Figure 10-28 Holographic image recording through a random medium.

across the medium are rapid enough, $\Psi_d(x, y) \neq \Psi_d(x - a, y - b)$ and the cancellation does not occur. A more detailed discussion of the restrictions in terms of distortion-caused deflections of the rays from the object has been made with essentially the same results [10-97].

If the distorting medium is very near the object, either a holographic or a conventional imaging system will produce an acceptable image. This can be illustrated by placing a distorting medium, such as a piece of vellum paper, on a photograph. The object can be seen as long as the medium is close to the object, but as they are separated, the image is degraded. One way of expressing the effect is by the use of an effective aperture diameter:

$$d_{\text{eff}} \approx \frac{z_0}{z_0 - z_d} \Delta_d. \tag{10-24}$$

The variable d_{eff} is, in essence, the projection of Δ_d onto the entrance pupil of the imaging system [10-92]. The effective aperture is limited by the distortion because it limits the size of the imaged point; a distorted wave cannot be focused to give a sharp point.

Even though an image can be obtained when the distortion is near either the object or the hologram, acceptable resolution may not be possible if a strong phase distorter is midway between them.

The impetus for much of this work is the desire to form an image through a turbulent medium. Both short-and long-range experiments have been conducted that demonstrate the compensation for atmospheric turbulence [10-98 through 10-102]. The experimental results showed some improvement using holographic imaging, but not so much as was hoped. One of the difficulties is that the turbulence causes large variations in the wave intensity. The dynamic range of the recorder (photographic film in the experiments) is then exceeded. Consequently, some regions of the hologram are heavily overexposed, some are underexposed, and only a relatively small portion are optimally exposed. It has been estimated that this effect reduces the S/N of the image by a factor of 10^3 compared with what could be obtained with a uniform reference wave [10-99]. In one set of experiments, the dynamic range of the film was increased by using the low-contrast developer POTA. This gave better resolution, but the S/N was still poor.

In recording holograms through turbulent media, a corner reflector is often used to provide the point reference. The speckle reference technique can also be used [10-103]. In this case, a portion of the illuminating beam is focused to a spot on the object, providing an approximation to a point reference. The turbulence distorts the spot, reducing the resolution of the final image, but distortions in the image are reduced. Figure 10-29*a* shows a spoke pattern photographed through turbulent water. The bright spot to the side is the speckle reference for the hologram. Notice the blurred regions and the regions where the spokes are bent. Figure 10-29*b* shows a photograph of the image from a hologram recorded at the same time. The resolution is improved and the distortions of the pattern reduced considerably. The blotches in the image are due to nonuniform illumination caused by the turbulence in the illumination path and cannot be removed.

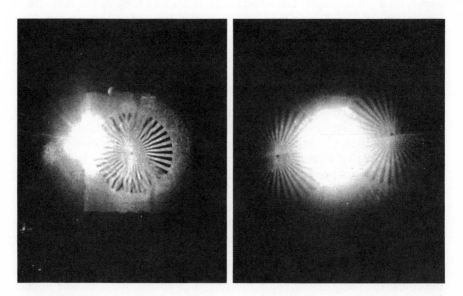

Figure 10-29 (*a*) Photograph of a spoke pattern and reference point through turbulent water. (*b*) Photograph of the image from a speckle reference hologram through the same medium at the same time. [Courtesy Ball Bros. Research Corp.]

A form of speckle interferometry can be used to measure stellar diameters and separation of double stars. The resolution limit is then determined by the aperture diameter, not the atmospheric effects [10-104, 10-105]. A speckle pattern is produced in the image because of the wave front perturbations due to the atmosphere. If the object is a double star, the speckle patterns are identical but displaced. The fringes formed in the Fourier transform plane then give the separation of the two stars. Alternatively, one star can be thought of as the reference point for the other star in a holographic recording.

10.10   Bandwidth Reduction and Three-Dimensional Television

The idea of three-dimensional television has intrigued many people. Aside from the problem of a suitable material for a real-time hologram, the bandwidth needed to transmit all the information on a hologram is much too large. If all of the information on a 4×5 in., high-resolution plate is to be transmitted for a typical scene, more than 10^{11} samples/sec must be transmitted to give 30 frames/sec [10-106]. This is compared with approximately 8×10^6 samples/sec for present television systems. Various techniques for reducing the required bandwidth are presented and possible combinations are discussed.

Spatial carrier frequency elimination.   The spatial carrier, necessary for separation of the primary and conjugate waves from the on-axis distribution, requires a higher resolution in the detector than would be needed otherwise. The on-axis distribution is twice as wide as the object distribution and overlap must be avoided. Hence to transmit the on-axis distribution, the primary wave, and the secondary wave, at least four times the bandwidth is needed. A heterodyne detecting system with a scanned reference beam can be used to reduce this requirement [10-107, 10-108].

Assume that the frequency of the reference wave is different from that of the object wave so that the intensity distribution on the hologram is

$$I(x,y,t) = U^2(x,y) + V^2(x,y)$$
$$+ 2U(x,y)V(x,y) \cos[(\omega_1 - \omega_2)t + 2\zeta x + \Psi(x,y) - \Phi(x,y)].$$

$$(10\text{-}25)$$

If the reference wave is a scanned beam of very small dimension, it can be represented by

$$V(x,y) = \delta(x - \mathbf{s}_x t)\delta(y - \mathbf{s}_y t) \qquad (10\text{-}26)$$

where s_x and s_y are the scan velocities in the x and y directions. Because the reference beam is small, a single, large detector can be used. Use of (10-26) in (10-25) and integration over the detector area yields the time-varying component of the current from the detector as

$$\mathbf{i}(t) = 2U(\mathbf{s}_x t, \mathbf{s}_y t) \cos[(\omega_1 - \omega_2 + 2\zeta \mathbf{s}_x)t + \Psi(\mathbf{s}_x t, \mathbf{s}_y t) - \Phi(\mathbf{s}_x t, \mathbf{s}_y t)]$$

$$(10\text{-}27)$$

where the constant related to the detector sensitivity has been dropped. We now have the information concerning the object wave alone, on an intermediate frequency carrier, and have traded the resolution require-ments on an array of detectors or a scanning detector for a small reference beam. The transmitted information is then used to form a hologram at the receiver [10-108].

Of course, the scanning reference need not be used if only a frequency offset is desired. A reduction in required resolution of conventional detec-tors is possible. Three in-line holograms of the same scene can be recorded with the phase of the reference set at 0, 120, and 240° for the first, second, and third holograms, respectively. It has been shown that an off-axis hologram can be synthesized using these three recordings [10-109]. Al-ternatively, an in-line system can be used with temporal frequency offset on the reference wave [10-110].

Parallax reduction.   A hologram can be thought of as the equivalent of a number of photographs of an object, taken from many viewing angles. This concept helps in appreciating some of the means of reducing the information, hence the bandwidth required to transmit the information stored on a hologram. Vertical parallax is due to the different views possible when one raises or lowers his head when viewing a hologram. The three-dimensional effect is produced primarily by seeing two views with each of the horizontally set eyes. Consequently, vertical parallax could be dispensed with and its absence noticed only by those who prefer to view television while reclining on their sides and devotees of the evening dress competition in the Miss America contests.

Two different approaches have been used in removing much of the vertical information. One is to record only a thin horizontal strip of the hologram and then replicate it at the receiver [10-111]. The other is to record a hologram in the horizontal direction and an image in the vertical direction [10-112].

A 1-mm high hologram strip replicated 100 times in the vertical direc-tion always gives the same view as the viewer moves up or down, but it preserves the horizontal parallax. The choice of 1 mm was made because smaller strips cause diffraction effects and larger ones produce periodic horizontal bars large enough to be easily seen. The bars appear because each strip hologram produces an image that extends beyond its width. Consequently, the strips must be modified to prevent duplication of the images. Figure 10-30 shows the problem encountered. Let the object consist of a regular spacing of points 0, 1, 2,... All of the points can be seen, depending on the angle at which the strip hologram is viewed. If the strip is replicated and the copies placed contiguously, the image points are duplicated as shown. For example, the points 2, 3, and 4 are imaged through the top strip and again by the second strip. Point 4 imaged by one strip overlaps point 2 imaged by the next, and points 3 and 4 show up an extra time. This can be removed by reducing the size of the replicated hologram strips or by removing alternate strips. Either procedure produces a striped hologram, but if the strips are small, it is not noticeable. The other effect is a stretching of the image in the vertical direction. This effect can be partially corrected by illuminating the hologram with cylindrical wave, but then points 0 through 4, for example, are too close. Some distortion is unavoidable, but is not always noticeable.

In recording a one-dimensional hologram to remove vertical parallax, a cylindrical lens is used to form an image in the vertical direction. The hologram is formed in the horizontal direction. By selecting the cylindrical lens to have a small aperture in the vertical direction, the depth of focus of the imaging system can be made large enough to record an image of an

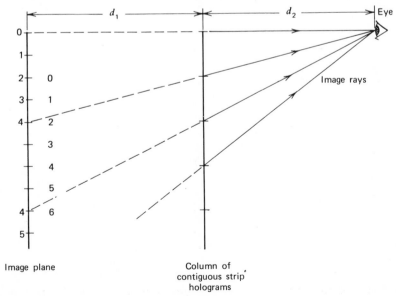

Figure 10-30 A column of strip holograms showing the duplicate images produced [10-111].

object with extended depth. The width of the lens in the nonimaging direction determines the field of view of the hologram in the horizontal coordinate. If the hologram is viewed through a cylindrical lens, a realistic virtual image is seen with parallax in the horizontal plane [10-112].

Hologram coding.   The high resolution possible with holographic imaging is not needed for many display and viewing applications. Consequently, the hologram size could be reduced. However, this has the undesirable effect of reducing the field of view. One technique that allows the field of view to be retained and conserve bandwidth by sacrificing resolution is illustrated in Fig. 10-31. It relies on the ability of a hologram to image through a distorting or diffusing medium. The wave from the diffusely reflecting object illuminates the entire diffusing plate, and a portion of the wave from every section of the diffusing plate is assumed to fall on the hologram. The reconstructed wave then goes through every portion of the diffuser to form a real image of the object. The field of view is determined by the extent of the diffuser [10-113, 10-114]. The same technique can be used to produce large-screen holographic movies even though the hologram is small [10-115]. Notice that in this illustration only a real image can be formed.

The reduction in the space-spatial frequency bandwidth product (SSFB) is given by

$$\frac{D_0}{d'_1} \frac{d''_1}{D_h} \qquad\qquad (10\text{-}28)$$

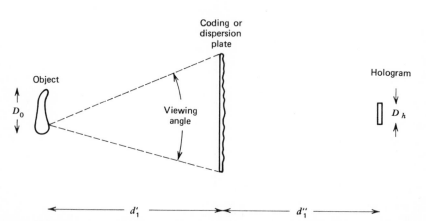

Figure 10-31 Imaging through a coding or dispersion plate to reduce the space-spatial frequency bandwidth product but preserve the viewing angle [10-114].

where the symbols are shown on Fig. 10-31. A decrease in hologram size decreases the space needed and an increase in distance to the hologram can reduce the resolution required. From the geometry of Fig. 10-31, we can see that converse statements hold for the size of the diffuser and its distance to the object. The question is, "What do we give up in order to save SSFB?" The answer depends on the type of coding plate used. If a diffuser is used, the S/N ratio suffers [10-114]. However, if the coding plate is properly chosen, resolution can be sacrificed. Each point on the plate, rather than scattering light over the entire hologram, should collect it from a small region of the object and direct it to a definite place on the hologram. This has been compared to using a fly's eye lens with the lenslets having random orientation and focal length. The search for an optimum coding plate continues.

One problem with this approach is that a real image is produced. A lens could be used to form a virtual image, but the size of the lens would then restrict the field of view. One proposed solution is to use a conjugate of the coding plate and the original wave rather than a duplicate of the plate and the conjugate wave [10-116]. The conjugate would have to be easily duplicated and scaled.

Rather than insist that all of the coding be done in space, some can be done in time [10-116]. Storage of the coded hologram can be considered to be *spatial* multiplexing of the signal. *Temporal* multiplexing would then occur when different rays from the coding plate are present at different times. This approach is not pursued here, but it should be pointed out that if temporal coding is used, the code could be generated electronically, making duplication or conjugation with scale changes easy.

Hologram sampling. The elimination of vertical parallax is a form of sampling in that only one sample is made in the vertical direction. Sampling can be done in the horizontal direction as well. One method is to place a screen of transparent and opaque strips in front of the hologram while recording [10-117]. This reduces the area of the recorded hologram. The blank spots are filled in by replicating the recorded strips a sufficient number of times. The replication can be done by multiply exposing a second hologram with the virtual image of the first. The first hologram is moved between each exposure to fill in the blank places. In addition, the illuminating beam of the first hologram must be shifted in angle so that the virtual image points seen through adjacent exposures of the second hologram appear in the same place. This, of course, compensates only for the image points in one plane.

A sampling technique that eliminates the problem of image smearing when the replicated portions of the hologram are moved is to sample a Fourier transform or quasi Fourier transform hologram [10-118, 119].

Translation of subhologram elements does not move the portion of the image in the transform plane. In terms of spatial frequencies, groups of spatial frequencies from different regions of the spectrum are recorded and repeated to fill the entire space. If samples are taken in vertical and horizontal directions, a mosaic of small holograms can be made. Vertical parallax can be eliminated by taking only one sample in the vertical direction. The reduction in information can be as great as 10^3 with the field of view preserved. Resolution and to some extent, S/N suffers.

Additional research. Although the bandwidth required has been reduced considerably, additional reductions would be helpful. The S/N ratio must be kept at a high level. The use of coding plates and hologram sampling are closely related, and future work should show a convergence of these two approaches. Another technique that could be added is the transmission of changes in the hologram only. The idea of transmitting only changes in an image is presently used in picture phones. An image-plane hologram might lend itself to this technique.

PROBLEMS

10-1. Show that a near Fourier transform hologram, recorded as shown in Fig. P 10-1, can be illuminated so that a lateral change in the hologram position produces only a phase shift in the image.

10-2. In making a HoloTape, attention must be given to the effects of twisting of the tape during illumination. A motion of the image due to tape twisting could smear the television picture. Find the angle of the read-out wave ψ for which the angle of the image wave θ' does not change with changes in ψ. That is, find the read-out angle at which tape twists produce *no* motion in the image position. *Hint*:

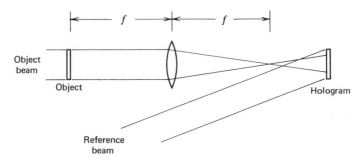

Figure P10-1

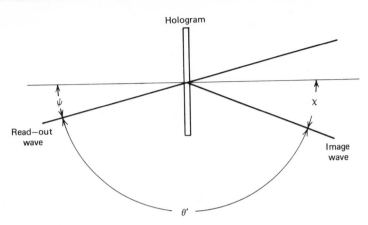

Figure P10-2

use the grating equation $\sin \psi + \sin \chi = \lambda/d$, where d is the spacing of the fringes on the hologram [10-12]. (Figure P 10-2.)

10-3. Describe the fringes formed and their location when a double-exposed hologram is made using the same object but with a rotation of 2 milliradians. Assume plane wave illumination and $\theta_i = 80°$, $\theta_s = 100°$.

10-4. Assume that a two-frequency laser is used to record a hologram of an object. Obtain (10-7) using a description of the temporal coherence of the source, and geometry.

10-5. Calculate a Fourier transform hologram distribution for the letter F, using a slide rule or computer. Draw on graph paper one of the formats used in computer-generated holograms, photograph it, and obtain an image from the resulting hologram. *Hint*: break the F into several slits.

10-6. Assume that the reference of (10-26) is described by a two-dimensional rectangle function. What is then the proper form for (10-27)?

REFERENCES

10-1. T. H. Jeong, "Cylindrical Holography and Some Proposed Applications," *J. Opt. Soc. Am.* **57**, 1396–1398 (1967).

10-2. M. C. King, A. M. Noll, and D. H. Berry, "A New Approach to Computer-Generated Holography," *Appl. Opt.* **9**, 471–475 (1970).

10-3. A. D. Jacobson, V. Evtuhov, and J. K. Neeland, "Motion Picture Holography," *Appl. Phys. Lett.* **14**, 120–122 (1969).

10-4.   D. J. DeBitetto, "A Front-Lighted 3-D Holographic Movie," *Appl. Opt.* **9**, 498–499 (1970).

10-5.   G. C. Sherman, "Hologram Copying by Gabor Holography of Transparencies," *Appl. Opt.* **6**, 1749–1753 (1967).

10-6.   E. N. Leith et al., "Holographic Data Storage in Three-Dimensional Media," *Appl. Opt.* **5**, 1303–1311 (1966).

10-7.   L. K. Anderson, "Holographic Optical Memory for Bulk Data Storage," *Bell Lab. Rec.* **46**, 319–325 (1968).

10-8.   B. Hill, "Some Aspects of a Large Capacity Holographic Memory," *Appl. Opt.* **11**, 182–191 (1972).

10-9.   A. Vander Lugt, "Design Relationships for Holographic Memories," *Appl. Opt.* **12**, 1675–1685 (1973).

10-10. P. Nisenson and S. Iwasa, "Real Time Optical Processing with $Bi_{12}SiO_{20}$ PROM," *Appl. Opt.* **11**, 2760–2767 (1972).

10-11. B. Hill and K. P. Schmidt, "New Page-Composer for Holographic Data Storage," *Appl. Opt.* **12**, 1193–1198 (1973).

10-12. R. Bartolini, W. Hannan, D. Karlsons, and M. Lurie, "Embossed Hologram Motion Pictures for Television Playback," *Appl. Opt.* **9**, 2283–2290 (1970).

10-13. D. H. McMahon, "Holographic Ultrafiche," *Appl. Opt.* **11**, 798–806 (1972).

10-14. C. B. Burckhardt, "Use of a Random Phase Mask for the Recording of Fourier Transform Holograms of Data Masks," *Appl. Opt.* **9**, 695–700 (1970).

10-15. W. C. Stewart, A. H. Firester, and E. C. Fox, "Random Phase Data Masks: Fabrication Tolerances and Advantages of Four Phase Level Masks," *Appl. Opt.* **11**, 604–608 (1972).

10-16. Y. Takeda, Y. Oshida, and Y. Miyamura, "Random Phase Shifters for Fourier Transformed Holograms," *Appl. Opt.* **11**, 818–822 (1972); Y. Takeda, "Hologram Memory with High Quality and High Information Storage Density," *Jap. J. Appl. Phys.* **11**, 656–665 (1972).

10-17. W. J. Dallas, "Deterministic Diffusers for Holography," *Appl. Opt.* **12**, 1179–1187 (1973).

10-18. A. H. Firester, E. C. Fox, T. Gayeski, W. J. Hannan, and M. Lurie, "Redundant Holograms," *RCA Rev.* **33**, 131–153 (1972).

10-19. R. A. Bartolini, J. Bordogna, and D. Karlsons, "Recording Considerations for RCA HoloTape," *RCA Rev.* **33**, 170–205 (1972).

10-20. R. E. Brooks, L. O. Heflinger, and R. F. Wuerker, "Pulsed Laser Holograms," *IEEE J. Quantum Electron.* **2**, 275–279 (1966).

10-21. F. M. Mottier, R. Dändliker, and B. Ineichen, "Relaxation of the Coherence Requirements in Holography," *Appl. Opt.* **12**, 243–248 (1973).

10-22. H. Nassenstein, H. Dedden, H. J. Metz, H. E. Rieck, and D. Schultze, "Physical Properties of Holographic Materials," *Photogr. Sci. Eng.* **13**, 194–199 (1969).

10-23. M. Hercher and B. Ruff, "High-Intensity Reciprocity Failure in Kodak 649-F Plates at 6943 Å," *J. Opt. Soc. Am.* **57**, 103–105 (1967).

10-24. F. J. McClung, A. D. Jacobson, and D. H. Close, "Some Experiments Performed with a Reflected-Light Pulsed-Laser Holography System," *Appl. Opt.* **9**, 103–106 (1970).

10-25. R. J. Collier, C. B. Burckhardt, and L. H. Lin, *Optical Holography,* (Academic Press, New York, 1971), Ch. 15.

10-26. W. G. Gottenberg, ed., *Applications of Holography in Mechanics*, (American Society of Mechanical Engineers, New York, 1971).

10-27. S. Walles, "Visibility and Localization of Fringes in Holographic Interferometry of Diffusely Reflecting Surfaces," *Arkiv för Fysik (Stockholm)* **40**, 299–403 (1969).

10-28. W. G. Gottenberg, "Some Applications of Holographic Interferometry," *Exp. Mech.* **8**, 405–411 (1968).

10-29. K. A. Haines and B. P. Hildebrand, "Interferometric Measurements on Diffuse Surfaces by Holographic Techniques," *IEEE Trans. Instr. Meas.* **15**, 149–161 (1966).

10-30. J. E. Sollid, "Holographic Interferometry Applied to Measurements of Small Static Displacements of Diffusely Reflecting Surfaces," *Appl. Opt.* **8**, 1587–1595 (1969).

10-31. E. B. Aleksandrov and A. M. Bonch-Bruevich, "Investigation of Surface Strains by the Hologram Technique," *Sov. Phys.—Tech. Phys.* **12**, 258–265 (1967).

10-32. A. E. Ennos, "Measurement of in-Plane Surface Strain by Holographic Interferometry," *J. Sci. Instr. (J. Phys.E.)* **1**, 731–734 (1968).

10-33. K. Snow and R. Vandewarker, "On Using Holograms for Test Glasses," *Appl. Opt.* **9**, 822–827 (1970).

10-34. A. J. MacGovern and J. C. Wyant, "Computer Generated Holograms for Testing Optical Elements," *Appl. Opt.* **10**, 619–624 (1971).

10-35. R. L. Powell and K. A. Stetson, "Interferometric Vibration Analysis by Wavefront Reconstruction," *J. Opt. Soc. Am.* **55**, 1593–1598 (1965).

10-36. C. C. Aleksoff, "Temporally Modulated Holography," *Appl. Opt.* **10**, 1329–1341 (1971).

10-37. K. Haines and B. P. Hildebrand, "Contour Generation by Wavefront Reconstruction," *Phys. Lett.* **19**, 10-11 (1965).

10-38. B. P. Hildebrand and K. A. Haines, "Multiple-Wavelength and Multiple-Source Holography Applied to Contour Generation," *J. Opt. Soc. Am.* **57**, 155–162 (1967).

10-39. T. Tsuruta and N. Shiotake, "Holographic Generation of Contour Map of Diffusely Reflecting Surface by Using Immersion Method," *Jap. J. Appl. Phys.* **6**, 661–662 (1967).

10-40. J. S. Zelenka and J. R. Varner, "A New Method for Generating Depth Contours Holographically," *Appl. Opt.* **7**, 2107–2110 (1968).

10-41. J. S. Zelenka and J. R. Varner, "Multiple Index Holographic Contouring," *Appl. Opt.* **8**, 1431–1434 (1969).

10-42. L. O. Heflinger and R. F. Wuerker, "Holographic Contouring Via Multifrequency Lasers," *Appl. Phys. Lett.* **15**, 28–30, (1969).

10-43. J. R. Varner, "Desentitized Hologram Interferometry," *Appl. Opt.* **9**, 2098–2100 (1970).

10-44. E. Archbold, J. M. Burch, A. E. Ennos, and P. A. Taylor, "Visual Observation of Surface Vibration Patterns," *Nature* **222**, No. 5190, 263–265 (1969).

10-45. H. J. Tiziani, "Application of Speckling for In-Plane Vibration Analysis," *Optica Acta* **18**, 891–902 (1971).

10-46. E. Archbold and A. E. Ennos, "Displacement Measurement from Double-Exposure Laser Photographs," *Optica Acta* **19**, 253–271 (1972).

10-47. H. J. Tiziani, "A Study of the Use of Laser Speckle to Measure Small Tilts of Optically Rough Surfaces Accurately," *Opt. Commun.* **5**, 271–276 (1972).

10-48. H. J. Tiziani, "Analysis of Mechanical Oscillations by Speckling," *Appl. Opt.* **11**, 2911–2917 (1972).

10-49. L. Hode and K. Biederman, "Observation of Surface Deformation in Real Time Using Laser Speckles," *Swed. Phys. Conf.*, Lund, Sweden (1972).

10-50. L. Ek and N.-E. Molin, "Detection of the Nodal Lines and the Amplitude of Vibration by Speckle Interferometry," *Opt. Commun.* **2**, 419–424 (1971).

10-51. D. Joyeux and S. Lowenthal, "Real Time Measurement of Angstrom Order Transverse Displacement or Vibrations, by Use of Laser Speckle," *Opt. Commun.* **4**, 108–112 (1971).

10-52. E. N. Leith and J. Upatnieks, "Microscopy by Wavefront Reconstruction," *J. Opt. Soc. Am.* **55**, 569–570 (1965).

10-53. M. E. Cox, R. G. Buckles, and D. Whitlow, "Cineholomicroscopy of Small Animal Microcirculation," *Appl. Opt.* **10**, 128–131 (1971).

10-54. R. F. van Ligten and H. Osterberg, "Holographic Microscopy," *Nature 211*, 282–283 (1966).

10-55. K. Snow and R. Vandewarker, "An Application of Holography to Interference Microscopy," *Appl. Opt.* **7**, 549–554 (1968).

10-56. E. N. Leith, "Quasi-Holographic Techniques in the Microwave Region," *Proc. IEEE* **59**, 1305–1318 (1971).

10-57. M. I. Skolnik, *Introduction to Radar Systems,* (McGraw-Hill Book Co., New York, 1962), Ch. 10.

10-58. E. N. Leith and A. L. Ingalls, "Synthetic Antenna Data Processing by Wavefront Reconstruction," *Appl. Opt.* **7**, 539–544 (1968).

10-59. A. Kozma, E. N. Leith, and N. G. Massey, "Tilted-Plane Optical Processor," *Appl. Opt.* **11**, 1766–1777 (1972).

10-60. L. J. Cutrona, E. N. Leith, L. J. Porcello, and W. E. Vivian, "On the Application of Coherent Optical Processing Techniques to Synthetic-Aperture Radar," *Proc. IEEE* **54**, 1026–1032 (1966).

10-61. E. N. Leith, "Range-Azimuth-Coupling Aberrations in Pulse-Scanned Imaging Systems," *J. Opt. Soc. Am.* **63**, 119–126 (1973).

10-62. B. A. Sichelstiel, W. M. Waters, and T. A. Wild, "Self-Focusing Array Research Model," *IEEE Trans. Antennas Propag.* **12**, 150–154 (1964).

10-63. K. Iizuka, "Microwave Holography, by Photoengraving," *Proc. IEEE* **57**, 813–814 (1969).

10-64. R. P. Dooley, "X-Band Holography," *IEEE Proc.* **53**, 1733–1735 (1965).

10-65. C. F. Augustine, C. Deutsch, D. Fritzler, and E. Marom, "Microwave Holography Using Liquid Crystal Area Detectors," *Proc. IEEE* **57**, 1333–1334 (1969).

10-66. K. Iizuka, "In Situ Microwave Holography," *Appl. Opt.* **12**, 147–149 (1973).

10-67. N. H. Farhat, "Millimeter Wave Holographic Imaging of Concealed Weapons," *Proc. IEEE* **59**, 1383–1384 (1971).

10-68. K. Iizuka and L. G. Gregoris, "Application of Microwave Holography in the Study of the Field from a Radiating Source," *Appl. Phys. Lett.* **17**, 509–512 (1970).

10-69. G. Papi, V. Russo, and S. Sottini, "Microwave Holographic Interferometry," *IEEE Trans. Antennas Propag.* **19**, 740–746 (1971).

10-70. G. Papi, V. Russo, and S. Sottini, "Two-Frequency Microwave Holographic Interferometry," *Proc. IEEE* **60**, 1004–1005 (1972).

10-71. D. C. Winter, "Correction of Unequal Longitudinal and Lateral Magnification in Holography," *Appl. Opt.* **10**, 2551–2553 (1971).

10-72. R. K. Mueller, "Acoustic Holography," *Proc. IEEE* **59**, 1319–1335 (1971).

10-73. A. F. Metherell, H. M. A. El-Sum, J. J. Dreher, and L. Larmore, "Introduction to Acoustical Holography," *J. Acoust. Soc. Am.* **42**, 733–742 (1967).

10-74. A. F. Metherell, H. M. A. El-Sum, and L. Larmore, eds., *Acoustical Holography*, Vol. 1 (Plenum Press, New York, 1968).

10-75. A. F. Metherell and L. Larmore, eds., *Acoustical Holography*, Vol. 2, (Plenum Press, New York, 1969).

10-76. A. F. Metherell, ed., *Acoustical Holography*, Vol. 3 (Plenum Press, New York, 1971).

10-77. R. K. Mueller and N. K. Sheridon, "Sound Holograms and Optical Reconstruction," *Appl. Phys. Lett.* **9**, 328–329 (1966).

10-78. K. Preston, Jr., and J. L. Kreuzer, "Ultrasonic Imaging Using a Synthetic Holographic Technique," *Appl. Phys. Lett.* **10**, 150–152 (1967).

10-79. G. A. Massey, "Acoustic Imaging by Holography," *IEEE Trans. Sonics Ultrasonics* **15**, 141–146 (1968).

10-80. R. L. Whitman, A. Korpel, and M. Ahmed, "Novel Technique for Real-Time Depth-Gated Acoustic Image Holography," *Appl. Phys. Lett.* **20**, 370–371 (1972).

10-81. E. J. Feleppa, "Holography and Medicine," *IEEE Trans. Biomed. Eng.* **19**, 194-205 (1972).

10-82. R. F. van Ligten, B. Grolman, and K. Lawton, "The Hologram and its Ophthalmic Potential," *Am. J. Optom. Am. Acad. Optom.* **43**, 351–363 (1966).

10-83. J. W. Giles, Jr., "Image Reconstruction from a Fraunhofer X-Ray Hologram with Visible Light," *J. Opt. Soc. Am.* **59**,, 1179–1188 (1969).

10-84. H. H. Barrett, "Fresnel Zone Plate Imaging in Nuclear Medicine," *J. Nucl. Med.* **13**, 382–385 (1972).

10-85. H. H. Barrett, K. Garewal, and D. T. Wilson, "A Spatially-Coded X-Ray Source," *Radiology* **104**, 429–430 (1972).

10-86. H. C. Becker, P. H. Meyers, and C. M. Nice, Jr., "Laser Light Diffraction, Spatial Filtering, and Reconstruction of Medical Radiographic Images: A Pilot Study," *IEEE Trans. Biomed. Engr.* **15**, 186–195 (1968).

10-87. G. Groh and M. Kock, "3-D Display of X-Ray Images by Means of Holography," *Appl. Opt.* **9**, 775–777 (1970).

10-88. W. T. Cathey, D. Shander, and R. C. Bigelow, "Three-Dimensional Display of a Vector Cardiogram," in R. J. Gowen, C. E. Tucker, and J. K. Aikawa, eds., *Biomedical Sciences Instrumentation* (Instrument Society of America, Pittsburgh, 1970), pp. 30–33.

10-89. D. Shander and T. M. Duran, "A New Three-Dimensional Display of the Vectorcardiogram Employing Holography," *Chest* **59**, 438–440 (1971).

10-90. H. Kogelnik, "Holographic Image Projection through Inhomogeneous Media," *Bell Syst. Tech. J.* **44**, 2451–2455 (1965).

10-91. E. N. Leith and J. Upatnieks, "Holographic Imagery through Diffusing Media," *J. Opt. Soc. Am.* **56**, 523 (1966).

10-92. J. W. Goodman, W. H. Huntley, Jr., D. W. Jackson, and M. Lehmann, "Wavefront-

Reconstruction Imaging Through Random Media," *Appl. Phys. Lett.* **8**, 311–313 (1966).

10-93. J. Upatnieks, A. Vander Lugt, and E. Leith, "Correction of Lens Aberrations by Means of Holograms," *Appl. Opt.* **5**, 589–593 (1966).

10-94. J. E. Ward, D. C. Auth, and F. P. Carlson, "Lens Aberration Correction by Holography," *Appl. Opt.* **10**, 896–900 (1971).

10-95. W. T. Cathey, "Holographic Simulation of Compensation for Atmospheric Wavefront Distortion," *Proc. IEEE* **56**, 340 (1968).

10-96. W. T. Cathey, C. L. Hayes, W. C. Davis, and V. F. Pizzurro, "Compensation for Atmospheric Phase Effects at 10.6 μ," *Appl. Opt.* **9**, 701–707 (1960).

10-97. P. N. Everett and A. J. Cantor, "Long-Range Holography," *Appl. Opt.* **11**, 1697–1707 (1972).

10-98. J. D. Gaskill, "Imaging Through a Randomly Inhomogeneous Medium by Wavefront Reconstruction," *J. Opt. Soc. Am.* **58**, 600–608 (1968).

10-99. J. W. Goodman, D. W. Jackson, M. Lehmann, and J. Knotts," Experiments in Long-Distance Holographic Imagery," *Appl. Opt.* **8**, 1581–1586 (1969).

10-100. J. D. Gaskill, "Atmospheric Degradation of Holographic Images," *J. Opt. Soc. Am.* **59**, 308–318 (1969).

10-101. P. N. Everett and A. J. Cantor, "Long-Range Holography," *Appl. Opt.* **11**, 1697–1707 (1972).

10-102. D. A. Cain, C. Billings, C. D. Johnson, and G. M. Mayer, "Holography of Large Objects in a Turbulent Atmosphere with a CW Laser," *Appl. Phys. Lett.* **23**, 37–38 (1973).

10-103. W. T. Cathey, J. F. Hadwin, and J. D. Pace, "Imaging Through Turbulent Water Using Speckle Reference Holography," *Appl. Opt.* **12**, 2683–2685 (1973).

10-104. A. Labeyrie, "Attainment of Diffraction Limited Resolution in Large Telescopes by Fourier Analysing Speckle Patterns in Star Images," *Astron. Astrophys.* **6**, 85–87 (1970).

10-105. D. Y. Gezari, A. Labeyrie, and R. V. Stachnik, "Speckle Interferometry: Diffraction-Limited Measurements of Nine Stars with the 200-Inch Telescope,"*Astrophys. J.* **176**, L1–L5 (1972).

10-106. E. N. Leith, J. Upatnieks, B. P. Hildebrand, and K. Haines, "Requirements for a Wavefront Reconstruction Television Facsimile System," *J. SMPTE* **74**, 893–896 (1965).

10-107. L. H. Enloe, W. C. Jakes, Jr., and C. B. Rubinstein, "Hologram Heterodyne Scanners," *Bell Syst. Tech. J.* **47**, 1875–1882 (1968).

10-108. A. B. Larsen, "A Heterodyne Scanning System for Hologram Transmission," *Bell Syst. Tech. J.* **48**, 2507–2527 (1969).

10-109. C. B. Burckhardt and L. H. Enloe, "Television Transmission of Holograms with Reduced Resolution Requirements on the Camera Tube," *Bell Syst. Tech. J.* **48**, 1529–1535 (1969).

10-110. A. Macovski, "Considerations of Television Holography," *Opt. Acta* **18**, 31–39 (1971).

10-111. D. J. DeBitetto, "Bandwidth Reduction of Hologram Transmission Systems by Elimination of Vertical Parallax," *Appl. Phys. Lett.* **12**, 176–178 (1968).

10-112. D. Fritzler and E. Marom, "Reduction of Bandwidth Required for High Resolution Hologram Transmission," *Appl. Opt.* **8**, 1241–1243 (1969).

10-113. K. A. Haines and D. B. Brumm, "A Technique for Bandwidth Reduction in Holographic Systems," *Proc. IEEE* **55**, 1512–1513 (1967).

10-114. K. A. Haines and D. B. Brumm, "Holographic Data Reduction," *Appl. Opt.* **7**, 1185–1189 (1968).

10-115. E. N. Leith, D. B. Brumm, and S. S. H. Hsiao, "Holographic Cinematography," *Appl. Opt.* **11**, 2016–2023 (1972).

10-116. B. P. Hildebrand, "Hologram Bandwidth Reduction by Space-Time Multiplexing," *J. Opt. Soc. Am.* **60**, 259–264 (1970).

10-117. C. B. Burckhardt, "Information Reduction in Holograms for Visual Display," *J. Opt. Soc. Am.* **58**, 241–246 (1968).

10-118. J.-C. Viénot, J. Duvernoy, G. Tribillon, and J.-L. Tribillon, "Three Methods of Information Assessment for Optical Data Processing," *Appl. Opt.* **12**, 950–960 (1973).

10-119. L. H. Lin, "A Method of Hologram Information Reduction by Spatial Frequency Sampling," *Appl. Opt.* **7**, 545–548 (1968).

Appendix 1

The Fourier Transform
of III(x/σ)

The Fourier transform of the sampling or shah function is another shah function of a different spacing:

$$\mathscr{F}\left[\text{III}\left(\frac{x}{\sigma}\right)\right] = \sigma\,\text{III}(\sigma u). \qquad \text{(A1-1)}$$

The two functions are

$$\text{III}\left(\frac{x}{\sigma}\right) = \sigma \sum_{m=-\infty}^{\infty} \delta(x - m\sigma) \qquad \text{(A1-2)}$$

and

$$\sigma\,\text{III}(\sigma u) = \sum_{m=-\infty}^{\infty} \delta\left(u - \frac{m}{\sigma}\right). \qquad \text{(A1-3)}$$

Equation (A1-1) can be written as

$$\sigma \int \sum_{m=-\infty}^{\infty} \delta(x - m\sigma)\, e^{-i2\pi xu}\, dx = \sigma \sum_{m=-\infty}^{\infty} e^{-i2\pi m\sigma u}. \qquad \text{(A1-4)}$$

Therefore, it only remains to show that

$$\text{III}(\sigma u) = \sum_{m=-\infty}^{\infty} e^{-i2\pi m\sigma u}, \qquad \text{(A1-5)}$$

or

$$\frac{1}{\sigma} \sum_{m=-\infty}^{\infty} \delta\left(u - \frac{m}{\sigma}\right) = \sum_{m=-\infty}^{\infty} e^{-i2\pi m\sigma u}. \qquad \text{(A1-5)}$$

Consider a Fourier series representation of $1/\sigma\sum_{m=-\infty}^{\infty}\delta(u-m/\sigma)$. The formula for the Fourier series representation of a function $G(u)$ is

$$G(u) = \sum_{m=-\infty}^{\infty} b_m e^{-im2\pi u/\mathcal{T}} \tag{A1-6}$$

where $\mathcal{T}$ is the period of the function. The coefficient b_m is found by

$$b_m = \frac{1}{\mathcal{T}} \int_{-\mathcal{T}/2}^{\mathcal{T}/2} G(u) e^{im2\pi u/\mathcal{T}} du. \tag{A1-7}$$

Substitution of the given function into (A1-7) gives

$$b_m = \sigma \int_{-1/2\sigma}^{1/2\sigma} \frac{1}{\sigma} \sum_{m=-\infty}^{\infty} \delta\left(u - \frac{m}{\sigma}\right) e^{im2\pi u\sigma} du \tag{A1-8}$$

where the period $\mathcal{T}$ is, in this case, $1/\sigma$. The only delta function contributing to the integral (A1-8) is the one associated with $m=0$. Therefore, $b_m = 1$. The formula (A1-6) then yields

$$G(u) = \sum_{m=-\infty}^{\infty} e^{-i2\pi m\sigma u}, \tag{A1-9}$$

the Fourier series representation of

$$\frac{1}{\sigma} \sum_{m=-\infty}^{\infty} \delta\left(u - \frac{m}{\sigma}\right).$$

An alternative proof of (A1-1) is given by Bracewell [A1-1].

REFERENCES

A1-1. Ron Bracewell, *The Fourier Transform and Its Applications* (McGraw-Hill Book Co., New York, 1965), p. 214.

Appendix 2

Operations Involving $\mathbf{h}(x,y;d)$

The function $\mathbf{h}(x,y;d)$ is defined as

$$\mathbf{h}(x,y;d) = \exp\left[\frac{ik}{2d}(x^2+y^2)\right]. \qquad \text{(A2-1)}$$

Integral

$$\int\int \mathbf{h}(x,y;d) = i\lambda d. \qquad \text{(A2-2)}$$

This is shown by the change of variables

$$\int\int \exp\left[\frac{ik}{2d}(x^2+y^2)\right]dx\,dy = \frac{2d}{k}\int\int \exp[i(s^2+w^2)]\,ds\,dw$$

where

$$s = \sqrt{\frac{k}{2d}}\,x, \qquad w = \sqrt{\frac{k}{2d}}\,y.$$

Evaluation of the integral gives $i\pi$, which, when combined with $2d/k$, gives the result of (A2-2).

Transform

$$\mathcal{F}[\mathbf{h}(x,y;d)] = i\lambda d\mathbf{h}\left(u,v;-\frac{1}{\lambda^2 d}\right). \qquad \text{(A2-3)}$$

Writing out the transform gives

$$\mathcal{F}[\mathbf{h}(x,y;d)] = \int\int \exp\left[\frac{ik}{2d}(x^2+y^2)\right]\exp[-i2\pi(xu+yv)]dx\,dy.$$

The square of the exponent can be completed to yield

$$\mathcal{F}[\mathbf{h}(x,y;d)]$$

$$= \exp\left[\frac{-ik\lambda^2 d}{2}(u^2+v^2)\right]\int\int\exp\left\{\frac{ik}{2d}\left[(x-\lambda ud)^2+(y-\lambda vd)^2\right]\right\}dx\,dy$$

$$= i\lambda d\mathbf{h}\left(u,v;-\frac{1}{\lambda^2 d}\right).$$

Convolution of g(x,y) with h(x,y;d)

$$\mathbf{h}(x,y;-d)[\mathbf{g}(x,y)\otimes\mathbf{h}(x,y;d)]=\mathbf{G}\left(\frac{x}{\lambda d},\frac{y}{\lambda d}\right)\otimes\tilde{\mathbf{h}}\left(\frac{x}{\lambda d},\frac{y}{\lambda d};-\frac{1}{\lambda^2 d}\right)$$

$$= i\lambda d\mathbf{G}\left(\frac{x}{\lambda d},\frac{y}{\lambda d}\right)\otimes\mathbf{h}\left(\frac{x}{\lambda d},\frac{y}{\lambda d};-\frac{1}{\lambda^2 d}\right). \quad \text{(A2-4)}$$

Let $\mathbf{G}$ be defined as the Fourier transform of $\mathbf{g}$ and $\tilde{\mathbf{h}}$ the transform of $\mathbf{h}$. Writing the convolution $\mathbf{h}(x,y;-d)[\mathbf{g}(x,y)\otimes\mathbf{h}(x,y;d)]$ in integral form gives

$$\mathbf{h}(x,y;-d)\int\int\mathbf{g}(s,w)\mathbf{h}(s-x,w-y;d)\,ds\,dw.$$

Expansion of $\mathbf{h}(s-x,w-y;d)$ causes the convolution to take the form

$$\mathbf{h}(x,y;-d)\mathbf{h}(x,y;d)\int\mathbf{g}(s,w)\mathbf{h}(s,w;d)\exp\left[-i2\pi\left(\frac{sx}{\lambda d}+\frac{wy}{\lambda d}\right)\right]ds\,dw$$

$$= \mathbf{G}\left(\frac{x}{\lambda d},\frac{y}{\lambda d}\right)\otimes\tilde{\mathbf{h}}\left(\frac{x}{\lambda d},\frac{y}{\lambda d};-\frac{1}{\lambda^2 d}\right)$$

$$= i\lambda d\mathbf{G}\left(\frac{x}{\lambda d},\frac{y}{\lambda d}\right)\otimes\mathbf{h}\left(\frac{x}{\lambda d},\frac{y}{\lambda d};-\frac{1}{\lambda^2 d}\right).$$

A change of variables indicates that (A2-4) can also be written as

$$\mathbf{G}(x,y)\otimes\tilde{\mathbf{h}}(x,y;-d)$$

$$= \mathbf{h}(\lambda\,dx,\lambda\,dy;-\lambda^2 d^3)[\mathbf{g}(\lambda\,dx,\lambda\,dy)\otimes\mathbf{h}(\lambda\,dx,\lambda\,dy;\lambda^2 d^3)].$$

Convolution of h with h

$$\mathbf{h}(x,y;d_1) \otimes \mathbf{h}(x,y;d_2) = \frac{i\lambda d_1 d_2}{d_1 + d_2} \mathbf{h}(x,y;d_1+d_2), d_1 \neq -d_2. \quad (A2\text{-}5)$$

Writing the convolution in integral form gives

$$\mathbf{h}(x,y;d_1) \otimes \mathbf{h}(x,y;d_2)$$

$$= \int \int \exp\left[\frac{ik}{2d_1}(s^2+w^2) \right] \exp\left\{ \frac{ik}{2d_2}\left[(x-s)^2 + (y-w)^2 \right] \right\} ds\,dw$$

where s and w are dummy variables.

$$\mathbf{h}(x,y;d_1) \otimes \mathbf{h}(x,y;d_2) = \exp\left[\frac{ik}{2d_2}(x^2+y^2) \right]$$

$$\int\int \exp\left[\frac{ik}{2}\left(\frac{1}{d_1} + \frac{1}{d_2}\right)(s^2+w^2) \right] \exp\left\{ -i2\pi\left[s\left(\frac{x}{\lambda d_2}\right) + w\left(\frac{y}{\lambda d_2}\right) \right] \right\} ds\,dw$$

$$= \mathbf{h}(x,y;d_2) \mathcal{F}\left[\mathbf{h}\left(s,w; \frac{d_1 d_2}{d_1+d_2} \right) \right]\Bigg|_{\left(\frac{x}{\lambda d_2}, \frac{y}{\lambda d_2}\right)}$$

$$= \mathbf{h}(x,y;d_2) \frac{i\lambda d_1 d_2}{d_1+d_2} \mathbf{h}\left[\frac{x}{\lambda d_2}, \frac{y}{\lambda d_2}; \frac{-(d_1+d_2)}{\lambda^2 d_1 d_2} \right]$$

$$= \frac{i\lambda d_1 d_2}{d_1+d_2} \mathbf{h}(x,y;d_2) \mathbf{h}\left[x,y; \frac{-(d_1+d_2)d_2}{d_1} \right]$$

$$= \frac{i\lambda d_1 d_2}{d_1+d_2} \mathbf{h}(x,y;d_1+d_2). \quad (A2\text{-}6)$$

Impulse function

$$\mathbf{h}(x,y;0)=\mathbf{h}(x,y;d)\otimes\mathbf{h}(x,y;-d)=|\lambda d|^2\delta(x,y). \qquad \text{(A2-7)}$$

Starting with (A2-6), let $d_1=-d_2$. Because

$$\mathbf{h}(x,y;\infty)=1,$$

$$\mathbf{h}(x,y;d)\otimes\mathbf{h}(x,y;-d)=\mathbf{h}(x,y;d)\mathcal{F}[1]\big|_{\left(\frac{x}{\lambda d},\frac{y}{\lambda d}\right)}$$

$$=\mathbf{h}(x,y;d)\delta\left(\frac{x}{\lambda d},\frac{y}{\lambda d}\right)=\delta\left(\frac{x}{\lambda d},\frac{y}{\lambda d}\right)$$

and

$$\delta\left(\frac{x}{\lambda d},\frac{y}{\lambda d}\right)=|\lambda d|^2\delta(x,y).$$

Appendix 3

Making Holograms
and Spatial Filters

The techniques involved in making spatial filters and holograms are essentially the same except for the selection of the proper reference-to-object wave intensity ratio. Consequently, this discussion is usually in terms of making holograms.

A3.1 Illumination

Many of the illumination principles of photography apply in holography. Illumination from more than one direction eliminates heavy shadows in the scene. One possible arrangement is shown in Fig. A3-1. An indication of how the image will appear can be obtained by looking at the object from the location of the hologram. One of the simplest hologram recording configurations is to lay the object on a table, the film over it, and illuminate the object through the film. The reference wave is the illumination wave going through the film, and the object wave is the reflected wave coming from the other direction. A white light hologram is obtained. Development with a low-contrast developer allows a large latitude in exposure.

Reference-to-object wave intensity ratio. In choosing the ratio of the reference wave to object wave intensities, a tradeoff must be made between hologram efficiency and image distortions due to recorder nonlinearities. The greatest efficiency is obtained when the reference and object wave intensities are equal. As shown in Chapter 6, however, the effects of nonlinearities can produce intolerable amounts of noise in the images of some scenes. A rule-of-thumb intensity ratio is from 2.5 to 5.

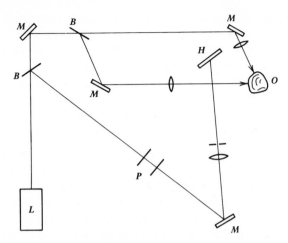

Figure A3-1 One arrangement that provides two-beam illumination of the object with nearly equal object and reference wave paths. The crossed polarizers allow easy adjustment of the intensity of the reference beam. *B*—beamsplitters, *M*—mirrors, *P*—polarizers, *H*—hologram, *O*—object, *L*—laser.

Effects of polarization. In making the intensity ratio measurements, attention must be given to the polarization of the waves. Only the components of the two waves having the same polarization interfere to produce a hologram. For example, the wave reflected from a diffusely reflecting object is depolarized. A polarized reference wave interferes with only that portion of the object wave which has the same polarization. The portion of the object wave having orthogonal polarization only contributes to the fogging or biasing of the recorder.

Neutral density filters can be used to control the relative wave intensities, but they may introduce unwanted distortions of the wave. An alternative method is the adjustment of the size of the beam illuminating the object. More precise control without changing the recording arrangement can be obtained by using two polarizers in the object or reference wave before the pinhole. The pinhole removes any distortions introduced by nonflat polarizers. The polarizer closest to the hologram recording position should be fixed with the polarization of the other wave. Rotation of the polarizer closest to the laser will vary the power transmitted through the pair of polarizers. This is illustrated in Fig. A3-1. If the laser output is polarized, it is usually vertically polarized. If the output has both vertical and horizontal components, the polarization ratio changes at the beamsplitters due to the different reflection coefficients for the two polarizations.

If a choice between vertical and horizontal polarization is to be made, the vertical one is preferable. When the polarization vector of two interfering waves is in the plane defined by the two directions of propagation, the visibility of the fringes formed is reduced by the cosine of the angle between the two waves. Consequently, if the reference and object waves intersect near 90°, no fringes are formed. This is not the case when the polarization is vertical because, if the layout is on a tabletop, the angle between the two polarization vectors is always zero.

Complex spatial filters. In the case of spatial filters, more care must be taken in adjusting the reference-to-object (or signal) intensity ratio. This is because the intensity distribution varies with the spatial frequency. If one can determine the region of spatial frequencies of greatest interest, the reference intensity could be selected to correspond to the optimum recording of that region. The reference wave amplitude could be tapered across the spectral plane. In spatial heterodyning, this tapering occurs naturally.

The reference wave intensity can be adjusted by eye by viewing the two waves on ground glass placed in the filter recording position. Rapid alternate blocking of the signal and reference beams gives a good indication of the relative intensities at any location in the spectral plane.

Exposure. In determining the optimum exposure time, the intensity distribution is measured with both object and reference waves present. In the case of spatial filters, an average reading of power in the spectral plane is not accurate. It may indicate a shorter exposure time than needed because of the intense zero-frequency spot or a longer one if the pattern has large spaces without energy. Obviously, the size of the detector is an important consideration. A few trial shots with exposures above and below that indicated by the power measurement is the best approach.

A3.2 Vibration and Temporal Coherence

Temporal coherence. Unless axial mode control is used in the laser, multiple frequencies will limit the temporal coherence of the source. The difference in the path lengths for the reference and object waves should be limited to a fraction of the length of the laser cavity. See Chapter 4 for a discussion of the reasons. Figure A3-1 shows one arrangement providing nearly equal paths. This is much less of a problem with local reference beam or speckle reference beam holograms.

Vibration. Vibration of mirrors or the object produces relative phase changes that can reduce the fringe visibility. The stability of the table used can be checked by using the optical components to assemble a Mach-

Zender or Michelson interferometers with coarse fringes. Observation of the fringes will show the effects of any vibration as well as any fringe changes caused by thermal or acoustical effects.

Holograms can be made in spite of vibrations if a pulsed laser is used, reducing the exposure time to less than the period of the highest frequency vibration. Alternatively, servo controls can be applied to the position of one of the mirrors in the reference path [A3-1]. The control system maintains stable fringes on the hologram in spite of motions of the object or other optical components.

A3.3   Selection of a Recording Medium

The medium chosen depends on the planned use of the hologram (permanent data storage, short-term erasable recordings) and the illumination energy available. Photographic film is the most commonly used and its properties are given in Chapter 6. Kodak 649-F and Agfa 10E70 or 10E75 are the most popular films for holographic work.

A glass substrate provides stability in holographic interferometry and is often useful in displays. Acetate or polyester film substrates can be obtained in 35 or 70 mm strips, which are often useful in making holographic displays.

A3.4   Film Development and Bleaching

The type developer and the process used affects the efficiency of the hologram and the S/N ratio in the image. The techniques described here are only a few of several approaches, but they have produced good results.

Normal development of high-resolution recordings.   The general process is as described in brochures produced by the photographic supply houses [A3-2, A3-3]. Kodak recommends that 649-F be developed in D-19 for 6–8 min at 20°C and fixed in Kodak Rapid Fixer for 5 min. Both baths should be continuously agitated and followed by a short rinse in water. The residual fixer is then removed by a $1\frac{1}{2}$-min bath in Kodak Hypo Clearing Agent followed by a 5-min wash in flowing water. Alternatively, a 20 to 25-min wash without the clearing agent can be used. The wash is followed by drying in a dustfree place. If precise control of the developing time is desired, Kodak Stop Bath SB-5 can be used after development to immediately stop the development. The stop bath should be followed by a rinse in water before the fixer is used.

Agfa suggests development of their plates in G3p or Metinol U at 20°C for 5 min and fixing in G334 for 4 min. For most purposes, D-19 provides satisfactory results with both types of emulsions.

Low contrast developers. If the normal dynamic range of the film is greatly exceeded by the range of exposures, a low-contrast developer may be used. This is the case, for example, in speckle reference holography. Only two such developers are described here, POTA and D-165.

POTA is a special developer that produces results with a very low photographic gamma [A3-4]. For example, the gamma of one high-resolution film is 2.6 when developed in D-19 for 5 min and 0.6 when developed in POTA for $2\frac{1}{2}$ min. The formula for POTA is:

1-Phenyl-3-pyrazolidone, 1.5 grams
Sodium sulfite, 30 grams
Water to make, 1 liter

A 1:4 dilution of Kodak D-165 in water can be used to develop 649-F plates to a gamma of 1.3 to 1.45, depending on the developing time [A3-5]. The formula for D-165 is given in reference [A3-5].

Bleaching after development. There are many bleaches that can be used on photographic emulsions after development. Many, such as Kodak's Chromium Intensifier, are readily accessible and easy to use, but lack permanence and produce noise. Many bleached holograms become dark when exposed to light, which makes them unsuitable for permanent storage. The problem of the chemical stability of bleached emulsions has been studied and formulas for various bleaches published [A3-6, A3-7]. One simple bleach consists of:

Potassium ferricyanide, 45 grams
Potassium iodide, 25 grams
Distilled water to make, 1 liter

In using this bleach, a previously developed plate is bleached until cleared, washed 1 min in water, 1 min in isopropyl alcohol, and dried.

Reversal bleaching. Much of the noise in bleached holograms is produced by *flare light*. This is caused by low spatial frequency fringes on the hologram, which in turn are caused by light from one part of a diffusely reflecting object interfering with light from another part of the object. The thickness variations of the emulsion are more pronounced at low spatial frequencies, producing excessive scattered light. In the process of reversal bleaching described in Chapter 6, the effects of the surface relief tend to cancel the effects of the index changes at low spatial frequencies, thereby reducing the flare light [A3-8].

The bleach process is given in Table A3-1 with the formulas for the developer, bleach, and clearing solution in Table A3-2. Care should be taken to avoid contact with the developer, especially the carcinogenic

TABLE A3-1   REVERSAL BLEACHING PROCESS

1.	Develop—Kodak SD-48	5 min	24°C
2.	Stop bath—Kodak SB-1	15 sec	21–27°C
3.	Rinse—running water	1 min	21–27°C
4.	Bleach—Kodak R-9	3 min	21–27°C
5.	Wash—running water	5 min	21–27°C
6.	Stain removal—Kodak S-13, Solution A	1 min	21–27°C
7.	Clear—Kodak S-13, Solution B	1 min	21–27°C
8.	Wash—running water	5–10 min	21–27°C
9.	Rinse—1 : 1 solution methanol and water	1 min	21–27°C
10.	Wash—isopropyl alcholol	1 min	21–27°C
11.	Dry		

TABLE A3-2   DEVELOPER, BLEACH, AND STAIN REMOVER FORMULAS

KODAK SPECIAL DEVELOPER SD-48

Solution A

Water	750 ml
Sodium sulfite, desiccated	8 grams
Pyrocatechol (Eastman P 604)	40 grams
Sodium sulfate, desiccated	100 grams
Water to make	1 liter

Solution B

Water	750 ml
Sodium hydroxide	20 grams
Sodium sulfate, desiccated	100 grams
Water to make	1 liter

KODAK BLEACH BATH R-9

Distilled water	1.0 liter
Potassium dichromate	9.5 grams
Sulfuric acid, concentrated	12.0 ml

KODAK STAIN REMOVER S-13

Solution A

Water	750.0 ml
Potassium permanganate	2.5 grams
Sulfuric acid, concentrated	8.0 ml
Water to make	1.0 liter

Solution B

Water	750 ml
Sodium bisulfite	10 grams
Water to make	1 liter

pyrocatechol. The dust or vapor of the pyrocatechol should not be breathed. Another precaution is that the sulfuric acid of the bleach and stain remover should be added to the solution slowly with constant stirring to avoid boiling and spattering of the acid. If you still want to proceed, you should get low-noise, bleached holograms.

An index matching liquid gate to reduce the effect of surface variations also suppresses the flare light.

A3.5 Emulsion Shrinkage

The emulsion shrinks during the fixing process because the unexposed silver halide grains are removed. Omission of the fixing step, however, causes the continued exposure and photodevelopment of the emulsion, darkening it. Emulsion shrinkage reduces the spacing between the reflecting planes of a color or white-light hologram, resulting in the reflection of a shorter wavelength light than was used in the recording. Transmission holograms suffer very little when the emulsion shrinks if the normal to the emulsion surface bisects the angle between the direction of propagation of the object and reference waves. This means that the diffracting planes are normal to the surface of the emulsion and shrinkage does not change their spacing.

The emulsion can be expanded back to its original size by soaking it in an 8% solution of triethanolamine [A3-9].

REFERENCES

A3-1. D. B. Neumann and H. W. Rose, "Improvement of Recorded Holographic Fringes by Feedback Control," *Appl. Opt.* **6**, 1097–1104 (1967).

A3-2. *Kodak Plates and Films for Science and Industry*, 2nd ed. (Eastman Kodak Co., 1972).

A3-3. *Photographic Materials for Holography* (Agfa-Gevaert Co., 1971).

A3-4. M. Levy, "Wide Latitude Photography," *Photogr. Sci. Eng.* **11**, 46–53 (1967).

A3-5. D. P. Jablonowski, R. A. Heinz, and J. O. Artman, "Achieving Low Gamma (Γ) Characteristics in Kodak 649-F Plates," *Appl. Opt.* **10**, 1988–1989 (1971).

A3-6. D. H. McMahon and W. T. Maloney, "Measurements of the Stability of Bleached Photographic Phase Holograms," *Appl. Opt.* **9**, 1363–1368 (1970).

A3-7. K. S. Pennington and J. S. Harper, "Techniques for Producing Low-Noise, Improved Efficiency Holograms," *Appl. Opt.* **9**, 1643–1650 (1970).

A3-8. R. L. Lamberts and C. N. Kurtz, "Reversal Bleaching for Low Flare Light in Holograms," *Appl. Opt.* **10**, 1342–1347 (1971).

A3-9. N. Nishida, "Correction of the Shrinkage of a Photographic Emulsion with Triethanolamine," *Appl. Opt.* **9**, 238–240 (1970).

Glossary

a	Aperture size; dummy variable
b	Aperture size
b_m	Coefficient of Fourier series
c	Velocity of light
d	Separation
d_i	Distance from image to lens or hologram
d_o	Distance from object to lens or hologram
d_1	Distance in front of lens or hologram
d_2	Distance behind lens or hologram
e	Base of natural (naperian) logarithms
$\mathbf{e}(x,y)$	Illumination wave
f	Focal length
g	Arbitrary function
h	Arbitrary function
$\mathbf{h}$	Impulse response
$\mathbf{h}(x,y;d)$	$\mathrm{Exp}[ik(x^2+y^2)/2d]$
$\mathbf{h}_1(x,y;d)$	$\mathrm{Exp}[ik_1(x^2+y^2)/2d]$
i	$\sqrt{-1}$
$\mathbf{i}$	Electrical current
j	$-\sqrt{-1}$
k	$2\pi/\lambda$
$\mathbf{k}$	Propagation vector
k_g	Grating vector normal to the fringe planes of a thick hologram
ℓ	Overlap in the image of a pair of waves representing a spatial frequency; line scan rate of a vidicon
m	Integral number
$\mathbf{n}$	Index of refraction
p	Phase modulation index
$p(x)$	Convolution
$\mathbf{p}_l$	Phase effect of lens
$\mathbf{p}_p$	Phase effect of parabola
q	Distance plane wave travels

383

$q(x)$	Correlation
r	Radial coordinate; distance
s	Dummy variable
$\mathbf{s}$	Scan velocity
$s(t)$	Signal
s_R	Separation of two point images at which they are just resolved according to the Rayleigh criterion
t	Time
$\mathbf{t}(x,y)$	Amplitude transmittance
u	Spatial frequency in x direction
u_c	Cutoff spatial frequency of an imaging system
v	Spatial frequency in y direction
$\mathbf{v}$	Velocity
w	Dummy variable
x	Rectangular coordinate
y	Rectangular coordinate
z	Rectangular coordinate
$\mathcal{C}$	Angular spectrum of an intensity distribution
A	Constant
$\mathbf{A}$	Angular spectrum of an amplitude distribution
B	Beamsplitter
C	Constant
D	Constant; diameter
$\mathfrak{D}$	Photographic density
$\mathfrak{D}_0$	Average density of a grating
$\mathfrak{D}_1$	Amplitude of the density modulation of a grating
$\mathbf{E}$	Electric field
$\mathbf{E}(u,v)$	Fourier transform of $e(x,y)$
$\mathcal{E}$	Exposure of photorecording material
F	Focal point
$F^{\#}$	Focal length divided by the diameter of an imaging system
$\mathcal{F}$	Denotes Fourier transform
G	$\mathcal{F}$ transform of g
H	$\mathcal{F}$ transform of h
$\mathbf{H}$	Spatial frequency transfer function for amplitude distributions
$\mathcal{H}$	Spatial frequency transfer function for intensity distributions
$\mathcal{H}$	Magnetic field
I	Intensity
I_i	Intensity (irradiance) of image distribution
I_o	Intensity (radiance) of object distribution
J	Bessel function

K	Constant; ratio of reference wave intensity to object wave intensity
L	Linear dimension of a hologram
M	Mirror; magnification factor
M_c	Magnification of hologram during copying process
N	Number
P	Point P
$\mathbf{P}(x,y)$	Pupil function
R	Distance; reference point
S	Source
$\mathbf{S}$	Signal, or image, wave from a hologram
s	Scattered power normalized to incident power and a two-dimensional spatial bandwidth
$\mathbf{T}(u,v)$	Fourier transform of $\mathbf{t}(x,y)$
T	Thickness
$T_l(x,y)$	Thickness of lens
$T(x,y)$	Thickness variation
$\mathcal{T}$	Period of function
$\mathbf{U}$	Complex wave field (object)
U	$|\mathbf{U}|$
$\mathbf{V}$	Complex wave field (reference)
V	$|\mathbf{V}|$
υ	Visibility of fringes
$\mathbf{W}$	Complex wave field (read-out)
W	$|\mathbf{W}|$
α	Propagation constant of the object wave in the x direction; argument of mutual coherence functions Γ,γ
α_0	Average attenuation constant
α_1	Modulated portion of attenuation of absorption grating
β	Propagation constant of the object wave in the y direction
β_0	Phase shift per unit wavelength propagated (associated with the general propagation constant γ_g)
γ	Normalized mutual coherence function
γ	Slope of H & D curve (photographic gamma)
γ	Propagation constant of the object wave in the z direction
γ_g	General propagation constant for the coupled wave in a thick hologram
δ	Discrete interval; geometric phase of mutual coherence function
δ_p	Phase difference
$\delta(x)$	Dirac delta function
ϵ	Dielectric permeability of material

ϵ	Phase of plane wave
ϵ_r	Relative dielectric constant
ζ	Propagation constant of the reference wave in the x direction
η	Intrinsic impedance of medium
η	Diffraction efficiency of a hologram
θ	Angle between object wave direction of propagation and x axis; polar coordinate
ι	$\pi n_1 T/\lambda_a \cos\theta_0$—phase-modulation index for thick holograms
κ	Coupling constant between read-out and reconstructed image waves in thick holograms
λ	Wavelength
λ_g	Period of holographic grating measured normal to the diffracting surfaces
μ	Magnetic permittivity of medium
μ	Propagation constant of the reference wave in the z direction
ν	Optical frequency
ξ	Angle between the object wave direction of propagation and the y axis
π	3.1415
ρ	Radial coordinate
ρ_V	Radius of curvature of a spherical reference wave
ρ_W	Radius of curvature of a spherical read-out wave
σ	Sample spacing; grating period
σ	Conductivity
τ	Time interval
τ	Intensity transmittance
ϕ	Angle between the reference wave direction of propagation and the z axis
ϕ_0	Bragg angle in a thick hologram, measured between the reflecting surface and the direction of propagation of the wave in the medium
φ	Polar coordinate
χ	Angle between the object wave direction of propagation and the z axis
ψ	Angle between the read-out wave direction of propagation and the z axis
ψ_0	Angle in a thick hologram between the hologram surface and the direction of propagation of the wave in the medium
ω	Angular frequency
Γ	Mutual coherence function
Δ	Continuous interval

Δ_d Average linear dimension over which the phase of a wavefront is constant to within $\lambda/8$

Δ Measure of deviation from optimum illumination for thick hologram; the subscripts t and r refer to transmission and reflection holograms

Θ_s Source angular size

Λ Triangle function

$\Lambda(x/a)$ $\{1-|X|,|X|<a;0,\text{ otherwise}\}$

Ξ Propagation constant of the read-out wave in the x direction

Σ Summation

Υ Normalized deviation of illumination from the optimum for thick holograms; the subscripts t and r refer to transmission and reflection holograms

Φ Phase of the reference wave

Ψ Phase of the object wave

Ω Phase of the read-out wave

Π Rectangle function: $\Pi(x/a)=\begin{cases} 1, & -a/2<x<a/2 \\ 0, & \text{otherwise} \end{cases}$

III Shah, comb, or sampling function: $\mathrm{III}(x/\sigma)=|\sigma|\sum_m \delta(x-m\sigma)$

$*$ Denotes complex conjugate

$\otimes$ Denotes convolution

$\star$ Denotes correlation

$\tilde{}$ Denotes spatial Fourier transform

$\hat{}$ Denotes temporal Fourier transform

Author Index

Subject Index